KV-414-656

WITHDRAWN

VOLUME FIFTY NINE

ADVANCES IN ECOLOGICAL RESEARCH

Next Generation Biomonitoring: Part 2

ADVANCES IN ECOLOGICAL RESEARCH

Series Editors

DAVID A. BOHAN
Directeur de Recherche
UMR 1347 Agroécologie
AgroSup/UB/INRA
Pôle GESTAD, Dijon, France

ALEX J. DUMBRELL
School of Biological Sciences
University of Essex
Wivenhoe Park, Colchester
Essex, United Kingdom

VOLUME FIFTY NINE

Advances in Ecological Research

Next Generation Biomonitoring: Part 2

Edited by

DAVID A. BOHAN

Directeur de Recherche
UMR 1347 Agroécologie
AgroSup/UB/INRA
Pôle GESTAD, Dijon, France

ALEX J. DUMBRELL

School of Biological Sciences
University of Essex
Wivenhoe Park, Colchester, Essex,
United Kingdom

GUY WOODWARD

Imperial College London, Ascot, Berkshire,
United Kingdom

MICHELLE JACKSON

Imperial College London, Ascot, Berkshire,
United Kingdom

Academic Press is an imprint of Elsevier
125 London Wall, London, EC2Y 5AS, United Kingdom
The Boulevard, Langford Lane, Kidlington, Oxford OX5 1GB, United Kingdom
525 B Street, Suite 1650, San Diego, CA 92101, United States
50 Hampshire Street, 5th Floor, Cambridge, MA 02139, United States

First edition 2018

ISBN: 978-0-12-814317-9
ISSN: 0065-2504

For information on all Academic Press publications
visit our website at https://www.elsevier.com/books-and-journals

Publisher: Zoe Kruze
Acquisition Editor: Jason Mitchell
Editorial Project Manager: Joanna Collett
Production Project Manager: Abdulla Sait
Cover Designer: Alan Studholme

Typeset by SPi Global, India

CONTENTS

Contributors *ix*
Preface *xiii*
Acknowledgements *xvii*

1. Bioinformatics for Biomonitoring: Species Detection and Diversity Estimates Across Next-Generation Sequencing Platforms **1**

Isaac M.K. Eckert, Joanne E. Littlefair, Guang K. Zhang, Frédéric J.J. Chain, Teresa J. Crease, and Melania E. Cristescu

1. Introduction 2
2. Materials and Methods 9
3. Results 15
4. Discussion 22
5. Conclusions 28

Acknowledgements 29
References 29

2. Linking DNA Metabarcoding and Text Mining to Create Network-Based Biomonitoring Tools: A Case Study on Boreal Wetland Macroinvertebrate Communities **33**

Zacchaeus G. Compson, Wendy A. Monk, Colin J. Curry, Dominique Gravel, Alex Bush, Christopher J.O. Baker, Mohammad Sadnan Al Manir, Alexandre Riazanov, Mehrdad Hajibabaei, Shadi Shokralla, Joel F. Gibson, Sonja Stefani, Michael T.G. Wright, and Donald J. Baird

1. Introduction 35
2. Building Heuristic Food Webs for Wetland Biomonitoring: A Case Study 40
3. The Future of Text Mining: New Tools for Rapid Expansion of Food Web Databases and Challenges to Their Widespread Adoption for Searching Ecological Literature 53
4. Towards an Open-Source Pipeline for the Rapid Construction and Assessment of Trait-Based Food Webs for Biomonitoring: The Promise, Challenges, and Next Steps 55
5. Perspectives: The Future of Biomonitoring 61

Acknowledgements 62
Appendix 62
Glossary 65
References 66
Further Reading 74

3. Volatile Biomarkers for Aquatic Ecological Research 75

Michael Steinke, Luli Randell, Alex J. Dumbrell, and Mahasweta Saha

1. Introduction 76
2. Principal Techniques for Measuring Biogenic Volatiles 79
3. Medical Volatilomics Provides a Blueprint for Ecological Research 80
4. Role of Volatiles in Aquatic Ecological Interactions 81
5. Application of Volatilomics to Ecological Research: Using Volatilomics to 'Direct' Environmental Management 85

Acknowledgements 89

References 89

4. Noninvasive Analysis of the Soil Microbiome: Biomonitoring Strategies Using the Volatilome, Community Analysis, and Environmental Data 93

Kelly R. Redeker, Leda L. Cai, Alex J. Dumbrell, Alex Bardill, James P.J. Chong, and Thorunn Helgason

1. Introduction 94
2. An Overview of the Soil Volatilome 101
3. Understanding Essex UK Salt Marsh Sediments Through the Volatilome 108
4. Conclusion 125

Acknowledgements 127

References 127

Further Reading 132

5. Using Social Media for Biomonitoring: How Facebook, Twitter, Flickr and Other Social Networking Platforms Can Provide Large-Scale Biodiversity Data 133

Jon Chamberlain

1. Introduction 134
2. Related Work 135
3. Examples of Biomonitoring Using Social Networking Platforms 141
4. Analysis of Posts About Wildlife on Social Networking Platforms 144
5. Discussion 155
6. Applications 161
7. Future Directions 163
8. Conclusions 165

Acknowledgements 165

References 166

6. A Vision for Global Biodiversity Monitoring With Citizen Science **169**

Michael J.O. Pocock, Mark Chandler, Rick Bonney, Ian Thornhill, Anna Albin, Tom August, Steven Bachman, Peter M.J. Brown, Davi Gasparini Fernandes Cunha, Audrey Grez, Colin Jackson, Monica Peters, Narindra Romer Rabarijaon, Helen E. Roy, Tania Zaviezo, and Finn Danielsen

1. Introduction 170
2. Citizen Science for Biodiversity Monitoring 171
3. The Global Need for Biodiversity Monitoring 177
4. The Global Potential for Biodiversity Monitoring With Citizen Science 179
5. Approaches for Biodiversity Monitoring With Citizen Science: Who, What and How? 183
6. Case Studies of Steps Towards Global Biodiversity Monitoring With Citizen Science 194
7. Conclusion 210

Acknowledgements 211
References 211

7. A Replicated Network Approach to 'Big Data' in Ecology **225**

Athen Ma, David A. Bohan, Elsa Canard, Stéphane A.P. Derocles, Clare Gray, Xueke Lu, Sarina Macfadyen, Gustavo Q. Romero, and Pavel Kratina

1. Introduction: A Need to Detect Ecosystem Change 226
2. Historical Perspective on Network Analysis 232
3. Promising Future Avenues to 'Big Data', Network Analyses of Change 243
4. Conclusions 252

Acknowledgements 253
References 253
Further Reading 263

Cumulative List of Titles *265*

CONTRIBUTORS

Anna Albin
NORDECO, Copenhagen, Denmark

Tom August
Centre for Ecology & Hydrology, Wallingford, United Kingdom

Steven Bachman
Royal Botanic Gardens, Kew, Richmond, United Kingdom

Donald J. Baird
Canadian Rivers Institute, Department of Biology; Environment and Climate Change Canada @ Canadian Rivers Institute, Department of Biology, University of New Brunswick, Fredericton, NB, Canada

Christopher J.O. Baker
Department of Computer Science, University of New Brunswick; IPSNP Computing Inc., Saint John, NB, Canada

Alex Bardill
Department of Biology, University of York, York, United Kingdom

David A. Bohan
UMR 1347 Agroécologie. AgroSup/UB/INRA, Pôle GESTion durable des ADventices, Dijon Cedex, France

Rick Bonney
Cornell Lab of Ornithology, Ithaca, NY, United States

Peter M.J. Brown
Applied Ecology Research Group, Department of Biology, Anglia Ruskin University, Cambridge, United Kingdom

Alex Bush
Canadian Rivers Institute, Department of Biology; Environment and Climate Change Canada @ Canadian Rivers Institute, Department of Biology, University of New Brunswick, Fredericton, NB, Canada

Leda L. Cai
Department of Biology, University of York, York, United Kingdom

Elsa Canard
INRA, Agrocampus-Ouest, Université de Rennes 1, UMR1349 IGEPP, Rennes Cedex, France

Frédéric J.J. Chain
Department of Biology, McGill University, Montréal, QC, Canada; Department of Biological Sciences, University of Massachusetts Lowell, Lowell, MA, United States

Jon Chamberlain
School of Computer Science and Electronic Engineering, University of Essex, Essex, United Kingdom

Mark Chandler
Earthwatch Institute, Boston, MA, United States

James P.J. Chong
Department of Biology, University of York, York, United Kingdom

Zacchaeus G. Compson
Canadian Rivers Institute, Department of Biology; Environment and Climate Change Canada @ Canadian Rivers Institute, Department of Biology, University of New Brunswick, Fredericton, NB, Canada

Teresa J. Crease
Department of Integrative Biology, University of Guelph, Guelph, ON, Canada

Melania E. Cristescu
Department of Biology, McGill University, Montréal, QC, Canada

Davi Gasparini Fernandes Cunha
Departamento de Hidráulica e Saneamento, Escola de Engenharia de São Carlos, Universidade de São Paulo, Avenida Trabalhador São-Carlense, São Carlos, Brazil

Colin J. Curry
Wolastoqey Nation in New Brunswick, Fredericton, NB, Canada

Finn Danielsen
NORDECO, Copenhagen, Denmark

Stéphane A.P. Derocles
UMR 1347 Agroécologie. AgroSup/UB/INRA, Pôle GESTion durable des ADventices, Dijon Cedex, France

Alex J. Dumbrell
School of Biological Sciences, University of Essex, Colchester, United Kingdom

Isaac M.K. Eckert
Department of Biology, McGill University, Montréal, QC, Canada

Joel F. Gibson
Entomology, Royal BC Museum, Victoria, BC, Canada

Dominique Gravel
Département de Biologie, Université de Sherbrooke, Sherbrooke, QC, Canada

Clare Gray
School of Biological and Chemical Sciences, Queen Mary University of London, London; Department of Life Sciences, Imperial College London, Ascot, Berkshire, United Kingdom

Audrey Grez
Facultad de Ciencias Veterinarias y Pecuarias, Universidad de Chile and Director of Kauyeken, Santiago, Chile

Mehrdad Hajibabaei
Biodiversity Institute of Ontario, Department of Integrative Biology, University of Guelph, Guelph, ON, Canada

Thorunn Helgason
Department of Biology, University of York, York, United Kingdom

Colin Jackson
A Rocha Kenya, Watamu, Kenya

Pavel Kratina
School of Biological and Chemical Sciences, Queen Mary University of London, London, United Kingdom

Joanne E. Littlefair
Department of Biology, McGill University, Montréal, QC, Canada

Xueke Lu
School of Electronic Engineering and Computer Science, Queen Mary University of London, London; School of Engineering, The University of Warwick, Coventry, United Kingdom

Athen Ma
School of Electronic Engineering and Computer Science, Queen Mary University of London, London, United Kingdom

Sarina Macfadyen
CSIRO, Black Mountain, Acton, ACT, Australia

Mohammad Sadnan Al Manir
Department of Computer Science, University of New Brunswick, Saint John, NB, Canada

Wendy A. Monk
Environment and Climate Change Canada @ Canadian Rivers Institute, Department of Biology; Faculty of Forestry and Environmental Management, University of New Brunswick, Fredericton, NB, Canada

Monica Peters
People + science, Hamilton, New Zealand

Michael J.O. Pocock
Centre for Ecology & Hydrology, Wallingford, United Kingdom

Narindra Romer Rabarijaon
Kew Madagascar Conservation Center, Antananarivo, Madagascar

Luli Randell
School of Biological Sciences, University of Essex, Colchester, United Kingdom

Kelly R. Redeker
Department of Biology, University of York, York, United Kingdom

Alexandre Riazanov
Department of Computer Science, University of New Brunswick, Saint John, NB, Canada

Gustavo Q. Romero
Departamento de Biologia Animal, University of Campinas, Campinas, Brazil

Helen E. Roy
Centre for Ecology & Hydrology, Wallingford, United Kingdom

Mahasweta Saha
School of Biological Sciences, University of Essex, Colchester, United Kingdom

Shadi Shokralla
Biodiversity Institute of Ontario, Department of Integrative Biology, University of Guelph, Guelph, ON, Canada

Sonja Stefani
Dresden University of Technology, Dresden, Germany

Michael Steinke
School of Biological Sciences, University of Essex, Colchester, United Kingdom

Ian Thornhill
Earthwatch Institute, Oxford; College of Liberal Arts (CoLA), Bath Spa University, Bath, United Kingdom

Michael T.G. Wright
Biodiversity Institute of Ontario, Department of Integrative Biology, University of Guelph, Guelph, ON, Canada

Tania Zaviezo
Facultad de Agronomía e Ingeniería Forestal, Pontificia Universidad Católica de Chile, Santiago, Chile

Guang K. Zhang
Department of Biology, McGill University, Montréal, QC, Canada

PREFACE

Biomonitoring the Earth's ecosystems and their attendant communities, functions and ecoservices underpins decision making in many areas of policy and can have considerable value for the public, particularly in the case of species with high conservation value. In almost all cases, however, current biomonitoring approaches suffer from problems of accuracy, high costs that restrict coverage and limited generality. Biomonitoring schemes are also based upon methods developed in the early or middle part of the last century and have largely ignored subsequent advances in ecological theory and techniques, especially those derived from molecular ecology, remote sensing, network science and ecoinformatics. Consequently, the full diversity of functions and species in an ecosystem has rarely been evaluated. This is problematic because it only provides a partial view of the greater whole and cannot account for—or predict—the 'ecological surprises' that commonly arise through indirect food web effects in nature. In this two-volume Thematic Issue of Advances in Ecological Research focusing on Ecological Biomonitoring, we showcase some of the new biomonitoring approaches that have begun to appear in the last 15 years and that have started to tackle these problems directly; to generate the more sophisticated Next-Generation Biomonitoring (NGB) approaches, we will need to cope with our rapidly changing environment. Potentially, NGB could, even within the next decade, revolutionise our understanding of the functioning of Earth's major ecosystems, allowing us to both measure and predict the effects of a range of abiotic stressors as well as those from the biotic sphere (e.g. species invasion and extinction), which will lead to better-informed and more effective management. Moreover, as they are often rooted in standardised, functional metrics, these approaches could potentially be applied at local to global scales, both accurately and cheaply.

The first volume covered a range of new monitoring methodologies designed to understand the complexities of change in ecosystems, most notably using DNA (Derocles et al., 2018) and remote sensing (Perennou et al., 2018), and also placed the use of these methods into real-world biomonitoring situations (e.g. Leese et al., 2018; Bramer et al., 2018; De Palma et al., 2018; Purvis et al. 2018).

This second volume opens with a paper comparing species detection and diversity estimates across next generation sequencing platforms (NGS;

Eckert et al., this issue). Sequencing platforms are frequently updated, creating challenges when it comes to comparing diversity estimates which were made at a different time, or using a different platform. Eckert et al. (this issue) assembled two mock plankton communities and compared accuracy of species detection using the NGS platforms Roche 454 and Illumina MiSeq. They found that the platforms were comparable with only slight differences in accuracy. However, they caution that OTU clustering as a proxy for genetic diversity must be used with caution, as it led to overestimates of diversity when using both platforms. This has implications for the next generation of biomonitoring tools—ensuring that past (both traditional taxonomic identification and earlier sequencing platforms) and future technologies are comparable is essential.

Compson et al. (this issue) then examine the construction of ecological networks for a boreal wetland macroinvertebrate community, using a combination of NGS metabarcoding and text mining. In the paper, the authors examine how standardizable analytical pipelines can be constructed to generate 'heuristic food webs' based upon organismal traits that are harvested from online databases using a General Architecture for Text Engineering (GATE) system of hybrid text mining. They then tested the networks constructed to examine whether this combination of metabarcoding and text mining could prove a powerful tool for rapid bioassessment, which could be made available to land managers and conservation biologists.

In the third paper of the issue, Steinke et al. (this issue) move biomonitoring away from DNA methodologies to explore volatile organic compounds for the biomonitoring, particularly in aquatic environments. The authors note that all organisms and ecosystems emit and consume volatile organic compounds (VOCs). In traditional biomonitoring, these compounds have been evaluated in isolation without full consideration of the full VOC 'signatures' produced in the system. They argue that volatilomics could provide a relatively fast diagnostic tool to investigate taxonomic and possibly functional diversity in aquatic systems, and show that the approach can differentiate between four different algal genera in a case study.

Redeker and colleagues (this issue) examine soil volatiles and metabolism in soil microbe communities. Soil metabolism, which reflects the soil functionality of a variety of microorganisms (including saprophytes, mycorrhizal fungi, nitrogen-fixing bacteria and parasitic bacteria and fungi), remains challenging to accurately quantify. Redeker et al. (this issue) argue that using non-invasive, volatile signature methods the tightly coupled, 'net' metabolism of soil communities could be studied and biomonitored. They propose a 'fingerprint' to describe this complex community that uses trace gas fluxes combined with environmental data and describe the promising outcomes from an initial case study.

The chapter by Chamberlain (this issue) is the first of two papers exploring citizen-sourced biomonitoring. Chamberlain (this issue) examines social networking platforms as a source of biomonitoring information. From a corpus of 39,039 Facebook conversation threads, the 'ease of access' to and the reliability of the biodiversity data they contained was evaluated by analysing and understanding how groups of people solve image classification problems. In principle, social network technologies offer researchers and managers a new opportunity to gather biodiversity data. However, there are considerable resource overheads and the promise of these methods will only be achieved with the development of methods for automatic processing of social network data.

Pocock et al. (this issue) develops the theme of citizen-sourced biomonitoring using Citizen Science by explicitly detailing the diversity approaches from people monitoring the environment in a voluntary capacity up to participatory monitoring in which people work collaboratively with scientists in developing monitoring. While there is great unrealised potential for citizen science in biomonitoring, there is currently relatively low use of these approaches because they often fail to the dual needs in the locality (for participants, communities and decision-makers, including people's own use of the data and their motivations to participate) and support global needs for biodiversity monitoring (including the United Nations' Sustainable Development Goals and the Aichi Biodiversity Targets). Using examples from around the world that demonstrate that monitoring can engage different types of participants, Pocock et al. (this issue) demonstrate that Citizen Science has great potential that will only be achieved if activities are feasible and useful, for all users, from local to global scales.

In the final paper of this issue, Ma et al. (this issue) outline some of the statistical landscape we will need to develop and understand if we are to biomonitor ecosystems using complex ecological data such as networks. The authors imagine global-scale network data will become available, at high spatial and temporal resolution, with the development of next-generation sequencing approaches. They then tackle the question of how we will detect and explain such changes, and ascribe ecological importance to any change, using classical ecological network statistics (low-level metrics such as connectance) and newer, high-level structural metrics that have recently come into Ecology from domains including the social sciences and telecommunications. Ma et al. (this issue) advocate that higher level, more generic network metrics, will be necessary to allow biomonitoring at extremely high replication, both to facilitate detection of change and generic comparison of change and effect between ecosystems.

These two volumes present a snapshot of some of the work currently being done in biomonitoring. The combination of papers across them reveals the huge value in using novel NGS, sensing and informatics approaches and better fusions of pure and applied disciplines to monitor and model how natural ecosystems will respond to the accelerating rates and increasing magnitude of environmental change we are already seeing across the globe. There is clearly plenty of exciting and challenging work still to be done, but this Thematic Issue illustrates some of the most important steps being taken towards developing the NGB approaches we will need to achieve a more sustainable future.

Alex J. Dumbrell
Guy Woodward
Michelle C. Jackson
David A. Bohan

REFERENCES

Bramer, I., Anderson, B.J., Bennie, J., Bladon, A.J., De Frenne, P., Hemming, D., et al., 2018. Advances in monitoring and modelling climate at ecologically relevant scales. Adv. Ecol. Res. 58, 101–161. https://doi.org/10.1016/bs.aecr.2017.12.005.

De Palma, A., Sanchez-Ortiz, K., Martin, P.A., Chadwick, A., Gilbert, G., Bates, A.E., et al., 2018. Challenges with inferring how land-use affects terrestrial biodiversity: study design, time, space and synthesis. Adv. Ecol. Res. 58, 163–199. https://doi.org/10.1016/bs.aecr.2017.12.004.

Derocles, S.A.P., Bohan, D.A., Dumbrell, A.J., Kitson, J.J.N., Massol, F., Pauvert, C., et al., 2018. Biomonitoring for the 21st Century: Integrating Next-Generation Sequencing Into Ecological Network Analysis. Elsevier. https://doi.org/10.1016/bs.aecr.2017.12.001.

Leese, F., Bouchez, A., Abarenkov, K., Altermatt, F., Borja, Á., Bruce, K., et al., 2018. Why we need sustainable networks bridging countries, disciplines, cultures and generations for aquatic biomonitoring 2.0: a perspective derived from the DNAqua-Net COST action. Adv. Ecol. Res. 58, 63–99. https://doi.org/10.1016/bs.aecr.2018.01.001.

Perennou, C., Guelmami, A., Paganini, M., Philipson, P., Poulin, B., Strauch, A., et al., 2018. Mapping Mediterranean wetlands with remote sensing: a good-looking map is not always a good map. Adv. Ecol. Res. 58, 243–277. https://doi.org/10.1016/bs.aecr.2017.12.002.

Purvis, A., Newbold, T., De Palma, A., Contu, S., Hill, S.L.L., Sanchez-Ortiz, K., et al., 2018. Modelling and projecting the response of local terrestrial biodiversity worldwide to land use and related pressures: the PREDICTS project. Adv. Ecol. Res. 58, 201–241. https://doi.org/10.1016/bs.aecr.2017.12.003.

ACKNOWLEDGEMENTS

David A. Bohan would like to acknowledge the support of the French Agence Nationale de la Recherche project *NGB* (ANR-17-CE32-0011) and FACCE SURPLUS project *PREAR* (ANR-15-SUSF-0002-03).

CHAPTER ONE

Bioinformatics for Biomonitoring: Species Detection and Diversity Estimates Across Next-Generation Sequencing Platforms

Isaac M.K. Eckert*, Joanne E. Littlefair*[,1], Guang K. Zhang*, Frédéric J.J. Chain*[,†], Teresa J. Crease[‡], Melania E. Cristescu*

*Department of Biology, McGill University, Montréal, QC, Canada
[†]Department of Biological Sciences, University of Massachusetts Lowell, Lowell, MA, United States
[‡]Department of Integrative Biology, University of Guelph, Guelph, ON, Canada
[1]Corresponding author: e-mail address: joanne.littlefair@mail.mcgill.ca

Contents

1. Introduction 2
2. Materials and Methods 9
   2.1 Mock Communities 9
   2.2 Library Preparation for Roche 454 10
   2.3 Library Preparation for Illumina MiSeq 11
   2.4 Bioinformatics and Data Analysis 11
   2.5 Assigning Taxonomy to OTUs and Dereplicated Sequences 13
3. Results 15
   3.1 Sequence and Read Depth 15
   3.2 OTU Clustering and the Effect of Singletons 16
   3.3 Species Detection 16
   3.4 OTU Precision 20
   3.5 Impact of Merging and Appending MiSeq Reads on Species Detection and OTU Estimates 20
4. Discussion 22
   4.1 Read Depth and Singletons 23
   4.2 OTU Clustering 24
   4.3 Experimental Design 25
   4.4 Implications for Biomonitoring 27
5. Conclusions 28
Acknowledgements 29
References 29

Advances in Ecological Research, Volume 59
ISSN 0065-2504
https://doi.org/10.1016/bs.aecr.2018.06.002

Abstract

As a fast-growing area of technology, sequencing platforms are updated frequently and this rapid technical revolution poses not only great advances but also challenges. To be effective, biomonitoring programmes need to deliver comparable results across research groups and time. Understanding the sources of bias in bioinformatics promotes reliable results that accurately reflect biodiversity. We assembled two mock communities of planktonic organisms to assess the accuracy of species recovery based on sequencing the 18S rRNA V4 region using two NGS platforms, Roche 454 (the platform of choice for early metabarcoding studies), and Illumina MiSeq (employed frequently in recent metabarcoding studies). Our findings suggest that the two platforms have comparable performance on metabarcoding datasets. When singletons (sequences represented by a single read) were excluded from analyses, Illumina MiSeq had a slightly better operational taxonomic unit (OTU) precision score than Roche 454 (calculated as the number of species detected divided by the number of OTUs generated) but only in one bioinformatics workflow (when paired reads were appended, not merged). Roche 454 performed slightly better than Illumina MiSeq in terms of species detection but only when simple mock communities with a single individual per species were analysed. When singleton sequences were included, both platforms detected more than 75% of species with a slightly higher detection achieved by Illumina MiSeq. The OTU clustering of both datasets resulted in a gross overestimation of species richness. This finding suggests that studies employing OTU clustering as a proxy for genetic diversity must carefully perform read processing, such as singleton exclusion, to avoid overestimates. Finally, this study provides insight into technical bioinformatic strategies that should accompany such transitions. In a field such as metabarcoding, where advances in sequencing technology constantly drive the discipline, ensuring the comparability of past and future technologies, and the derived ecological conclusions is important.

1. INTRODUCTION

Metabarcoding has the potential to become a powerful tool for rapid biodiversity assessment, describing long-term biodiversity trends, studying the ecology and evolution of natural communities, and developing new, rapid, and efficient techniques for biomonitoring (Cristescu, 2014; Littlefair and Clare, 2016). At the core of metabarcoding is next-generation sequencing (NGS) technology, which allows the sequencing of mixed, complex environmental samples (Pompanon et al., 2012; Taberlet et al., 2012). The main advantage of NGS over the traditional Sanger sequencing is that it provides greater sequencing depth, high-resolution analyses, and eliminates the need to generate single-individual libraries prior to sequencing. These advantages make NGS approaches attractive for biomonitoring studies. However, the high depth comes with the cost of reduced length of sequenced reads (Leigh et al., 2015) and a relatively high error rate (Goodwin et al., 2016). Metabarcoding approaches using NGS allow the parallel examination of multiple taxonomic groups through the use of universal barcodes (specified short fragments of DNA), without a great deal of a priori knowledge about the targeted organisms. Species assignments are often conducted by matching the retrieved barcodes against large open-access reference libraries of DNA sequences that are constantly being populated with reference barcodes (e.g., BOLD, an informatics workbench containing open-access COI barcode records; SILVA, an open-access database which curates small (e.g., 16S/18S) and large (e.g., 23S/28S) subunit ribosomal RNA sequences).

Within the scope of biomonitoring, metabarcoding approaches have the potential to save time and alleviate the problem of deficient taxonomic expertise in the identification of more obscure groups (Deiner et al., 2017); as such expertise is often distributed unevenly around the globe (Ji et al., 2013). It has also been shown that metabarcoding approaches can provide taxonomic information with increased resolution in relation to existing monitoring protocols. For example, NGS techniques provided higher taxonomic resolution at lower cost than morphological identification using Environment Canada's Canadian Aquatic Biomonitoring Network protocols (Gibson et al., 2015). This "big data" approach to biodiversity science allows us to focus on multiple taxonomic groups, rather than monitoring with indicator species, the use of which can be problematic if not carefully linked to ecosystem functioning and measures of true diversity (Moonen and Bàrberi, 2008). If combined with almost real-time DNA sequencers currently in development (e.g., MinION, GridION), metabarcoding has the potential to provide very fine scale temporal and spatial monitoring data from around the globe (Bohan et al., 2017). The power derived from greater amounts of data will allow us to monitor ecological networks and their properties, from which we can infer or model information about ecosystem structure and stability (Evans et al., 2016). Governments are starting to consider integrating molecular methods into existing monitoring programmes (Darling and Mahon, 2011; Kelly et al., 2014). For example, in 2014 the UK government approved the use of environmental DNA as an alternative to conventional surveys to monitor the great crested newt, whose habitats are protected from development by European and UK law. The European Union has begun the integration of DNA-based tools into European ecological monitoring programmes by developing a series of work packages known as DNAqua-NET (Leese et al., 2016; see chapter "Next generation biomonitoring of aquatic ecosystems" by Leese). However, to integrate NGS into existing management strategies, we need to evaluate the consistency of biodiversity estimates based on metabarcoding datasets as sequencing platforms and bioinformatic tools are replaced with ever-evolving technology.

While the tremendous amount of data produced by NGS is advantageous, the accompanying need for stringent and specific filtering of data is paramount, with appropriate recording of processing steps when it comes to producing repeatable and reliable results and confirming species detection (Clare et al., 2016). Additionally, bioinformatic pipelines associated with metabarcoding continue to change rapidly with evolving NGS technology. Rapid change can be problematic when such technologies become integrated into long-term biomonitoring programmes, where consistency of, and comparison between results is valued (Coissac et al., 2012). In particular, NGS platforms are constantly updated as technological innovations become available, providing more accurate and in-depth analyses (Glenn, 2011; Goodwin et al., 2016; Zhou et al., 2013). This emphasizes the need to set appropriate guidelines on the use of sequencing platforms and bioinformatics.

To date, the major platforms that have been used in metabarcoding studies (Table 1) are Roche 454, Illumina HiSeq, and MiSeq, and Ion Torrent (Heather and Chain, 2016). The Roche 454 sequencing platform produces continuous single reads by employing pyrosequencing. This involves the addition of a nucleotide base onto a growing nucleotide chain coupled with the enzymatic release of light, which is monitored by a camera and recorded

Table 1 A Summary of the Next-Generation Sequencing (NGS) Technologies Discussed in This Chapter, Their Output Characteristics, and Their Main Types of Errors

NGS Technology	Output: Read Length	Output: Number of Reads	Error Rate %	Main Error Type
Roche 454	Single reads: 400–700 bp	>1 million	1	Insertion/deletion
Illumina MiSeq	Paired reads: 2 × 300 bp	25 million	>0.1%	Substitution
Illumina HiSeq	Paired reads: 2 × 150 bp	>300 million	≥0.1%	Substitution
Ion Proton 1	Single reads: 200 bp	60 million	1%	Insertion/deletion

Modified from Glenn, T.C., 2011. Field guide to next-generation DNA sequencers. *Mol. Ecol. Resour.* 11 (5), 759–769.

as one of the four DNA nucleotides (Balzer et al., 2010; Leamon et al., 2003). Illumina MiSeq and HiSeq both use the same method of dye labelling and sequencing by synthesis. Fluorescently labelled bases release fluorescence when incorporated into a growing nucleotide chain, which is detected by a laser and recorded. Unlike Roche 454, Illumina platforms can also produce paired reads that represent the start and end of a target amplicon (Bentley, 2008). These paired reads can be merged to produce continuous sequences (contigs) if they overlap with one another (e.g., the fragment size is shorter than the combined paired read length). HiSeq was designed to provide much higher data output compared to MiSeq, and can produce a much greater number of raw reads, but sacrifices short run time and long read length to do so. Ion Torrent sequencing relies on the monitored release of hydrogen atoms from a growing strand of nucleotides (hydrogen atoms released into the reaction solution change the pH, causing a signal to be recorded by an ion-sensitive field-effect transmitter). This platform can produce continuous reads at a much lower cost compared to other NGS technologies (Rothberg et al., 2011).

Despite the importance of understanding sources of potential bias in the lab and bioinformatics steps within metabarcoding (Alberdi et al., 2017; Brown et al., 2015; Clare et al., 2016; Flynn et al., 2015), the taxonomic bias introduced through the use of different sequencing technologies has received relatively little attention, particularly within the scope of metazoan communities. Conclusions from studies using genomics and transcriptomics applications suggest a variety of platform-specific biases due to the differing technologies (Table 2). For example, Quince et al. (2009) found a higher error rates associated with Roche 454 when sequencing homopolymer regions. Although Illumina improves on this problem, it suffers from its own limitations. For example, reads of different quality can be produced depending on their location on the sequencing plate (Erlich et al., 2008). While Illumina reads are shorter than Roche reads, the ability to use 300 bp paired reads to sequence both ends of a fragment gives the option of targeting the longer amplicons that can be analysed with Roche. Long continuous regions of the genome assembled from Illumina reads have been shown to be more accurate and complete when compared to Roche assemblies, due to the increased read depth (Luo et al., 2012). Despite these differences, Luo et al. (2012) found that Illumina and Roche are comparable in terms of the ability to assess the diversity of a microbial community from a pooled genetic sample. Furthermore, while Tremblay et al. (2015) identified a taxonomic bias associated with NGS platforms and the ability to detect

Table 2 A Summary of Comparative Studies of Sequencing Platforms

Publication	Platforms Investigated	Methods and Applications	Conclusions
Clooney et al. (2016)	Illumina MiSeq, Illumina HiSeq, Ion PGM	Microbiome analysis of stool samples for bacterial species detection	• HiSeq shotgun libraries had the greatest read depth • HiSeq shotgun sequences identified the highest number of species • MiSeq and Ion PGM provided a better basis for identifying functional genes due to longer read length
Mahé et al. (2015)	Illumina MiSeq and Roche 454	Comparison of quality and quantity of reads in an environmental diversity assessment	• MiSeq produced an order of magnitude more reads • More amplicons of different taxonomic identity (above the genus level) were found with MiSeq due to deeper sequencing capabilities • Roche 454 produced reads of better quality; therefore, more raw reads were retained after filtering
Tremblay et al. (2015)	Illumina MiSeq and Roche 454	Analysis of amplified V4, V6–V8, and V7–V8 reads from a microbial mock community, sequenced with both platforms	• Paired end MiSeq reads produced higher quality data and allowed for more aggressive quality control parameters, resulting in a higher retention rate of reads for further analysis • The impact of sequencing platform bias was relatively minor compared to the bias introduced by primer selection
Luo et al. (2012)	Illumina Genome Analyzer II and Roche 454	Compared base-call error, frameshift frequency and contig length between Illumina and Roche 454 sequencing data produced from a complex freshwater planktonic community	• Illumina yielded longer and more accurate contigs (fewer truncated genes due to frameshifts) • Roche 454 produced assemblies that contained a significantly higher proportion of frameshift errors compared to Illumina assemblies from the same genome • Both sequencing platforms were reliable for quantitatively assessing genetic diversity within natural communities

Li et al. (2014)	Illumina and Roche 454	Sequencing identical libraries composed of plasma from patients failing antiretroviral therapy for HIV	• Illumina data resulted in higher coverage as well as increased sensitivity for detecting HIV-1 minority variants • Illumina also produced fewer false-positive variant calls compared to Roche 454
Salipante et al. (2014)	Illumina MiSeq and Ion Torrent PGM	Comparison of sequencers for 16S rRNA-based bacterial community profiling in terms of differences in error rates, read truncation, and species detection	• Both platforms were comparable for species detection with only minor differences • Ion Torrent PGM exhibited much higher error rates as well as a pattern of premature sequence truncation • The suggested cause of premature sequence truncation by Ion PGM was the secondary structure of the sequences, as no identifiable primary sequence pattern was found to exist among the truncated sequences
Divoll et al. (2018)	Illumina MiSeq and Ion Torrent PGM	Compared the ability of both sequencers to resolve a list of prey species from bat faecal matter, using similar library preparation and identical analytical workflows	• 104 prey OTUs were detected by both platforms • 176 prey OTUs were detected by MiSeq only • 17 prey OTUs were detected by Ion Torrent only • Results suggest that Illumina MiSeq greatly outperformed Ion Torrent in terms of species resolution from a sample of community DNA

There is a paucity of eukaryotic metabarcoding studies that directly compare Illumina MiSeq and Roche 454 in terms of species detection, biomonitoring, and diversity estimates.

species within a mock community, the bias was relatively minor, compared to the taxonomic bias introduced through primer selection (Tremblay et al., 2015). The authors conclude that Illumina is advantageous due to the higher quality (lower insertion and deletion error rates) of data produced, resulting in a higher retention rate of reads for final analysis.

Early biomonitoring studies based on metabarcoding used Roche 454 technology (e.g., Geml et al., 2014; Hatzenbuhler et al., 2017; Lallias et al., 2015) due to its long read length and simplified bioinformatics, which did not require merging steps for forward and reverse reads. Later, Illumina MiSeq technologies were adopted due to the attraction of greater read depth (Evans et al., 2017; Hänfling et al., 2016). However, platform-specific differences exist in terms of read depth and error rate, as well as specific bioinformatics steps necessary to account for the inherent difference in read length. The recent transition from Roche to Illumina as the most commonly used NGS platform raises questions of comparability for long-term biomonitoring studies aiming to assess changes in community composition. If metabarcoding techniques are integrated into long-term monitoring schemes, it is important to examine how technological transitions will impact observational results obtained by examining experimental communities with defined species compositions. For example, increased read depth might influence the sensitivity for species detection in complex samples, and whether rare species and invasion fronts can be detected. Higher error rates can obscure taxonomic identities when interspecies divergence is low. These questions are not just important for the comparability of sequencing technologies (Mahé et al., 2015; Tremblay et al., 2015), but also for the field of metabarcoding as a whole. Standardization of a particular NGS platform for biomonitoring, while it has its advantages, could prevent researchers from using the most suitable or up-to-date platform for the focus of their study.

Once NGS libraries are sequenced, raw reads go through several bioinformatics quality control steps before species identity can be assigned with confidence. These include demultiplexing (assigning mixed sequences back to their original samples), read length filtering, quality control filtering, and clustering into operational taxonomic units (OTUs) (Floyd et al., 2002). OTUs can then be used to identify species based on alignment with a species reference sequence. Species abundance has been correlated to both sequenced read counts and OTU richness (Lim et al., 2016). Transforming the hundreds of thousands of raw reads produced in a sequencing run into a conclusive list of taxa is a daunting task, and also provides many opportunities for bias

(Schmidt et al., 2013). The analytical procedures need to be tailored to the study and can, if misapplied, result in an overestimation of biodiversity (Brown et al., 2015; Flynn et al., 2015). Understanding the possible introductions of bias at each step in a metabarcoding workflow is essential when it comes to generating reliable results including sample collection, molecular techniques, and bioinformatics tools (Alberdi et al., 2017; Clare et al., 2016; Flynn et al., 2015; Pawluczyk et al., 2015).

In this study, we compare two NGS technologies in the context of biomonitoring of zooplankton communities. As NGS technologies are steadily evolving, it is important to consider whether we can achieve consistent results over long-term studies that use molecular methods. One major recent transition involved the shift from the discontinued Roche 454 to Illumina MiSeq as the platform of choice for metabarcoding studies. Here, we use Roche 454 and Illumina MiSeq to sequence two mock zooplankton communities (artificial communities in which the composition of species is known) targeting the hypervariable V4 region of the 18S rRNA gene (18S-V4 region). We analyse the data using very similar bioinformatics pipelines for both datasets. We used workflows and pipelines originally tested by Flynn et al. (2015) with one additional step to deal with merging or appending paired reads from the MiSeq. This study compares and contrasts the number of reads, species detection results, and precision scores generated from both NGS platforms with the aim of qualifying the transition from Roche to Illumina sequencing for biomonitoring studies using metabarcoding.

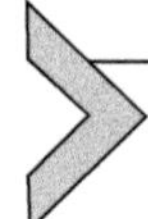

2. MATERIALS AND METHODS

2.1 Mock Communities

We used two mock communities assembled by Brown et al. (2015). The simple community 1 was represented by a single individual per species and consisted of 56 species of zooplankton: 46 arthropods, 2 chordates, and 8 molluscs. The complex community 2 was represented by populations and consisted of a total of 76 individuals belonging to 14 species: 12 crustaceans and 2 molluscs (Table S1 in the online version at https://doi.org/10.1016/bs.aecr.2018.06.002). Mock communities were assembled to avoid congeneric species, with some exceptions (Balanus, Daphnia, Hyallela, and Gammarus). Individuals included in the mock communities were identified to species or genus level based on morphology and Sanger sequencing. Although eight individuals could only be identified to the family level, we nevertheless included them in the community since these individuals were taxonomically and genetically divergent from other community members. The individual specimens, preserved in ethanol, were washed with distilled water prior to assembly. To increase the efficiency of the DNA extraction involving a large number of individuals, both mock communities were assembled in four microcentrifuge tubes. Each tube, containing approximately equal numbers of individuals, was then centrifuged to remove any liquid that remained after the specimens were washed. Genomic DNA was extracted from each of the four tubes using DNeasy Blood and Tissue Kit (Qiagen, Venlo, Limburg, Netherlands) following the manufacturer's protocol. The DNA extractions were then used to prepare NGS libraries to be run on two separate sequencing platforms to profile the zooplankton mock community, the Roche 454 and the Illumina MiSeq.

2.2 Library Preparation for Roche 454

DNA amplification of the 18S-V4 region was performed, and the amplicons were sequenced using the Roche 454 (Brown et al., 2015). PCR was performed using universal 18S-V4 primers to generate a 400–700 bp fragment (F: AGGGCAAKYCTGGTGCCAGC, R: GRCGGTATCTRATCGYCTT) (Zhan et al., 2013). To reduce PCR amplification bias, eight reactions were performed on each independent DNA extraction, and equimolar aliquots of each replicate were pooled for sequencing. PCR reactions consisted of 100 ng of genomic DNA, 1 × PCR buffer, 2 mmol/L of Mg^{2+}, 0.2 mmol/L of dNTPs, 0.4 μmol/L of each primer, and 2 U of Taq DNA polymerase (Genscript, Piscataway, NJ, USA), in a final reaction volume of 25 μL.

PCR cycling parameters consisted of an initial denaturation step at 95°C for 5 min, followed by 25 amplification cycles of 95°C for 30 s, 50°C for 30 s, 72°C for 90 s, and a final elongation step at 72°C for 10 min. PCR products were cleaned using the solid-phase reversible immobilization paramagnetic bead method (ChargeSwitch, Invitrogen, Carlsbad, CA, USA). Quantity and quality were assessed using gel electrophoresis and the Quant-iT PicoGreen dsDNA assay kit (Invitrogen, Carlsbad, CA, USA). Cleaned PCR products were then pooled together in equimolar concentrations before pryosequencing at the 1/2 PicoTiter plate scale. Roche 454 flex adapters (A: CCATCTCATCCCTGCGTGTCTCCGACTCAG, B: CCTATCCCCTGTGTGCCTTGGCAGTCTCAG) were added to the 5′ end of the 18S-V4 amplicons. Sequencing was performed using 454 FLEX Adapter A on a GS-FLEX Titanium platform (454 Life Sciences, Branford, CT, USA) by Génome Québec (Montréal QC, Canada). The data generated include 483,986 reads for community 1 (accession SRX884895) and 204,922 reads for community 2 (accession SRX884904) with an average length of 518 bp and can be found in the Sequence Read Archive.

2.3 Library Preparation for Illumina MiSeq

Library preparation for the MiSeq was performed by first PCR amplifying the 18S-V4 region with gene-specific primers as in the Roche workflow. The Forward gene-specific primer was as follows: Adapter–Spacer[Gene-Specific Region]: 5′-TCGTCGGCAGCGTC-AGATGTGTATAAGAGACAG [AGGGCAAKYCTGGTGCCAGC]-3′. The Reverse gene-specific primer was as follows: Adapter–Spacer[Gene-Specific Region]: 5′-GTCTC GTGGGCTCGG-AGATGTGTATAAGAGACAG[GRCGGTATCTR ATCGYCTT]-3′. All the initial PCR products were visualized on a 1.5% agarose gel and submitted to the Genomics Facility at the University of Guelph for further processing. The initial PCR products were purified using AMPure beads and indexed by PCR amplification with primers containing the index sequences; Forward primers were (Flowcell Adapter–Index–Adapter): 5′-AATGATACGGCGACCACCGAGATCTACAC-INDEX-TCGTCGG CAGCGTC-3′ and Reverse primers were (Flowcell Adapter–Index–Adapter): 5′-CAAGCAGAAGACGGCATACGAGAT-INDEX-GTCT CGTGGGCTCGG-3′. An equal volume of the eight PCR amplicons from each of the four samples from each community was pooled before performing the index amplification. Two libraries were created for each community from the pooled templates for a total of four pooled libraries. The indexed libraries were again purified with AMPure beads, quantified, normalized, and pooled for sequencing using the paired-end 2× 300 bp cartridge on the MiSeq (Illumina, Inc., San Diego, CA, USA). The data generated include 1,005,261 paired raw reads and can be found in the Sequence Read Archive, National Center for Biotechnology Information NCBI ID: SRR6848116 (Community 1) and SRR6848115 (Community 2).

2.4 Bioinformatics and Data Analysis

The data were processed in Unix using a series of data filtering and analysis packages (Table 3), and were compared to evaluate the species detection and performance of the Roche and MiSeq platforms. For all pipelines, taxonomic assignment was performed using BLAST alignments against a local reference database that was constructed to contain an 18S-V4 reference sequence for all members of the mock community, which were sufficiently

Table 3 The Major Steps of Each Bioinformatics Workflow for the Two Sequencing Platforms (Roche 454 or Illumina MiSeq) and the Main Workflows: With Singletons Included (S) or Excluded (N), and Merging (M) or Appending (AP) Steps for Illumina MiSeq

Bioinformatic Steps	Illumina MiSeq				Roche 454	
	N, M	N, AP	S, M	S, AP	N	S
Adapter removal and trimming (Trimmomatic)	✓	✓	✓	✓	✓	✓
Merging (FLASH)	✓	×	✓	×	×	×
Appending (FASTX)	×	✓	×	✓	×	×
Primer Removal (FASTX)	✓	✓	✓	✓	✓	✓
Dereplication (USEARCH)	✓	✓	✓	✓	✓	✓
Chimera removal (UCHIME)	✓	✓	✓	✓	✓	✓
Singleton exclusion (USEARCH)	✓	✓	×	×	✓	×
Abundance sorting	✓	✓	✓	✓	✓	✓
Clustering (UPARSE)	✓	✓	✓	✓	✓	✓
Taxonomic assignment (BLAST)	✓	✓	✓	✓	✓	✓

divergent such that each species could be individually identified. The local reference database was the same database used by Brown et al. (2015), assembled using reference sequences downloaded from the NCBI nucleotide database and the SILVA database (Quast et al., 2013) as well as sequences generated for closely related species using Sanger sequencing of the 18S-V4 region.

We received demultiplexed reads from both Génome Québec (Roche) and the University of Guelph (MiSeq). Reads from the two replicate libraries, for each mock community, were pooled prior to analysis. For both Roche and MiSeq pipelines, adapters were removed and reads were trimmed using Trimmomatic v0.32 (Bolger et al., 2014) based on Phred quality scores (a method of assessing the probability of an incorrect base call). Leading (5′) and trailing (3′) bases were trimmed if they had a quality score of less than 10, representing a 90% base accuracy threshold. In addition, bases were removed from a sequence if the average quality of a sliding window of 20 base pairs (bp) fell beneath an average quality score of 10. Finally, sequences that were shorter than 100 bp were discarded.

Assembly of MiSeq paired end reads into contigs occurred using two distinct methods: merging reads (M) and blunt-end appending reads (AP).

The overlapping and merging of paired end reads were conducted in Flash v1.2.7 (Magoč and Salzberg, 2011). Reads that did not merge were discarded, based on the expectation that the target amplicon should be smaller than the combined read pairs (~570 bp), which would mean the read pairs would overlap. Since the merging step can introduce a bias against species with long 18S-V4 regions if read pairs do not overlap, the impact of merging MiSeq reads was investigated through analysis of blunt-end appended paired reads to recover species with long V4 regions that are filtered out due to insufficient read overlap during the merging step. Appended reads were created using FASTX-Toolkit 0.0.13.2 (http://hannonlab.cshl.edu/fastx_toolkit/) by reverse complementing the reverse read, so both paired reads were in the same orientation (5′-3′), and then joining them together in the middle without overlap.

The 18S-V4 amplicon primers were removed in both the Roche and MiSeq pipelines using the FASTX clipper. Reads that did not meet a minimum length of 200 bp were discarded prior to dereplication (collapsed into a set of unique sequences) and chimeras were filtered out using UCHIME (Table 4, Edgar et al., 2011). Singletons were either retained or discarded using USEARCH (Edgar, 2010). Datasets were analysed with OTU clustering using UPARSE with a 3% divergence threshold implemented in USEARCH (Edgar, 2013) following Brown et al. (2015). Direct taxonomic assessment of dereplicated sequences without OTU clustering was performed (Table 4) by aligning dereplicated sequences independently against reference sequences in the local reference database using BLAST. OTUs were taxonomically assigned using the same method against the same local reference database.

2.5 Assigning Taxonomy to OTUs and Dereplicated Sequences

The taxonomic classification of each dereplicated sequence and each OTU was based on the best BLAST hit against a local reference database used by Brown et al. (2015), which was defined as a hit with at least 90% identity and an alignment of at least 300 nucleotides with a reference sequence in the assembled local reference database. A threshold of 90% was chosen to accommodate congeneric reference sequences that represented species only identified to the family level, and a minimum alignment of 300 bp was chosen based on the methods used in Brown et al. (2015). We ensured that species-specific reference sequences for species belonging to these genera were included in our local database to allow identification

Table 4 A Summary of the Reads/Sequences Retained at Each Quality Filtering Step for the Mock Communities (1 and 2) Sequenced With the Roche 454 or Illumina MiSeq Platforms

Workflow	**Illumina MiSeq-1**		**Illumina MiSeq-2**		**Roche 454-1**	**Roche 454-2**
Merged (M) or Appended (AP)	**M**	**AP**	**M**	**AP**	**N/A**	**N/A**
Raw reads	543,024	543,024	462,237	462,237	483,986	204,922
Trimmed reads	348,216	538,871	298,597	458,611	151,196	46,764
Dereplicated sequences	212,701	523,656	179,562	438,918	58,941	19,518
Sequences after chimera filtering	199,303	517,828	173,718	436,371	58,717	19,483
Percentage of reads that are chimeras	4.6	1.1	2.2	0.6	0.2	0.1
Sequences after removing singletons	12,959	3171	8973	2956	7321	2411
Percentage singletons	56	97	57	95	34	37
Total filtered reads	332,074	533,003	292,055	456,063	150,942	46,722
Nonsingleton filtered reads	144,980	18,334	126,976	22,636	99,501	29,620

M = merged paired-end reads. AP = appended paired-end reads.

at the species level. We recorded the number of reads that matched a species, as well as the number of OTUs that returned each species (Table S1 in the online version at https://doi.org/10.1016/bs.aecr.2018.06.002). Finally, we compared the results from the two sequencing platforms in terms of numbers of OTUs produced, number of species detected, and assigned precision scores. Species detection was calculated as a percentage (number of species detected by a workflow, divided by number of species in the mock community). OTU precision was calculated as the number of species detected divided by the number of OTUs generated. An OTU precision score of 1.0 indicates that each OTU correctly corresponded 1:1 to the species included in the mock community with no extra or missing OTUs, while a low precision score indicates the presence of many additional OTUs. We examined read depth, OTU diversity estimates, singleton OTUs, and species detection for the Roche data and for each treatment of the MiSeq reads (merged or appended).

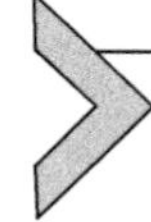

3. RESULTS

3.1 Sequence and Read Depth

The MiSeq runs produced greater numbers of retained reads and dereplicated sequences compared to the Roche runs, despite similar numbers of raw reads (Table 4). After filtering, dereplication, and excluding singletons, MiSeq produced 12,959 and 8973 (communities 1 and 2, respectively) successfully merged sequences compared to 7321 and 2411 filtered Roche sequences. However, fewer reads (3171 and 2956) were retained in the pipeline that involved appended MiSeq reads. Including singletons in the analysis produced a larger proportion of new sequences from MiSeq data, compared to Roche. Furthermore, final tallying of filtered merged reads shows that the MiSeq produced a greater proportion of singletons (56% and 57%) compared to Roche (34% and 37%) (Fig. 1). After assigning taxonomy, MiSeq produced an average read depth per species of 5000 and 15,101 (communities 1 and 2) excluding singletons, and 7532 and 25,013 including singletons. Roche produced an average read depth per species of 2584 and 2506 excluding singletons, and 3408 and 3678 including singletons (Table S1 in the online version at https://doi.org/10.1016/bs.aecr.2018.06.002). Excluding singletons, 144,961 (27% of raw reads, 99% of filtered reads) and 120,815 (26%, 95%) MiSeq reads successfully matched a reference sequence in our local BLAST database, while Roche only produced 98,203 (20%, 99%) and 25,065 (12%, 85%) reads that matched a reference sequence (Table 5). After including singletons, these numbers increased to 331,392 (61%, 99%) and 275,147 (59%, 94%) for MiSeq, and 146,566 (30%, 97%) and 40,456 (20%, 87%) for Roche 454.

3.2 OTU Clustering and the Effect of Singletons

Similar numbers of OTUs were identified from the Roche and MiSeq data when singletons were excluded (merged MiSeq: 53 and 26; Roche: 57 and 18; Table 5). However, the number of OTUs created during the clustering of dereplicated reads was significantly increased by the inclusion of singletons, which produced 18,120 and 16,269 OTUs (communities 1 and 2, respectively) from the merged MiSeq data and 459 and 154 OTUs from the Roche data. When singletons were included, the number of OTUs created from the MiSeq reads increased by three orders of magnitude, compared to an order of magnitude increase for Roche. With both platforms, the exclusion of singletons produced OTU estimates that are comparable to the number of species that were present in the mock community, while the inclusion of singletons created a large discrepancy between number of OTUs and species in the mock community. Upon taxonomical assignment, 83% and 42% (communities 1 and 2) of merged MiSeq OTUs successfully matched a species in the mock community when singletons were excluded, compared to 84% and 78% of Roche OTUs. When singletons were included, 99% and 94% of merged MiSeq OTUs matched a species in the mock community, compared to only 63% and 79% of Roche OTUs. The vast majority of singleton OTUs generated from MiSeq data matched a species in the mock community.

3.3 Species Detection

Only species included in the mock community were represented by a reference sequence in the constructed local BLAST database. The number of species detected in this study refers to the number of species present in the mock communities and detected by a best BLAST hit, excluding parasites, bacteria, or contaminants that were not included in the local database.

The large variability in OTU clustering of the Roche and MiSeq data was not reflected in the species detected mainly because we focused on false negatives (species present but not detected) and not false positives (species absent but detected). Detection scores were calculated as the number of species recovered divided by the number of species in the mock community (Table 5). Eleven of the 56 species in community 1 and 1 of the 14 species

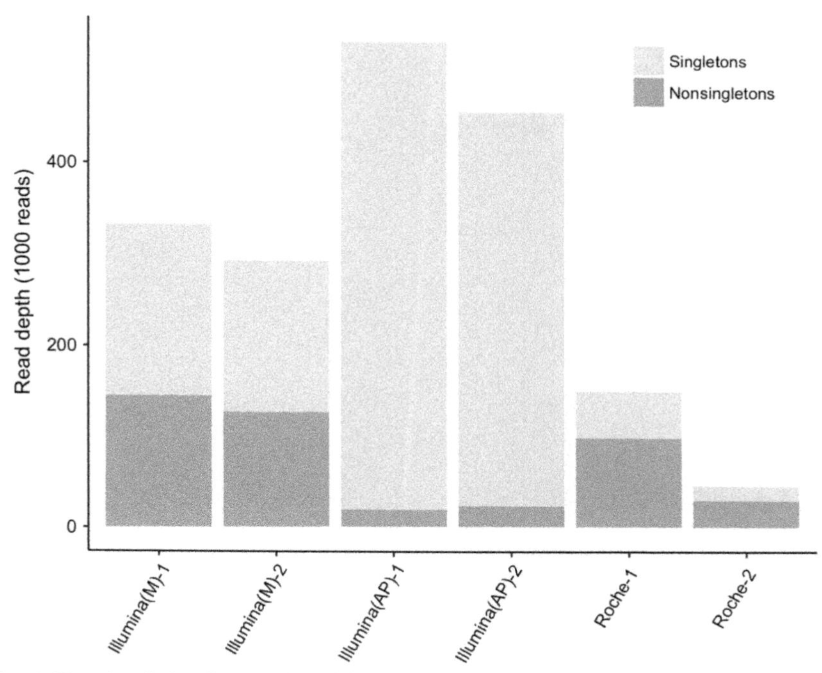

Fig. 1 The depth (in thousands) of filtered reads from Illumina MiSeq and Roche 454 sequencing of mock zooplankton communities. There is increased singleton generation in the MiSeq platform, especially in workflows that append reads (AP), instead of merging (M). The *x-axis labels* refer to the mock communities; simple community 1 with 56 species, each represented by a single individual and the complex community 2 with 14 species, represented by populations of individuals (Table S1 in the online version at https://doi.org/10.1016/bs.aecr.2018.06.002).

in community 2 were never detected using either platform, with or without including singletons. This could have been due to poor DNA extraction, but the undetected species from community 1 were spread across the four subcommunities (Table S1 in the online version at https://doi.org/10.1016/bs.aecr.2018.06.002) suggesting that the tissue for these species may have been of low quality.

When singletons were excluded, fewer species were detected in the merged MiSeq data (29 species; detection score [DS]=0.518 in community 1, and 8 species; DS=0.571 in community 2) compared to Roche (38 species; DS=0.679 in communities 1 and 10 species; DS=0.714 in community 2; Fig. 2A). The inclusion of singletons increased the number of species detected in both communities with the merged MiSeq data (44 species; DS=0.786 for communities 1 and 11 species; DS=0.786 for community 2)

Table 5 Comparison of Results From the Two NGS Platforms (Roche 454 or Illumina MiSeq)

Sequencing Platform	**Illumina MiSeq-1**				**Illumina MiSeq-2**				**Roche 454-1**		**Roche 454-2**	
S or N	**N**	**S**	**N**	**S**	**N**	**S**	**N**	**S**	**N**	**S**	**N**	**S**
Merged or Appended	**M**	**M**	**AP**	**AP**	**M**	**M**	**AP**	**AP**	**N/A**	**N/A**	**N/A**	**N/A**
Reads that matched a BLAST hit	144,961	331,392	18,315	502,435	120,815	275,147	22,240	385,658	98,203	146,566	25,065	40,456
Number of species detected	29	44	30	44	8	11	10	13	38	43	10	11
Number of OTUs	53	18,120	35	258,914	26	16,269	15	220,207	57	459	18	154
Percentage of singleton OTUs	0	85	0	91	0	88	0	93	0	48	0	44
Number of OTUs that matched a species in the mock community	44	17,924	34	243,640	11	15,283	12	161,706	48	287	14	122
Percentage of raw reads that match a reference sequence	27	61	3.4	93	26	60	4.8	83	20	30	12	20
Percentage of filtered reads that match a reference sequence	99	99	99	94	95	94	98	85	99	97	85	87
Percentage of OTUs that matched a species in the mock community	83	99	97	94	42	94	80	73	84	63	78	79
Detection score	0.518	0.786	0.536	0.786	0.571	0.786	0.714	0.929	0.679	0.769	0.714	0.786
OTU precision	0.547	0.003	0.857	0	0.308	0.001	0.667	0	0.667	0.094	0.556	0.071

Singletons were either included in (S) or excluded from (N) the analyses. Illumina MiSeq reads were either merged traditionally (M) or appended (AP) to form contigs. Precision score refers to the number of species recovered divided by the number of operational taxonomic units (OTU) generated. Detection score was calculated as the number of species recovered divided by the number of species in the mock community. Precision was calculated as the number of species recovered divided by the number of OTUs generated.

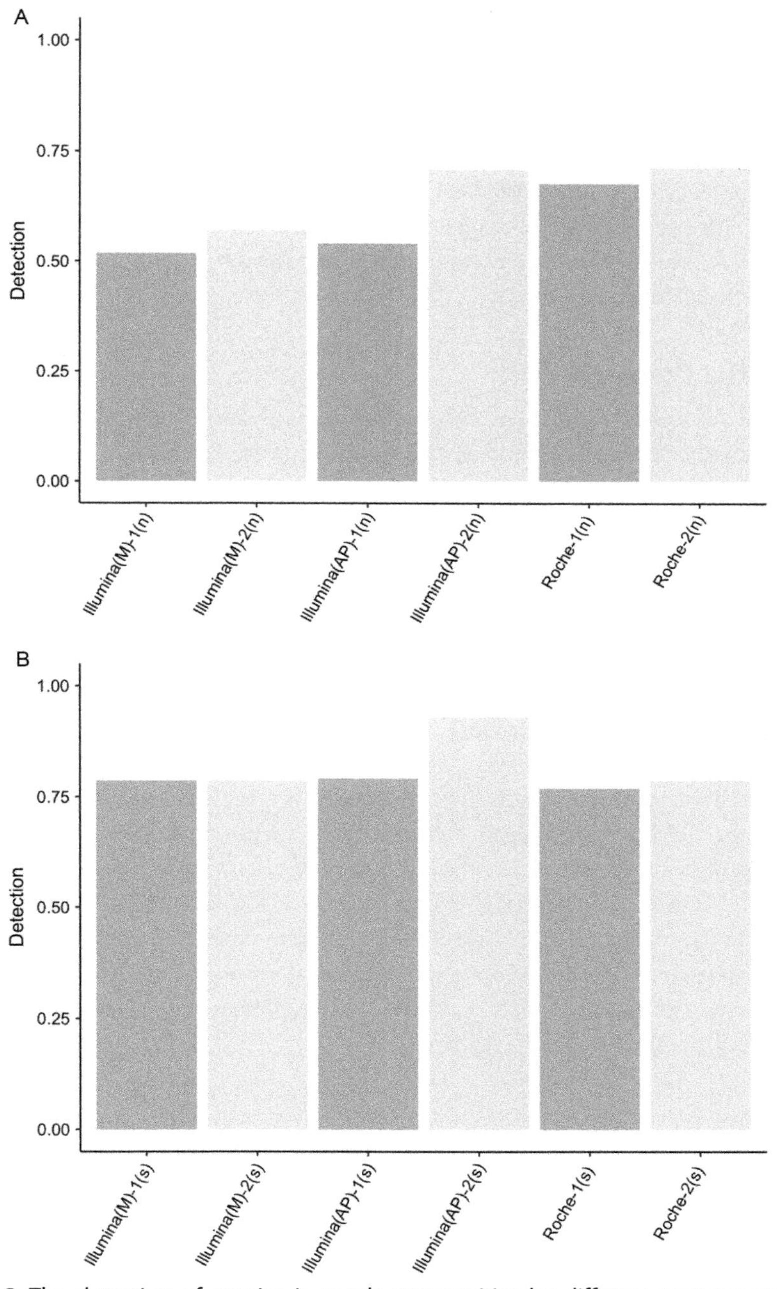

Fig. 2 The detection of species in mock communities by different next-generation sequencing platforms and bioinformatic workflows. Values were calculated by dividing the number of species recovered by the number of species in the mock community. (A) Workflows that exclude singletons. (B) Workflows that include singletons. Species detection varied with choice of platform and workflow between the simple community 1 and the complex community 2. The *x-axis labels* refer to the choice of platform, the workflow (M = merged MiSeq reads, AP = appended MiSeq reads), the mock community (1 in *dark grey* or 2 in *light grey*) and the inclusion(s) or exclusion(n) of singletons.

and Roche (43 species; DS = 0.769 for communities 1 and 11 species; DS = 0.786 for community 2; Fig. 2B). We recovered one species (*Hyalella* clade 8) with the MiSeq data that was never detected using the Roche data, and one species (*Hyperoche mediterranea*) that was only recovered with Roche data. With the inclusion of singletons, we recovered 15 and 3 (communities 1 and 2) additional species with merged MiSeq data compared to 5 and 1 additional species with Roche data.

3.4 OTU Precision

The increased ability to detect species with singleton inclusion was accompanied by a large decrease in precision (Fig. 3, Table 5), which was calculated as the number of species recovered divided by the total number of OTUs generated (including OTUs that did not match a species in the mock community). The precision value represents the inverse of the average number of OTUs created per species, where a smaller precision score means more OTUs per species, or more OTUs that do not match a species in the mock community. Merged MiSeq data produced consistently lower OTU precision scores (0.547 and 0.308, communities 1 and 2 without singletons; 0.002 and 0.001 with singletons) compared to Roche data (0.667 and 0.556 without singletons; 0.094 and 0.071 with singletons). After excluding singletons from the analysis, Roche produced higher precision scores than the merged MiSeq data, but lower scores than the appended MiSeq data (Fig. 3). The appended MiSeq data including singletons produced the lowest precision scores. Low precision scores are due to multiple OTUs being generated for a single species (Table S1 in the online version at https://doi.org/10.1016/bs.aecr.2018.06.002) and are prevalent even when a stringent filtering/clustering approach (exclusion of singletons and a 3% divergence threshold for OTU clustering) is used.

3.5 Impact of Merging and Appending MiSeq Reads on Species Detection and OTU Estimates

Merging MiSeq reads requires overlapping sequences, which will generally not occur with long amplicon sequences due to read length limitations imposed by the sequencing platform. Even so, some species with long 18S-V4 regions were detected with the merged data (Table S1 in the online version at https://doi.org/10.1016/bs.aecr.2018.06.002). However, MiSeq read pairs can be appended to one another to create potentially discontinuous, but informative sequences. The appending technique joins paired

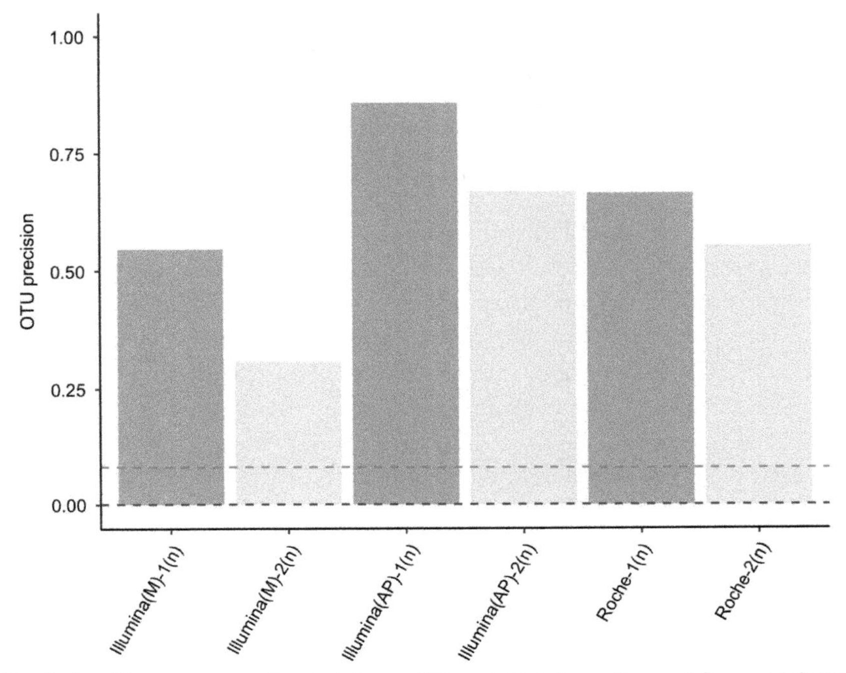

Fig. 3 Precision scores estimated from different bioinformatic workflows analysing next-generation sequence data from two mock zooplankton communities. Singletons were excluded from the analysis. Precision is defined as the number of species recovered divided by the number of operational taxonomic units generated. The *x-axis labels* refer to the choice of platform, the bioinformatic workflow (M = merged MiSeq reads; AP = appended MiSeq reads) and the mock community (simple community 1 in *dark grey* and complex community 2 in *light grey*). The *red and blue lines* indicate the average precision including singletons for Roche and MiSeq workflows, respectively.

reads together without overlap to generate a contig. In the appending process, reads that would overlap and merge contain a repeated sequence in the middle where the overlap would have occurred, and the length of the overlap can vary. In addition, paired-end reads that do not overlap may not always end at the same nucleotide position. As a result, there is substantial variation among appended sequences and they do not dereplicate into as few sequences as merged sequences. For this reason, appending pipelines generate a much higher number of singletons (Table 4).

There was little difference in the number of species detected with merged and appended reads, but more species were detected when singletons were included. When singletons were excluded, 30 species (DS = 0.536) were detected in community 1 with appended reads compared to 29 species

(DS = 0.518) with merged reads (Table 5). Ten species (DS = 0.714) were detected in community 2 with appended reads compared to 8 species (DS = 0.571) with merged reads. When singletons were included, 44 species (DS = 0.786) were detected in community 1 with both merged and appended reads. Thirteen species (DS = 0.929) were detected in community 2 with appended reads compared to 11 species (DS = 0.786) with merged reads.

Clustering of appended reads into OTUs resulted in the generation of very high numbers of OTUs when singletons were included in the analysis, even more so than merged reads. Whereas appended reads produced 35 and 15 OTUs (communities 1 and 2) when singletons were excluded, 258,914 and 220,207 OTUs were produced when singletons were included (Table 5). Owing to the variation generated by the appending process, very low precision scores (1.69×10^{-4} and 5.90×10^{-5} for communities 1 and 2) were obtained when singletons were included compared to exclusion of singletons (0.857 and 0.667).

4. DISCUSSION

The Illumina MiSeq and the Roche 454 NGS platforms were used to sequence two mock communities, one simple community containing 56 species represented by a single individual per species and a more complex community containing 14 species represented by individuals or populations. The MiSeq runs produced more filtered reads compared to the Roche runs, and a larger percentage of singleton reads, from a similar amount of raw data. When singletons were excluded from the analysis, the MiSeq and Roche datasets generated similar numbers of OTUs. Moreover, the two platforms generated comparable species detection and OTU precision for both mock communities, although the species detection rate for community 1 was slightly higher with the Roche data. When singletons were included, one more species in community 1 and two more species in community 2 were detected with the appended MiSeq reads compared to the Roche data. However, when singletons were included, Roche produced much higher number of OTUs than species included in the communities (459 for community 1 and 154 for community 2). Even higher numbers of OTUs were generated with the MiSeq data when singletons were included resulting in very low precision scores. Even so, the majority of OTUs matched a species in the mock community (99% and 94% for merged and appended reads in community 1, and 94% and 73% in community 2).

With one exception, these values are considerably higher than those obtained for Roche OTUs (63% and 79% for communities 1 and 2).

Merging vs appending MiSeq paired reads had a small impact on species detection but a large impact on OTU precision as the precision scores for appended reads were orders of magnitude lower than those for merged reads. This is a consequence of the variation generated by the appending process in which forward and reverse reads are joined together even if they overlap (which is required for merging), and regardless of their length, as long as it meets the length threshold. This results in highly variable contigs due to the loss of quality control that occurs during merging because reads that do not meet an overlap threshold (a combination of length of overlap and sequence similarity of overlapped region) are discarded in merged workflows. Other bioinformatics studies have confirmed that merging Illumina reads helps decrease the numbers of spurious OTUs, as it improves predictions of the overlapping sequence when forward and reverse reads are aligned (Edgar and Flyvbjerg, 2015). While merging reads did substantially decrease the number of OTUs in our study, it also slightly decreased species detection, compared to appending reads.

4.1 Read Depth and Singletons

Despite a greater read depth, the MiSeq data did not provide an increase in species detection ability compared to the Roche. The inclusion of singletons, which can represent rare authentic sequences or sequencing artefacts, did aid species detection with both platforms. However, it also caused a large decrease in precision. Singletons provided detection of 15 and 3 (communities 1 and 2) additional species with merged MiSeq data, and 5 and 1 additional species with Roche 454 data that were not detected in workflows excluding singletons (Table 5). Most singletons detected a species already recovered when included in taxonomic analysis. The relationship between inclusion of singletons, an increase in number of species detected, and a decrease in precision is conserved across sequencing platforms, although it is more pronounced with the MiSeq data. This suggests that the MiSeq platform is more susceptible to singleton generation (it generated a higher percentage of singletons), especially when appending reads. Most studies do exclude singleton sequences (Abad et al., 2016; Chain et al., 2016; although see Gibson et al., 2015). However, studies focused on the detection of rare species such as aquatic invaders often include singletons in their analysis.

4.2 OTU Clustering

The number of OTUs generated was extremely variable between sequencing platforms and between the workflows involving the inclusion and exclusion of singletons. However, the number remained relatively consistent within each platform in terms of order of magnitude, and for results from communities 1 and 2. A greater number of OTUs was consistently produced from the MiSeq data compared to Roche, which resulted in lower precision scores. Many of the additional MiSeq OTUs matched the same species in the mock community. For example, when reads from community 1 (excluding singletons) were clustered into OTUs and taxonomically assigned, four OTUs matching *Artemia* spp. were identified in the MiSeq data, while only one was identified in the Roche data.

The generation of more than one OTU per species is due to increased sequence variation produced by either genuine intragenomic variation (genetic variation within the genome of a single individual/species) or sequencing artefacts. OTU inflation needs to be taken into consideration when designing a metabarcoding workflow and when generating conclusions. This study found that with singletons included, MiSeq produced a much higher number of OTUs compared to Roche. Flynn et al. (2015) reported very high OTU estimates with Roche data and suggested that terminal gaps are partially responsible. These gaps occur when the single read falls short of the reverse primer, which can contribute to divergence during OTU clustering. Analysis of different gap treatments using clustering algorithms that excluded and included terminal gaps supported this hypothesis (Flynn et al., 2015). However, merged and appended MiSeq reads do not contain terminal gaps. With paired read sequencing, the target amplicon is sequenced from both the 5′ and 3′ ends and the reads are merged or appended to produce a full contig. Therefore, it is likely that internal gaps generated during the merging and appending process are the source of OTU overestimation with MiSeq data. Studies that rely on OTU-based diversity estimates should consider excluding singletons to avoid gross overestimation.

When singletons were excluded from analysis, both platforms produced similar OTU estimates. However, some species were still represented by more than one OTU. For example, *Mytilus edulis* in community 1 was represented by five OTUs in the MiSeq data, but was only represented by one OTU in the Roche data. Conversely, *Leptodora kindtii* was represented by four OTUs in the Roche data for community 1, but only represented by one OTU in the MiSeq data. These results suggest that the sequencing platform seems to introduce some taxonomical bias with respect to inflating OTU diversity. One possible explanation for this phenomenon is the genetic variation between different copies of the marker sequence within a single genome. The 18S-V4 region is known for its hypervariability in terms of both nucleotide sequence and length (Wuyts et al., 2000, 2001). Moreover, the occurrence of variation at both the intraspecific and intra-individual levels has been well documented (Crease and Taylor, 1998; Hancock and Vogler, 2000). It has been suggested that slippage replication is an important mechanism causing length expansion in hypervariable regions (Crease and Taylor, 1998; Hancock and Vogler, 2000), and the maintenance of stem-loop structures in these regions may be a function of the mutational process itself, regardless of functional constraint to maintain function (Hancock and Vogler, 2000). Nevertheless, it is clear that at least some of the variation in the V4 sequences is of biological origin and must be taken into account when estimating OTU diversity.

4.3 Experimental Design

The construction of a mock community allows creation of a local reference BLAST database, which removes the need for OTU clustering. OTU clustering can reduce the time it takes to generate results when using large databases like NCBI or SILVA, where searching each dereplicated sequence against millions of reference sequences requires substantial computing time. In this study, the filtered dereplicated reads were aligned individually, using BLAST, against a local reference database and tallied after returning a hit. While the use of a local reference database is efficient and effective when studying mock communities of known composition, it is not applicable for metabarcoding projects, for example when there is no a priori knowledge of community composition.

The direct comparability of results from the MiSeq and Roche platforms depends on the similarity of the workflows used to analyse the sequence reads. Comparability of results between platforms is of high importance when considering the use of novel technologies in long-term studies. Complete contigs can be generated from MiSeq paired-end reads that are usually merged together when the amplicon is short enough that paired reads are expected to overlap. On the other hand, Roche data are always single ended. Although the use of appended MiSeq reads detected a slightly larger number of species compared to merged MiSeq reads, there are several reasons why these processing steps may not be suitable for some biodiversity studies.

When singletons were included, using appended reads overestimated the number of OTUs by over three orders of magnitude compared to the MiSeq and the Roche 454 estimates. Thus, the confidence of researchers in the species detection results generated from appended Illumina reads can also be questioned, especially if a local reference database cannot be used as in biodiversity surveys.

Excluding singletons from the workflow improves precision of appended reads to the point where appended workflows produce better precision scores than merged ones. However, these workflows sacrifice a considerable amount of read depth because there is considerable variation around the point where the paired reads are joined due to length variation among reads, and the fact that overlapping reads are joined instead of overlapped. This means that many appended sequences are unique and fail to cluster into OTUs. Therefore, the vast majority of appended reads exit the pipeline as singletons. For example, MiSeq data for community 1 contained 543,024 raw paired end reads. After appending and filtering, those raw reads produced 514,657 singleton sequences and only 3171 nonsingleton sequences compared to 12,959 non-singleton merged sequences. Thus, merging allowed the retention of four times as many nonsingleton sequences compared to appending. While use of appending techniques as an alternative to merging may provide slightly higher species detection, researchers must compensate for the increased read diversity and decreased OTU precision that accompanies appended MiSeq workflows involving singletons, as well as the decreased sequence retention in workflows that exclude singletons.

Although the 18S-V4 region is commonly used as a phylogenetic marker for eukaryotes, there are several shortcomings of this barcode as a marker for metabarcoding that should be recognized. Ribosomal sequences are present in multiple copies, providing the opportunity for variation within the genome of a single individual (Bik et al., 2012). Decelle et al. (2014) assessed the intragenomic variability of the V4 region between different nuclei in two species of radiolarian and detected up to five OTUs in a single individual. However, the authors concluded that the analytical procedure used to obtain the results had a larger impact than the contribution of intragenomic variation to the presence of multiple OTUs per species. Even so, variation in length between gene copies is particularly common in expansion segments such as the V4 region (James et al., 2009), and such length variation can introduce bias against species with long amplicons during the MiSeq merging process. Indeed, the V4 region is longer than 570 bp in 12 (80%) of the 15 species in community 1 that were not detected with merged MiSeq data when singletons were excluded, but were detected with singleton inclusion. Thus, use of the 18S-V4 region as a genetic barcode should be taken into consideration when designing future studies because alternatives, such as the mitochondrial CO1 gene, exist. The transition from Roche to Illumina as the accepted NGS platform for metabarcoding adds importance to this idea by providing increased opportunity for overestimation of biodiversity.

4.4 Implications for Biomonitoring

The application of metabarcoding procedures to biodiversity, community ecology, and invasive species projects will depend on many sampling decisions, including study design, sample capture, and molecular and bioinformatic methods. Such decisions will be made in conjunction with the aims of each individual study. Increasingly, molecular methods will be integrated into ongoing biomonitoring campaigns, or may be used to start new biodiversity surveys (Leese et al., 2016). Methodological consistency is key when designing studies which range over temporal and spatial scales. Here, we have shown that changes in sequencing platform will produce different outcomes for species detection and OTU diversity estimates. It is likely that sequencing platforms and molecular methods will continue to evolve, probably at a greater pace than traditional methods, given their recent and rapid development. Therefore, it is likely that over a decades-long monitoring campaign, several sequencing platforms may be used.

Some studies aim to generate diversity estimates only, and prefer to assign OTUs or sequences to higher taxonomic ranks such as Order or Family. This may be because the primer pair in use cannot discriminate between taxa at lower taxonomic ranks, or because species-level reference sequences are not available, for example when work is conducted on tropical invertebrates (Salinas-Ramos et al., 2015). OTUs without species-level assignment can still be included in estimates of alpha and beta diversity, patterns of distribution and seasonality, and simple ecological analyses such as niche overlap (Clare et al., 2016). In this study, we demonstrated that the two sequencing platforms, Roche 454 and Illumina MiSeq produced comparable OTU values when singletons where excluded. Conversely, when singletons were included, vastly inflated OTU numbers were generated from the Illumina MiSeq data; sometimes several orders of magnitude greater than estimates based on Roche 454 data, as well as the true number of species included in the mock community. We therefore do not recommend comparisons of data produced by these platforms when singletons are included.

Many researchers choose to remove singletons from analysis (e.g., Port et al., 2016), but we detected some species only when including them. It is often emphasized that the inclusion of singletons could facilitate the detection of very rare species or invasion fronts (Brown et al., 2015; Jousset et al., 2017; Leray and Knowlton, 2017). Even so, Scott et al. (2018) recommend removal of singletons to reduce false-positive errors at the cost of slightly reduced sensitivity. The authors found that retaining singletons can reduce false negatives (species not detected) only when clustering is used prior to species assignment.

Many biomonitoring studies are focused on species detection as a starting point before exploring community richness and diversity. There may also be particular species of interest, for which detection as well as temporal and spatial distribution is important. While singleton inclusion might initially seem advantageous, researchers may not be able to use a local reference database to assign species identity in a biomonitoring study, but instead BLAST their results against a universal database such as BOLD or NCBI. It is likely that many of these additional OTUs might match to closely related species, or even OTUs of different taxa such as parasites or fungi, that while not intentionally included in the mock community, are inevitably present, and will likely be amplified by any barcoding study that uses universal primers.

5. CONCLUSIONS

Overall, the two NGS platforms used in this study generated comparable species detection and OTU precision scores. However, we detected unexpectedly high variation when using appended MiSeq reads, and when including singletons in the analysis. While the inclusion of singletons can aid in species detection, reducing false negatives, as demonstrated by both platforms, the generation of spurious OTU clusters that accompanies their inclusion can increase false positives and needs to be taken into account. The results from this study support the trends identified by Flynn et al. (2015) in their analysis of Roche data, as well as the impact of singletons identified in other studies (Clare et al., 2016). Specific analysis of the trade-off between accurate biodiversity estimation (singleton exclusion) and increased detection ability (singleton inclusion) should be considered when performing metabarcoding analyses. Furthermore, while the use of NGS technologies for metabarcoding shows great promise for the study of community composition, species detection, and biomonitoring; this study illustrates the need for tailored and cautious analytical procedures. While these results illustrate how the transition from the Roche to an Illumina platform is accompanied by increases in technological ability, and greater genomic resolution, Illumina sequencing is not without limitations since the short reads generated require additional bioinformatics manipulations that impact species detection. Future sequencing technologies and analytical tools may enable researchers to overcome the shortcomings of Illumina sequencing, and make further progress in the application and reliability of NGS platforms to the field of metabarcoding.

ACKNOWLEDGEMENTS

The authors would like to acknowledge Jullien M. Flynn and Emily A. Brown for their advice with analysis and for preparing libraries for sequencing. This study was supported by NSERC Discovery grants to T.J.C. and M.E.C.

REFERENCES

Abad, D., et al., 2016. Is metabarcoding suitable for estuarine plankton monitoring? A comparative study with microscopy. Mar. Biol. 163 (7), 1–13.

Alberdi, A., Aizpurua, O., Bohmann, K., 2017. Scrutinizing key steps for reliable metabarcoding of environmental samples. Methods Ecol. Evol. 9 (1), 134–147.

Balzer, S., et al., 2010. Characteristics of 454 pyrosequencing data—enabling realistic simulation with flowsim. Bioinformatics 26, 420–425.

Bentley, D.R., 2008. Accurate whole human genome sequencing using reversible terminator chemistry. Nature 456, 53–59.

Bik, H.M., et al., 2012. Sequencing our way towards understanding global eukaryotic biodiversity. Trends Ecol. Evol. 27 (4), 233–243.

Bohan, D.A., et al., 2017. Next-generation global biomonitoring: large-scale, automated reconstruction of ecological networks. Trends Ecol. Evol. 32 (7), 477–487.

Bolger, A.M., Lohse, M., Usadel, B., 2014. Trimmomatic: a flexible trimmer for Illumina sequence data. Bioinformatics 30 (15), 2114–2120.

Brown, E.A., et al., 2015. Divergence thresholds and divergent biodiversity estimates: can metabarcoding reliably describe zooplankton communities? Ecol. Evol. 5 (11), 2234–2251.

Chain, F.J.J., et al., 2016. Metabarcoding reveals strong spatial structure and temporal turnover of zooplankton communities among marine and freshwater ports. Divers. Distrib. 22 (5), 493–504.

Clare, E.L., et al., 2016. The effects of parameter choice on defining molecular operational taxonomic units and resulting ecological analyses of metabarcoding data. Genome 59 (11), 981–990.

Clooney, A.G., et al., 2016. Comparing apples and oranges?: next generation sequencing and its impact on microbiome analysis. PLoS One 11 (2), e0148028. Available at: https://doi.org/10.1371/journal.pone.0148028.

Coissac, E., Riaz, T., Puillandre, N., 2012. Bioinformatic challenges for DNA metabarcoding of plants and animals. Mol. Ecol. 21, 1834–1847.

Crease, T.J., Taylor, D.J., 1998. The origin and evolution of variable-region helices in V4 and V7 of the small-subunit ribosomal RNA of branchiopod crustaceans. Mol. Biol. Evol. 15 (11), 1430–1446.
Cristescu, M.E., 2014. From barcoding single individuals to metabarcoding biological communities: towards an integrative approach to the study of global biodiversity. Trends Ecol. Evol. 29 (10), 566–571. Available at: http://linkinghub.elsevier.com/retrieve/pii/S016953471400175X.
Darling, J.A., Mahon, A.R., 2011. From molecules to management: adopting DNA-based methods for monitoring biological invasions in aquatic environments. Environ. Res. 111, 978–988.
Decelle, J., et al., 2014. Intracellular diversity of the V4 and V9 regions of the 18S rRNA in marine protists (radiolarians) assessed by high-throughput sequencing. PLoS One 9 (8), e104297.
Deiner, K., et al., 2017. Environmental DNA metabarcoding: transforming how we survey animal and plant communities. Mol. Ecol. 26 (21), 5872–5895.
Divoll, T.J., et al., 2018. Disparities in second-generation DNA metabarcoding results exposed with accessible and repeatable workflows. Mol. Ecol. Resour. 18 (3), 1–12. Available at: http://doi.wiley.com/10.1111/1755-0998.12770.
Edgar, R.C., 2010. Search and clustering orders of magnitude faster than BLAST. Bioinformatics 26 (19), 2460–2461.
Edgar, R.C., 2013. UPARSE: highly accurate OTU sequences from microbial amplicon reads. Nat. Methods 10 (10), 996–998.
Edgar, R.C., Flyvbjerg, H., 2015. Error filtering, pair assembly and error correction for next-generation sequencing reads. Bioinformatics 31 (21), 3476–3482.
Edgar, R.C., et al., 2011. UCHIME improves sensitivity and speed of chimera detection. Bioinformatics 27 (16), 2194–2200.
Erlich, Y., et al., 2008. Alta-cyclic: a self-optimizing base caller for next-generation sequencing. Nat. Methods 5 (8), 679–682.
Evans, D.M., et al., 2016. Merging DNA metabarcoding and ecological network analysis to understand and build resilient terrestrial ecosystems. Funct. Ecol. 30 (12), 1904–1916.
Evans, N.T., et al., 2017. Fish community assessment with eDNA metabarcoding: effects of sampling design and bioinformatic filtering. Can. J. Fish. Aquat. Sci. 74 (9), 1362–1374.
Floyd, R., et al., 2002. Molecular barcodes for soil nematode identification. Mol. Ecol. 11 (4), 839–850.
Flynn, J.M., et al., 2015. Toward accurate molecular identification of species in complex environmental samples: testing the performance of sequence filtering and clustering methods. Ecol. Evol. 5 (11), 2252–2266.
Geml, J., et al., 2014. The contribution of DNA metabarcoding to fungal conservation: diversity assessment, habitat partitioning and mapping red-listed fungi in protected coastal *Salix repens* communities in the Netherlands. PLoS One 9 (6), e99852.
Gibson, J.F., et al., 2015. Large-scale biomonitoring of remote and threatened ecosystems via high-throughput sequencing. PLoS One 10 (10), e0138432. Available at: https://doi.org/10.1371/journal.pone.0138432.
Glenn, T.C., 2011. Field guide to next-generation DNA sequencers. Mol. Ecol. Resour. 11 (5), 759–769.
Goodwin, S., McPherson, J.D., McCombie, W.R., 2016. Coming of age: ten years of next-generation sequencing technologies. Nat. Rev. Genet. 17 (6), 333–351.
Hancock, J.M., Vogler, A.P., 2000. How slippage-derived sequences are incorporated into rRNA variable-region secondary structure: implications for phylogeny reconstruction. Mol. Phylogenet. Evol. 14 (3), 366–374.
Hänfling, B., et al., 2016. Environmental DNA metabarcoding of lake fish communities reflects long-term data from established survey methods. Mol. Ecol. 25, 3101–3119.

Hatzenbuhler, C., et al., 2017. Sensitivity and accuracy of high-throughput metabarcoding methods for early detection of invasive fish species. Sci. Rep. 7, 46393. Available at: https://doi.org/10.1038/srep46393.

Heather, J.M., Chain, B., 2016. The sequence of sequencers: the history of sequencing DNA. Genomics 107, 1–8.

James, S.A., et al., 2009. Repetitive sequence variation and dynamics in the ribosomal DNA array of *Saccharomyces cerevisiae* as revealed by whole-genome resequencing. Genome Res. 19, 626–635.

Ji, Y., et al., 2013. Reliable, verifiable and efficient monitoring of biodiversity via metabarcoding. Ecol. Lett. 16 (10), 1245–1257. Available at: http://doi.wiley.com/10.1111/ele.12162.

Jousset, A., et al., 2017. Where less may be more: how the rare biosphere pulls ecosystems strings. ISME J. 11 (4), 853–862. Available at: http://www.nature.com/doifinder/10.1038/ismej.2016.174.

Kelly, R.P., Port, J.A., Yamahara, K.M., Martone, R.G., Lowell, N., et al., 2014. Harnessing DNA to improve environmental management. Science 344, 1455–1456.

Lallias, D., et al., 2015. Environmental metabarcoding reveals heterogeneous drivers of microbial eukaryote diversity in contrasting estuarine ecosystems. ISME J. 9 (5), 1208–1221. Available at: https://doi.org/10.1038/ismej.2014.213.

Leamon, J.H., et al., 2003. A massively parallel PicoTiterPlate based platform for discrete picoliter-scale polymerase chain. Electrophoresis 24, 3769–3777.

Leese, F., et al., 2016. DNAqua-Net: developing new genetic tools for bioassessment and monitoring of aquatic ecosystems in Europe. Res. Ideas Outcomes 2, e11321. https://doi.org/10.3897/rio.2.e11321.

Leigh, M.B., Taylor, L., Neufeld, J.D., 2015. Clone libraries of ribosomal RNA gene sequences for characterization of microbial communities. In: McGenity, T.J., Timmis, K.N., Nogales, B. (Eds.), Hydrocarbon and Lipid Microbiology Protocols. Springer, pp. 127–154.

Leray, M., Knowlton, N., 2017. Random sampling causes the low reproducibility of rare eukaryotic OTUs in Illumina COI metabarcoding. PeerJ 5, e3006. Available at: https://peerj.com/articles/3006.

Li, J.Z., et al., 2014. Comparison of Illumina and 454 deep sequencing in participants failing raltegravir-based antiretroviral therapy. PLoS One 9 (3), e90485.

Lim, N.K.M., et al., 2016. Next-generation freshwater bioassessment: eDNA metabarcoding with a conserved metazoan primer reveals species-rich and communities. R. Soc. Open Sci. 3, 160635.

Littlefair, J.E., Clare, E.L., 2016. Barcoding the food chain: sanger to high-throughput sequencing. Genome 59 (11), 946–958.

Luo, C., et al., 2012. Direct comparisons of Illumina vs. Roche 454 sequencing technologies on the same microbial community DNA sample. PLoS One 7 (2), e30087.

Magoč, T., Salzberg, S.L., 2011. FLASH: fast length adjustment of short reads to improve genome assemblies. Bioinformatics 27 (21), 2957–2963.

Mahé, F., et al., 2015. Comparing high-throughput platforms for sequencing the V4 region of SSU-rDNA in environmental microbial eukaryotic diversity surveys. J. Eukaryot. Microbiol. 62, 338–345.

Moonen, A.C., Bàrberi, P., 2008. Functional biodiversity: an agroecosystem approach. Agric. Ecosyst. Environ. 127, 7–21.

Pawluczyk, M., et al., 2015. Quantitative evaluation of bias in PCR amplification and next-generation sequencing derived from metabarcoding samples. Anal. Bioanal. Chem. 407 (7), 1841–1848.

Pompanon, F., et al., 2012. Who is eating what: diet assessment using next generation sequencing. Mol. Ecol. 21 (8), 1931–1950.

Port, J.A., et al., 2016. Assessing vertebrate biodiversity in a kelp forest ecosystem using environmental DNA. Mol. Ecol. 25 (2), 527–541.
Quast, C., Pruesse, E., Yilmaz, P., Gerken, J., Schweer, T., Yarza, P., Peplies, J., Glöckner, F.O., 2013. The SILVA ribosomal RNA gene database project: improved data processing and web-based tools. Nucleic Acids Res. 41, D590–D596.
Quince, C., Lanzen, A., Curtis, T.P., Davenport, R.J., Hall, N., Head, I.M., Read, L.F., Sloan, W.T., 2009. Accurate determination of microbial diversity from 454 pyrosequencing data. Nat. Methods 6 (9), 639–641.
Rothberg, J.M., et al., 2011. An integrated semiconductor device enabling non-optical genome sequencing. Nature 475, 348–352.
Salinas-Ramos, V.B., et al., 2015. Dietary overlap and seasonality in three species of mormoopid bats from a tropical dry forest. Mol. Ecol. 24 (20), 5296–5307.
Salipante, S.J., et al., 2014. Performance comparison of Illumina and ion torrent next-generation sequencing platforms for 16S rRNA-based bacterial community profiling. Appl. Environ. Microbiol. 80 (24), 7583–7591.
Schmidt, P.-A., et al., 2013. Illumina metabarcoding of a soil fungal community. Soil Biol. Biochem. 65, 128–132. Available at: https://doi.org/10.1016/j.soilbio.2013.05.014.
Scott, R., Zhan, A., Brown, E.A., Chain, F.J.J., Cristescu, M.E., Gras, R., MacIsaac, H.J., 2018. Optimization and performance testing of a sequence processing pipeline applied to detection of nonindigenous species. Evol. Appl. 11 (6), 891–905.
Taberlet, P.P., et al., 2012. Towards next-generation biodiversity assessment using DNA metabarcoding. Mol. Ecol. 21 (8), 2045–2050.
Tremblay, J., et al., 2015. Primer and platform effects on 16S rRNA tag sequencing. Front. Microbiol. 6, 771.
Wuyts, J., et al., 2000. Comparative analysis of more than 3000 sequences reveals the existence of two pseudoknots in area V4 of eukaryotic small subunit ribosomal RNA. Nucleic Acids Res. 28 (23), 4698–4708.
Wuyts, J., Van De Peer, Y., De Wachter, R., 2001. Distribution of substitution rates and location of insertion sites in the tertiary structure of ribosomal RNA. Nucleic Acids Res. 29 (24), 5017–5028.
Zhan, A., et al., 2013. High sensitivity of 454 pyrosequencing for detection of rare species in aquatic communities. Methods Ecol. Evol. 4 (6), 558–565.
Zhou, X., et al., 2013. Ultra-deep sequencing enables high-fidelity recovery of biodiversity for bulk arthropod samples without PCR amplification. GigaScience 2 (4), 1–12.

CHAPTER TWO

Linking DNA Metabarcoding and Text Mining to Create Network-Based Biomonitoring Tools: A Case Study on Boreal Wetland Macroinvertebrate Communities

Zacchaeus G. Compson[*,†,1], Wendy A. Monk[†,‡], Colin J. Curry[§], Dominique Gravel[¶], Alex Bush[*,†], Christopher J.O. Baker[||,#], Mohammad Sadnan Al Manir[||], Alexandre Riazanov[||], Mehrdad Hajibabaei[**], Shadi Shokralla[**], Joel F. Gibson[††], Sonja Stefani[‡‡], Michael T.G. Wright[**], Donald J. Baird[*,†]

[*]Canadian Rivers Institute, Department of Biology, University of New Brunswick, Fredericton, NB, Canada
[†]Environment and Climate Change Canada @ Canadian Rivers Institute, Department of Biology, University of New Brunswick, Fredericton, NB, Canada
[‡]Faculty of Forestry and Environmental Management, University of New Brunswick, Fredericton, NB, Canada
[§]Wolastoqey Nation in New Brunswick, Fredericton, NB, Canada
[¶]Département de Biologie, Université de Sherbrooke, Sherbrooke, QC, Canada
[||]Department of Computer Science, University of New Brunswick, Saint John, NB, Canada
[#]IPSNP Computing Inc., Saint John, NB, Canada
[**]Biodiversity Institute of Ontario, Department of Integrative Biology, University of Guelph, Guelph, ON, Canada
[††]Entomology, Royal BC Museum, Victoria, BC, Canada
[‡‡]Dresden University of Technology, Dresden, Germany
[1]Corresponding author: e-mail address: zacchaeus.greg.compson@gmail.com

Contents

1. Introduction 35
   1.1 Ecological Networks as Biomonitoring Tools 35
   1.2 New Tools to Rapidly Assess Biodiversity and Annotate It With Ecological Information 37
   1.3 Freshwater Biodiversity Hotspots as Candidates for Exploring Novel Approaches for Biodiversity Assessment 39
   1.4 Exploring the Generation of DNA Ecological Networks for Aquatic Biomonitoring 39
2. Building Heuristic Food Webs for Wetland Biomonitoring: A Case Study 40
   2.1 DNA Sample Collection, Metabarcoding, and Bioinformatics Pipeline 40
   2.2 Development of Text-Mining Pipeline for Traits of Freshwater Organisms 41

Advances in Ecological Research, Volume 59
ISSN 0065-2504
https://doi.org/10.1016/bs.aecr.2018.09.001

2.3 Evaluation, and Iterative Refinements to Text-Mining Pipeline for Body Size 42
2.4 Rule-Based Procedure for Retrieving Trait Information for Heuristic Food Web Construction 46
2.5 Comparing Food Web Properties in Two River Deltas: A Spatiotemporal Comparison 48
3. The Future of Text Mining: New Tools for Rapid Expansion of Food Web Databases and Challenges to Their Widespread Adoption for Searching Ecological Literature 53
4. Towards an Open-Source Pipeline for the Rapid Construction and Assessment of Trait-Based Food Webs for Biomonitoring: The Promise, Challenges, and Next Steps 55
4.1 Advancing Network Approaches for Biomonitoring 55
4.2 The Need for Vastly Expanded Trait Databases 57
4.3 Assessing Uncertainty Within Ecological Networks and Heuristic Food Webs 58
4.4 Assessing the Effectiveness of Heuristic Food Webs at Detecting Environmental Change 59
5. Perspectives: The Future of Biomonitoring 61
Acknowledgements 62
Appendix 62
Glossary 65
References 66
Further Reading 74

Abstract

Ecological networks are powerful tools for visualizing biodiversity data and assessing ecosystem health and function. Constructing these networks requires considerable empirical efforts, and this remains highly challenging due to sampling limitations and the laborious and notoriously limited, error-prone process of traditional taxonomic identification. Recent advancements in high-throughput gene sequencing and high-performance computing provide new ways to address these challenges. DNA metabarcoding, a method of bulk taxonomic identification from DNA extracted from environmental samples, can generate detailed biodiversity information through a standardizable analytical pipeline for species detection. When this biodiversity information is annotated with prior knowledge on taxon interactions, body size, and trophic position, it is possible to generate trait-based networks, which we call "heuristic food webs". Although curating trait matrices for constructing heuristic food webs is a laborious, often intractable process using manual literature surveys, it can be greatly accelerated via text mining, allowing knowledge of relevant traits to be gathered across large databases. To explore this possibility, we employed a General Architecture for Text Engineering (GATE) system to create a hybrid text-mining pipeline combining rule-based and machine-learning modules. This pipeline was then used to query online repositories of published papers for missing data on a key trait, body size, that could not be gathered from existing trophic link libraries of freshwater benthic macroinvertebrates. Combining text-mined body size information with feeding information from existing sources allowed us to generate a database of over 20,000 pairwise trophic interactions. Next, we developed a pipeline that uses taxa lists generated from DNA metabarcoding and annotates this matrix with trophic information from existing databases and text-mined body size data. In this way, we generated heuristic food webs for wetland sites within a large delta complex formed by the confluence of the Peace and Athabasca rivers in northern Alberta: the Peace–Athabasca delta. Finally, we used these putative food webs and their network properties to resolve spatial and temporal differences between the benthic subwebs of wetlands in the Peace and Athabasca sectors of the delta complex. Specifically, we asked two questions. (1) How do food web properties (e.g. number of links, linkage density, trophic height) differ between the wetlands of the Peace and Athabasca deltas? (2) How do food web properties change temporally in wetlands of the two deltas? We discuss using DNA-generated, trait-based food webs as a powerful tool for rapid bioassessment, assess the limitations of our current approach, and outline a path forward to make this powerful tool more widely available for land managers and conservation biologists.

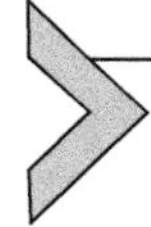

1. INTRODUCTION

1.1 Ecological Networks as Biomonitoring Tools

As we move deeper into the Anthropocene, Earth's biosphere is experiencing unprecedented, human-caused biodiversity loss (Dirzo et al., 2014; Steffen et al., 2015), and we now face the possibility of losing species before they are named and described (Costello et al., 2013a,b; Mora et al., 2013). As the delivery of ecosystem services can be positively related to biodiversity (Nelson et al., 2009; Zedler and Kercher, 2005), ecologists and resource managers need tools to facilitate the rapid assessment of the latter (Pecl et al., 2017). Despite considerable money and research allocated to develop biomonitoring, current efforts for biodiversity assessment fall short of adequately diagnosing perturbations to biological systems, largely because of lacking baseline data and a poor understanding of the ecological mechanisms responsible for mitigating or amplifying these perturbations (Friberg et al., 2011; Gray et al., 2014).

Biomonitoring can be particularly challenging in biodiversity hot spots, where samples must be extensively subsampled, and taxonomic resolution is limited by logistical and financial constraints, as well as taxonomic expertise (Curry et al., 2018; Orlofske and Baird, 2013; Woodward et al., 2013). Studies in complex, biologically diverse systems are thus often poorly replicated, limited to small, easily accessed areas, and focused on a narrow subset of the food web (e.g. macroinvertebrate assemblages in freshwaters). Over the last several decades, numerous methods have been used to assess ecosystem health and function (e.g. Bartoldus, 1999; Carletti et al., 2004), but these approaches suffer from being overly qualitative (i.e. assessing a suite of cultural, physical, and biological attributes) and restricted to short time periods (Gibson et al., 2015b; Woodward et al., 2013), limiting inferences on the degree to which key ecosystem functions or services are performed, and how these relate to biodiversity status. This is especially true given that ecosystem functions and services often result from processes occurring over large spatial and long temporal scales (Brose and Hillebrand, 2016) and that biodiversity loss at these scales could compromise many ecosystem properties (Isbell et al., 2017; Oliver et al., 2015; Wang and Loreau, 2016).

A leap forward in biomonitoring is clearly overdue (Bohan et al., 2017; Woodward et al., 2013), and recent interest has focused on the prospect of using ecological networks to improve mechanistic and diagnostic understanding of ecosystems (Gray et al., 2014, 2015a; Jordano, 2016). Ecological networks are potentially powerful biomonitoring tools because they enable visualization of complex biological interactions within diverse assemblages and reveal indirect interactions that cannot be predicted from more traditional reductionist approaches. While all food webs are essentially heuristic in nature, as it is impossible to observe or measure all trophic interactions in a given natural system, we use the term "heuristic food web" to refer to a trait-based network constructed using prior information (i.e. taxon interactions, body size, and trophic habit) gathered from the published literature. Food webs conceptually link individuals to populations, communities, and ecosystems (Delmas et al., 2018; Dunne, 2009; Ings et al., 2009), and their metrics, such as trophic level and linkage density, can be used for the assessment of ecosystem health and function. Consequently, many food web attributes provide a relatively direct link to ecosystem services (e.g. secondary production and fisheries yield) (Cohen et al., 2003; Jonsson et al., 2005). In this way, heuristic food webs can provide new information and generate insights regarding the structure and function of ecosystems when compared to current biomonitoring approaches, which generally focus on single taxa or functional groupings, or consider only limited aspects of the community (e.g. total abundance, species richness, diversity). Since ecosystem services are often best envisioned at landscape scales, where metacommunity and food web dynamics are important (Massol et al., 2011; Melián et al., 2011), network-based approaches constitute especially powerful tools because they also integrate ecosystem services across multiple scales (Dee et al., 2017). Species interaction networks, which support important functions and services of ecosystems (Gray et al., 2014), can reveal system dynamics and their resilience to environmental change or disturbance (Scheffer and Carpenter, 2003; Thompson et al., 2012; Tylianakis et al., 2007). This can be seen in cases where a system is altered despite no apparent change to the richness (i.e. the number of nodes) or community composition in a system (e.g. Spaak et al., 2017; Tylianakis et al., 2007). Moreover, in the occurrence of extreme environmental perturbations, such as droughts, loss of rare species that are often viewed as "noise" in traditional community bioassessments and are routinely trimmed from community data sets prior to analysis (McCune et al., 2002), can actually degrade a food web's resilience to future perturbations (Lu et al., 2016).

1.2 New Tools to Rapidly Assess Biodiversity and Annotate It With Ecological Information

New genomics and computational tools are revolutionizing the way we do biomonitoring research, and, together with growing trait and trophic interaction databases (e.g. Poelen et al., 2014; Poisot et al., 2016), can enable the rapid generation of trait-based networks for biomonitoring purposes. The advent of high-throughput sequencing (HTS) has enabled the rapid generation of DNA sequence information, facilitating biodiversity discovery across broad phylogenetic groups (Bik et al., 2012; Ercolini, 2013; Ushio et al., 2018) and providing orders of magnitude more taxa than traditional morphological sample assessment (Gibson et al., 2015a). Standardized DNA barcodes can then be matched to thousands of records for plant, animal, fungal, and microbial taxa in public databases (e.g. BOLD, GenBank) (Hajibabaei et al., 2007a,b). The use of metabarcoding for biodiversity assessment has been termed "metasystematics", and these advances, coupled with molecular methods for assessing functional gene regions for organisms (e.g. Mason et al., 2012; Pujolar et al., 2012), are called "ecogenomics". Ecogenomics has already moved biodiversity assessment forward in multiple ways, including assessment of biodiversity (Gibson et al., 2015a), rapid detection of species of interest (Zhan et al., 2013), direct evaluation of the effects of changing water quality (Gardham et al., 2014; Sims et al., 2013), development of advanced new biomonitoring tools accounting for taxa sensitivities to environmental stressors (e.g. Dafforn et al., 2014), measurement of the putatively active microbial community (e.g. Wilhelm et al., 2014), measurement of ecosystem function and biogeochemical processes (Eaton et al., 2011, 2012), and linking microbial community composition with ecosystem function (Peralta et al., 2010). When coupled with other information and analytical approaches, such as stable isotope analysis, ecogenomics can even be used to make inferences about food web structure (Gray et al., 2014).

Concurrently, advancements in computing, information management, and data discovery are making rapid, automatic text mining a reality (Baird et al., 2011; Bolón-Canedo et al., 2015; Tan et al., 2014), accelerating the annotation of biodiversity data with ecological information. Text mining offers a set of interoperable methods facilitating extraction of information, usually of very specific kinds in the form of structured data, from documents written in natural languages, such as scientific publications and technical documentation (Fleuren and Alkema, 2015; Inzalkar and Sharma, 2015; Spangler et al., 2014). Unlike research software attempting to implement general purpose natural language understanding (NLU), text-mining programs are always specialized for some well-defined type of extracted information, which makes them fully customizable, facilitating high degrees of accuracy in text matching. A typical text-mining program accepts a machine-readable text as input and returns extracted data in some simple structured (often simply tabular) form. Large repositories of the published literature are being updated from hard copies or raster scans to machine-readable text files. As databases like Google Scholar and Web of Science achieve deeper coverage of the historical literature, computational tools are being developed that can take advantage of modern computing clusters that are no longer limited by memory or processing capacity. The potential for text mining to extract targeted information effectively from the literature has been demonstrated in several domains, including medicine (Vandervalk et al., 2013), ecotoxicology (Riazanov et al., 2012), and lipidomics (Chepelev et al., 2011). Developing text-mining approaches for extracting species-trait information will likely be more challenging, however, as the language used to describe ecological traits is often imprecise when compared to the fields of medicine, microbiology, or toxicology. Yet, the tools are available, and when coupled with emerging machine-learning approaches (Bijalwan et al., 2014a,b; Witten et al., 2016), the potential to create robust tools for extracting ecological information from the literature exists. The combination of these advances could revolutionise the use of networks as the next-generation biomonitoring tools, by removing the previous logistical constraints in this field, which has for decades relied heavily on laborious and expensive empirical methods to construct networks from natural systems.

1.3 Freshwater Biodiversity Hotspots as Candidates for Exploring Novel Approaches for Biodiversity Assessment

Freshwater ecosystems are ideal candidates for exploring novel approaches to biomonitoring because the history of their management has been underpinned by biomonitoring (Friberg et al., 2011) and they harbour a disproportionate amount of imperilled biodiversity (Dudgeon et al., 2006). Wetlands, in particular, pose challenges to traditional biomonitoring, as they are hydrologically complex systems experiencing rapid change. An estimated 30%–90% of the wetlands in most countries have been destroyed or highly modified (Davidson, 2014; Junk et al., 2013). The unique properties of wetlands influence global biogeochemical and hydrological cycles (Mitsch and Gosselink, 2007), delivering ecosystem services at large scales (Barbier, 2011). Despite recognition of the wide-ranging ecosystem services wetlands provide, including fodder production, water flow and quality regulation, and climate regulation (Moor et al., 2015), wetland loss and impairment show no signs of waning (Junk et al., 2013). Water regulation and agriculture, industrial, and urban expansion pose growing threats to wetlands (Davis et al., 2010; Finlayson et al., 2013), and climate change is expected to add additional pressures to wetlands through hydrological changes, increasing temperature and sea-level rise (Junk et al., 2013). These pressures are expected to affect wetlands differently, requiring specific study and management to understand their particular conservation needs (Erwin, 2009). Wetlands pose particular challenges to the consistent observation of biodiversity because of their high spatiotemporal heterogeneity and fragility (Gibson et al., 2015b). Because wetlands are often highly productive and complex, consistent sampling of more than a narrow component of the food web is difficult, labor limited, and cost prohibitive. In summary, wetlands are ideal testbeds for developing novel informatics-based biomonitoring approaches.

1.4 Exploring the Generation of DNA Ecological Networks for Aquatic Biomonitoring

Most studies about "synthetic data sets" (sensu Poisot et al., 2016) and biomonitoring using DNA and network approaches (Bohan et al., 2017) have been conceptual, theoretical, or only employed tools for constructing networks based on existing trait or trophic interaction databases. Here, we take this a stage further by combining taxonomic list information derived from DNA metabarcoding with trait information derived from existing trophic link libraries and supplemented by text mining to construct heuristic food webs. Specifically, we explore the utility of heuristic food webs for comparing different sectors of the highly threatened Peace–Athabasca delta, an area of international conservation significance and a World Heritage site. Our example focuses on investigating two simple questions. (1) How do food web properties (e.g. number of links, linkage density, trophic height) differ between the Peace and Athabasca sectors of the delta complex? (2) Do food web characteristics change temporally in these two deltas? Our study is the first to examine the utility of DNA-based food webs for resolving structural differences in benthic wetland macroinvertebrate assemblages across space and time.

2. BUILDING HEURISTIC FOOD WEBS FOR WETLAND BIOMONITORING: A CASE STUDY

2.1 DNA Sample Collection, Metabarcoding, and Bioinformatics Pipeline

Samples were collected at two different time points (June and August) in both 2012 and 2013 in Wood Buffalo National Park (see Gibson et al., 2015a for detailed methods). This park encapsulates the largest inland wetland in Canada, which is recharged by two large rivers, collectively making up the Peace–Athabasca delta. We chose four sites in each river delta and sampled the benthic macroinvertebrate community ($n=3$ replicate samples per site, approximately 50 m apart) during each sampling event using a modified Canadian Aquatic Biomonitoring Network (CABIN; http://www.ec.gc.ca/rcba-cabin/) protocol for sampling wetland. The CABIN method for sampling macroinvertebrates was designed to be simple, robust, and readily adopted by resource managers and regulators. Because our samples were processed for DNA, we modified the existing CABIN protocol for wetland sampling to employ sterile technique at all stages of sample collection and processing. Sampling consisted of a 2-min travelling kick-net sample using a 400 μm mesh D-net with an attached collection cup. A clean, sterile net was used for harvesting each new sample to prevent contamination between samples. Samples were transferred into sterile 1L polyethylene jars filled with 95% EtOH, placed in a cooler, and taken to the lab where they were stored at −80°C until shipment to the University of Guelph for DNA extraction, PCR amplification, and HTS.

Each bulk sample was homogenized in ethanol, DNA was extracted, and a 230 bp and a 314 bp portion of the mitochondrial cytochrome *b* oxidase

I (COI) gene region was amplified. Amplified DNA from each sample was sequenced on an MiSeq platform. Complete DNA extraction, amplification, and sequencing data, including amplification primers, are given in Gibson et al. (2015a).

All DNA sequences generated were merged and filtered for length, quality, and chimeras. This produced an average of 59,820 sequences per sample for the longer segment of COI and 182,120 sequences per sample for the shorter segment. Sequences for each sample were pooled into OTUs at 98% similarity. All OTUs were compared to existing DNA barcode databases to obtain an unambiguous best taxonomic identification. Resulting identifications were used to generate taxonomic lists for sample. Complete bioinformatics pipeline information, including software used and total number of sequences generated, is given in Gibson et al. (2015a).

2.2 Development of Text-Mining Pipeline for Traits of Freshwater Organisms

The initial step in developing our text-mining pipeline was to compose a corpus of documents with unstructured text containing trophic information. Because of the importance of body size for understanding food web structure (Brose et al., 2017), we developed a corpus of literature for this trait, focusing on freshwater organisms. Briefly, we aggregated a corpus of 50 peer-reviewed documents from Google Scholar using a Boolean search including all of the following terms: "trophic", "food web", "aquatic", and "body size". While 50 is a relatively low number of documents for building text-mining applications that require a high degree of precision, our pipeline was not intended to produce only relevant text, but was instead used to rapidly generate a large list of potentially useful body size information that was then manually curated and incorporated into existing trait databases. Additionally, because of the high variation in the ecological language from paper to paper, our corpus contained a wide variety of body size instances, which allowed us to capture much of the variation in this language in a relatively small number of documents. We screened our initial Boolean search, from the most relevant citation first, to determine if each paper contained body size information related to aquatic macroinvertebrates. In order to be included, a paper needed to contain at least one instance where body size type (e.g. length, mass) and an associated measurement value and unit were associated with a specific taxa name. Once 50 documents were discovered that had the required information, we annotated all of them, noting all instances that body size information was found in the text, as well as one-, two-, and three-sentence windows containing that information.

Next, we constructed a gazetteer for key terms necessary for constructing our text-mining pipeline. Our gazetteer consisted of a list of terms (e.g. "body size", "length", "weight") and their forms and synonyms. Our gazetteer consisted of 14 terms and between 3 and 10 different homologues that occurred in our 50-paper corpus. This step allowed for the coding of sentences into consistent, useful directives for finding body size information for specific taxa.

After annotating our corpus and composing our gazetteer, we then developed a rule-based text-mining pipeline using GATE (Cunningham et al., 2011) to retrieve from our corpus important information related to organism body size. This pipeline was developed iteratively, with improvements made after each run on our annotated corpus. We made the a priori determination that four key pieces of information would be needed to make sense of body size traits: body size type, body size value, measurement unit, and organism name. In Box 1 we briefly describe the functionality and some implementation details of the components of our final text-mining pipeline for body size.

2.3 Evaluation, and Iterative Refinements to Text-Mining Pipeline for Body Size

We developed our final body size text-mining pipeline using an iterative process where new components were developed or refined (i.e. added or refined JAPE rules) after each run of the pipeline on our corpus. We evaluated the pipeline using standardized computer science performance metrics, including precision and recall, which are commonly used for assessing pattern recognition, information retrieval, and binary classification (Klein et al., 2014; Witte and Baker, 2007). Precision (*Pr*), which is the fraction of relevant elements among all retrieved elements, is mathematically defined as

$$Pr = \frac{tp}{tp + fp}, \tag{1}$$

where *tp* is true-positive elements and *fp* is false-positive elements. Recall (*Re*), or the fraction of true-positive relative elements among the total amount of relative elements, is defined as

$$Re = \frac{tp}{tp + fn}, \tag{2}$$

BOX 1 GATE-based text-mining pipeline for extracting sentences containing body size information.

The first section of the pipeline is designed to extract bulk text. The pipeline starts using **Document Reset PR** to remove named annotation sets or resets the default annotation set for a pdf that has been converted to text. This component enables the document to be reset to its original state, by removing all the annotation sets and their contents, apart from the one containing the document format analysis. Next, **Tokenizer** splits text into tokens, or minimal text segments such as numbers, punctuations, and words of different types. **Sentence Splitter** then identifies sentences and parses documents into a string of sequential sentences; this component uses a gazetteer list of abbreviations to help distinguish sentence-marking full stops from other kinds. Both the tokenization and sentence splitting are high-level processes that break up and annotate whole documents, enabling further text processing. **Organism Tagger**, an open-source, hybrid rule-based, and machine-learning system (CITE; http://www.semanticsoftware.info/organism-tagger), is then used to extract organism mentions from the document, normalize them to their scientific name, and map them to the NCBI Taxonomy database (https://www.ncbi.nlm.nih.gov/taxonomy).

The next section of the pipeline reassembles the bulk, annotated text and then extracts tagged information about body size. A JAPE-rule-based post-processor, **Split Organism Sentence Combiner**, reassembles consecutive sentences that were split due to the abbreviated representation of an organism. **Gazetteer-based Named Entity Extractor** then extracts body traits (e.g. dry mass, head length) and measurement units (e.g. mg, mm), and **Numbers Tagger** annotates numbers made up from numbers or numeric words.

The final part of the pipeline is a series of components that implements JAPE rules to refine the mined information to improve the precision of our results while retaining recall. **Large Number Remover** first filters out numbers larger than 1,000 because the scientific literature seldom reports numerical values larger than this for body size, using instead different measurement units for very large or very small organisms; additionally this rule eliminates many false-positive results due to the fact that values over 1000 often correspond to time values (year) that are found in papers, particularly the methods and literature cited sections. Next, **Body Size Value Determiner** uses a series of 12 JAPE-based rules to filter and blacklist sentence values that are not body size values (e.g. values related to temperature, sample size, statistical information). **Body Size Black List Filter** then removes blacklisted sentences from the list of annotated sentences. **Co-occurrence Matcher** subsequently annotates sentences having co-occurrences of annotations for organism names, body size types, body size values, and measurement units in one-, two-, and three-sentence text windows. Finally, **Black Listed Sentence Remover** eliminates blacklisted sentences, annotated by **Body Size Black List Filter**, from the sentences of co-occurring annotations derived by **Co-occurrence Matcher**.

Continued

BOX 1 GATE-based text-mining pipeline for extracting sentences containing body size information.—cont'd

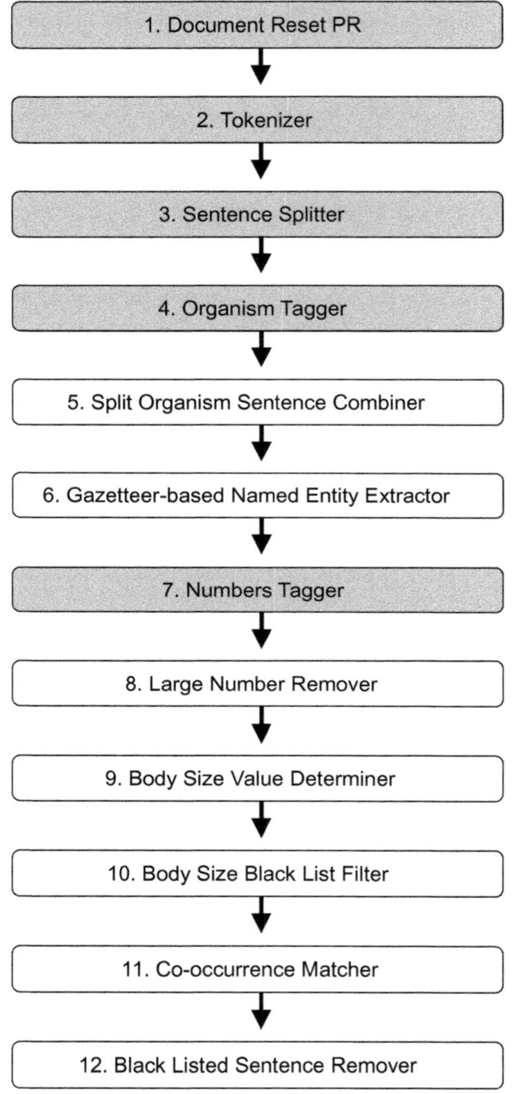

Gray boxes correspond to native GATE components or third-party components adapted for GATE, and white boxes correspond to components developed for this project.

where *fn* is false-negative elements. For the purposes of this assessment, we defined an element as one of the four key pieces of information needed to obtain body size trait information for a given species: (1) species name, (2) body size type, (3) body size value, and (4) body size measurement unit. In addition to overall assessments of precision and recall, we also assessed the performance of our pipeline in terms of these individual pieces of information and across sentence windows ranging from one to three sentences. Finally, we assessed Rijsbergen's *F*-measure (F_β) or the measure of effectiveness of retrieval (van Rijsbergen, 1979) as

$$F_\beta = \left(1 + \beta^2\right) \times \frac{Pr \times Re}{\beta^2 \times Pr + Re}, \tag{3}$$

where β is a factor that weights the importance a user places on precision as compared to recall. Because of the nature of our text-mining application, where multiple papers could have overlapping information on body size, we were less concerned with recall, and so used the special case of F_β, where precision was weighted more than recall ($F_{0.5}$).

We then developed a second corpus in order to test how our final, refined pipeline performed at extracting useful information on other, untested literature. This was completed following the same search query described in Section 2.1, except that we searched deeper into the search results until we retrieved 200 additional papers that contained useful body size information (i.e. instances that included body size type and measurement values and units for a given species). We then annotated all of the documents (see Section 2.1), ran our refined text-mining pipeline on this naïve corpus, and evaluated the results using the same approach described in Section 2.2.

In general, our text-mining pipeline was much better at precision than recall. When precision was defined broadly as the model's ability to retrieve any piece of useful information (i.e. one of the four key pieces of information for an organism's body size: species name, body size type, body size value, body size unit), precision was 0.98. This value was much lower (0.15), however, when we assessed sentences that contained all four pieces of information. Additionally, when we assessed precision on specific elements of the corpus, we found that there was a lot of variation based on the specific element we looked at (species name: 0.92; body size type: 0.67; body size number: 0.19; body size unit: 0.20). Recall, on the other hand, was much lower overall, with a total recall of 0.40. When we split recall up into one-, two-, and three-sentence search windows, we found that recall dropped off precipitously after the search window was widened from a one-sentence window (one-sentence window: 0.64; two-sentence window: 0.07; three-sentence window: 0.09). Finally, the measure of effectiveness of retrieval (van Rijsbergen, 1979), Rijsbergen's F-measure ($F_{0.5}$), was 0.59 for body size, 0.73 for species name, 0.21 for measurement number, 0.23 for measurement unit, and 0.17 for sentences containing all four pieces of information.

2.4 Rule-Based Procedure for Retrieving Trait Information for Heuristic Food Web Construction

Heuristic food webs were constructed in R using the package *cheddar* (Hudson et al., 2013). Generating these food webs requires input matrices of all consumer–resource interactions. We generated heuristic food webs for benthic macroinvertebrates from DNA-generated species lists by identifying all possible consumer–resource interactions using a rule-based approach that integrated both existing trophic trait information and information we mined from the literature (heuristic food web pipeline, Fig. 1). First, we assembled a DNA-based, presence–absence community matrix for each sample. Samples averaged 75.7 ± 0.13 (mean $\pm$ S.E.) genera, with 407 total genera across all samples. Once the species pool was assembled, we then followed our rule-based pipeline to determine all possible trophic interactions within each community (Fig. 1). Briefly, this involved constructing a pairwise trophic interaction matrix for all taxa in the species pool. To obtain this information, we queried online databases for existing trophic interactions at the genus level. Gaps in the interaction matrix were then filled using a series of filters, including functional feeding guilds, and linkages were further refined by limiting consumer–resource interactions based on a body size filter, which was partly developed using the body size information text mined from the literature (Box 1). When information on taxa was not available, information from the closest taxonomic relative was used (Fig. 1). In total, this process yielded a database of over 20,000 pairwise trophic interactions. This matrix was then filtered based on the taxa present for a given sample (for an individual community) or group of samples (for a metacommunity). Trophic interaction matrices were then imported into R, and *cheddar* routines were conducted to generate heuristic food webs and measure their properties, including trophic height, network density, and connectance.

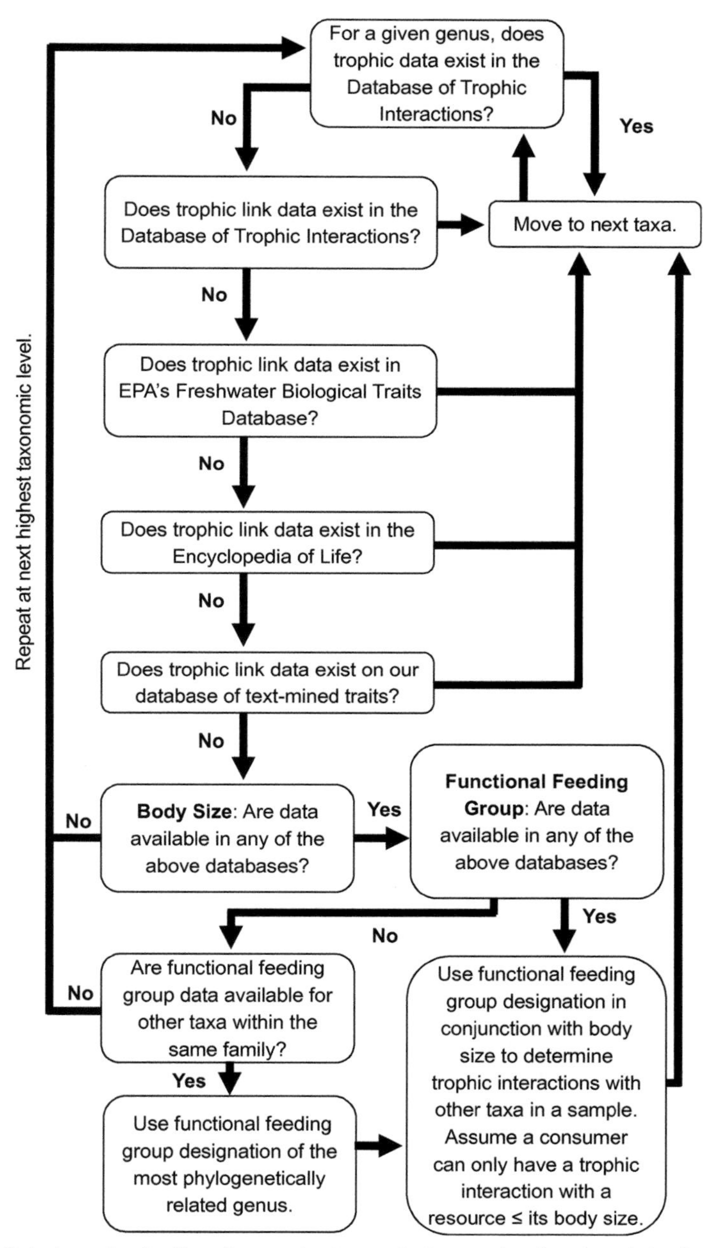

Fig. 1 Rule-based pipeline for gathering missing traits for heuristic food web construction.

2.5 Comparing Food Web Properties in Two River Deltas: A Spatiotemporal Comparison

We investigated how well our heuristic food webs could resolve spatial and temporal differences in the Peace–Athabasca delta located in the northern Canadian boreal region, downstream of the Alberta Oil Sands mining area. Specifically, we used linear mixed effects models to examine how important food web properties differed (a) between deltas, (b) across seasons, and (c) between two successive years. In total we examined 91 total food webs from 96 samples ($n = 2$ years $\times$ 2 seasons $\times$ 2 river deltas $\times$ 4 sites $\times$ 3 replicate samples $= 96$ total samples); 5 samples were not analysed because they were lost during transport or processing of DNA. The food web properties we examined were the number of nodes, trophic links, connectance and omnivory, the number of unique trophic species, prey-averaged trophic level, mean vulnerability, and the variation (standard deviation) in vulnerability (see Glossary). We examined these properties the *lmer* package in R to assess the full model

$$y = Delta^{*} Year^{*} Season + (1|Delta/Site/Sample).$$

Using a model comparison approach, we found that the nested term *Delta* did not significantly influence any model, so it was removed for the final analysis, and we used the following, most-parsimonious model:

$$y = Delta^{*} Year^{*} Season + (1|Site/Sample).$$

Random effects were assessed using REML.

Overall, differences in heuristic food web structure were evident, but difficult to visualize as entire networks due to the high number of species (nodes) retrieved from our DNA benthic samples (Fig. 2). Strong seasonal and interannual differences in the trophic structure, however, were apparent. August networks, for example, generally had more nodes and higher linkage density and trophic level than June networks (Fig. 2). Additionally, the trophic level was generally higher in 2012 than in 2013, especially for August (Fig. 3C, D, G, and, H). Finally, while differences in delta were less pronounced than seasonal differences, the Peace (Fig. 2A, C, E, and G) delta generally had more nodes and trophic links than the Athabasca delta (Fig. 2B, D, F, and H).

Differences in individual food web properties were much more revealing than simple visual comparison. Overall, food webs composed from individual samples, or subwebs, contained 75.7 ± 0.1 nodes and 712 ± 2.8 trophic

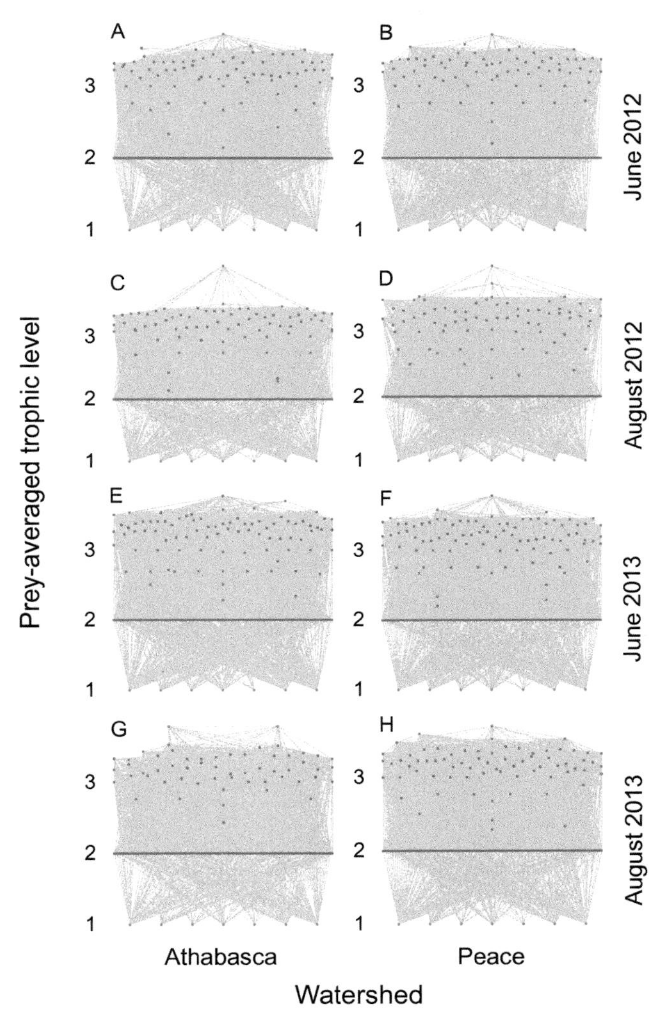

Fig. 2 Metawebs aggregated across all samples for the Athabasca (A, C, E, G) and Peace deltas (B, D, F, H) for June (A–B, E–F) and August (C–D, G–H) in 2012 (A–D) and 2013 (E–F).

links (mean ± S.E.). Connectance was 0.12 ± 0.00 and trophic height was 3.5 ± 0.0, with 68.5 ± 0.1 unique trophic species (mean ± S.E.). Temporal differences were more pronounced than spatial differences across most food web metrics (Fig. 3; the Appendix). Despite the fact that visual assessment of food webs seemed to indicate more nodes and links in August compared to June (Fig. 2), we observed significantly higher numbers of nodes and links

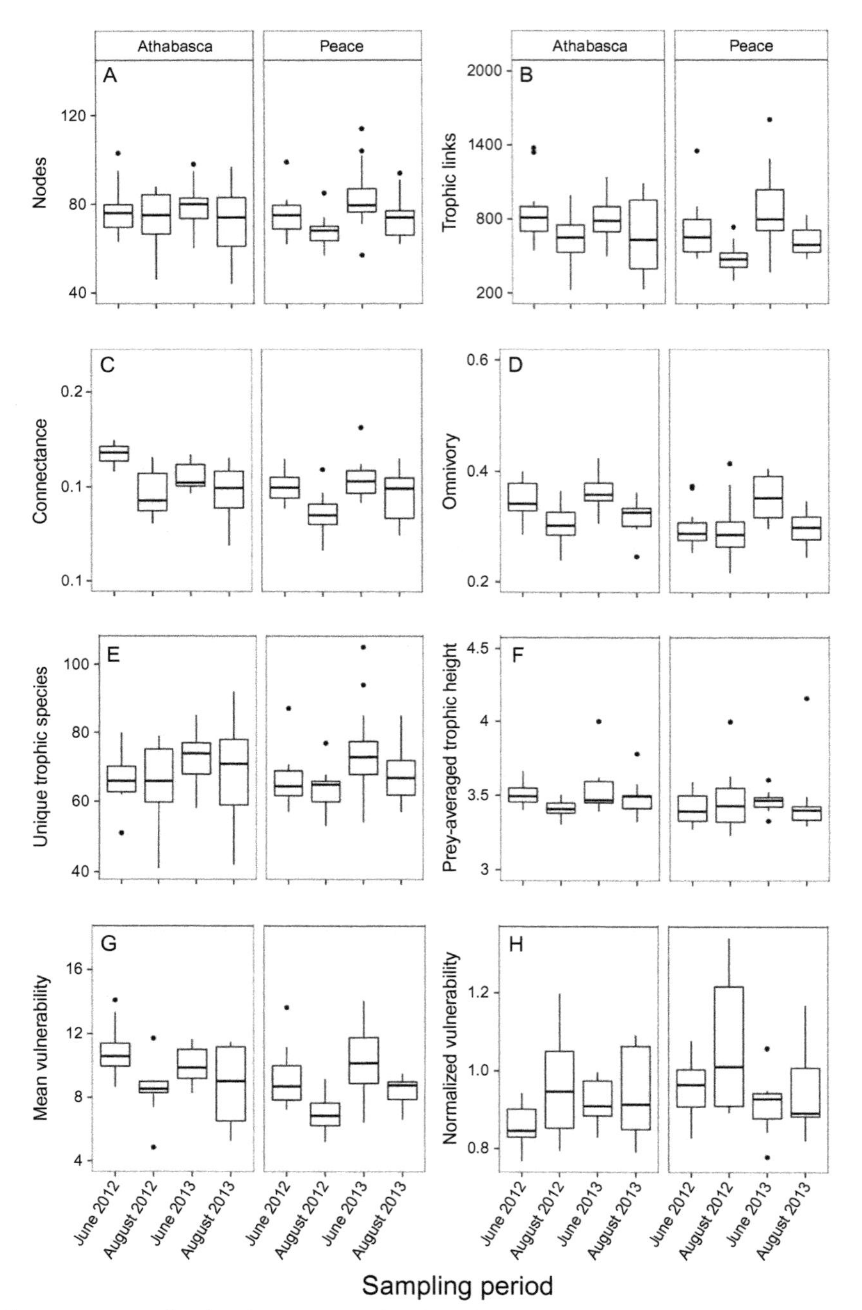

Fig. 3 Heuristic food web properties for the Athabasca and Peace deltas in June and August for 2012 and 2013: (A) number of nodes (or taxa), (B) trophic links, (C) directed connectance, (D) proportion of omnivory, (E) the number of unique trophic species, (F) prey-averaged trophic height, (G) mean vulnerability, and (H) normalized vulnerability.

for June compared to August across subwebs (Fig. 3A and B; the Appendix). Additionally, differences were more pronounced for *Season* than *Year*, with all models differing between June and August except for the number of unique trophic species and trophic level, but only omnivory and the number of unique trophic species differing between years (i.e. both were higher in 2013 compared to 2012) (Fig. 3; the Appendix).

Pronounced seasonal differences in food webs, or more precisely subwebs, have been observed in other systems (Carvalho et al., 2017; Closs and Lake, 1994; Warren, 1989). For example, in an intermittent stream in Australia, the number of nodes present in Lerderderg River subwebs was attributed to increases in the duration of baseflow (Closs and Lake, 1994). Similarly, the hydrological regime influenced temporal patterns in food webs of two floodplain lakes in Argentina by altering riverine inputs to these lakes and influencing food supply (Carvalho et al., 2017). In our system, seasonal differences in food web structure could have also been caused by seasonal flow variation, likely because flow can influence the ontology and phenology of aquatic insects (Bunn and Arthington, 2002). Interestingly, despite apparent higher trophic levels in the food web for August compared to June, we found higher number of nodes, trophic links, connectance, and omnivory in June compared to August, indicating a larger, more connected community with more diverse feeding links. The visibly higher trophic levels in August could potentially be explained by seasonal increases in predators, as this can lead to increased food chain length (Closs and Lake, 1994) and likely trophic height. The Broadstone Stream food web also displayed strong seasonal patterns in food chain length, with maximum chain length reaching its highest point in the summer, and then declining into the winter and spring (Woodward et al., 2005). However, because August subwebs in our study did not differ in trophic height from June, it is likely that while predators became more diverse, food webs in August merely contained different, redundant predators, an idea supported by the trend toward less unique trophic species in August compared to June (Fig. 3; the Appendix). Additionally, larger, more abundant predators later in the season could have explained the reduced number of nodes, trophic links, and connectance in August compared to June. Because predators have been shown to have a seasonal lag in colonization (Closs and Lake, 1994) and large predators can have strong top-down effects (Lemmens et al., 2018; Newsome et al., 2017; Worm and Myers, 2003), they can reduce the number of primary consumers, storing biomass at the top of the food web (i.e. inverted biomass pyramids; e.g. Trebilco et al., 2013, 2016; Wang et al., 2009).

This idea is also supported by a study of shallow lakes in Argentina, where higher connectance was found in a food web with lower numbers of predators compared to a food web with higher numbers of predators (Carvalho et al., 2017). However, this does not mean that organisms were extirpated from the system; rather, their numbers could have been reduced by predation to levels below detection. This is evident by seasonal cycles in food web properties across years despite few year-to-year differences in food web properties (Layer et al., 2011). Alternatively, the seasonal patterns observed in our system could have resulted from higher numbers of species, particularly predators, emerging later in the summer, decreasing our ability to detect them in August.

Spatial variation in food web metrics was not as pronounced as temporal variation. Normalized vulnerability was marginally different between river deltas, with the Peace river delta tending to be more variable than Athabasca (Fig. 3; the Appendix). Additionally, the Athabasca delta had a marginally higher connectance than the Peace delta (Fig. 3; the Appendix). There were also many significant *Delta * Year* interactions (i.e. for trophic links, connectance, mean vulnerability, and the variation in vulnerability; Fig. 3; the Appendix), indicating that many of the spatial patterns in food web properties changed through time.

The lack of many spatial differences in food web structure among localities could be due to the high degree of spatial connectedness in our wetland system: the Peace–Athabasca delta is a wetland of highly interconnected sites, with hydrological connections varying largely from year to year, driven by distance to open water and habitat structure (i.e. the degree of perched basin) (Peters et al., 2006a,b). Alternatively, the sites within this delta could simply be similar in habitat at scales relevant for most taxa. Evidence from other studies demonstrates mixed effects of spatial variation on food web structure. For example, similar to our study, seasonal variation in food web structure was much more pronounced than spatial variation in an intermittent stream (Closs and Lake, 1994); in their study, however, sites were only 1.5 km apart, and so the spatial autocorrelation in food webs could be much larger than temporal variation, especially as it relates to the intermittent streamflow seen in their system. Other studies of freshwater food webs have demonstrated strong spatial patterns (Thompson and Townsend, 1999; Warren, 1989), which can be more pronounced than temporal patterns (Winemiller, 1990). Spatial differences in food web structure and dynamics are likely to be more detectable in less hydrologically connected systems, in systems with pronounced disturbance gradients, and in systems that span greater spatial scales, have greater variation in energy availability, or exhibit large differences in ecosystem size (McHugh et al., 2010; Nyström and McIntosh, 2003; Thompson and Townsend, 2005).

Interestingly, the food web metrics that did differ by delta in our system could have arisen because of variation in predator abundance. For example, normalized vulnerability differed significantly among river deltas (the Appendix). Normalized vulnerability, which standardizes to linkage density (Hudson et al., 2013), is more comparable across webs of different sizes (Williams and Martinez, 2000). By comparing the variance around these estimates, one may be able to get a sense of the stability of a system (Williams and Martinez, 2000). Because vulnerability is the average number of predators per prey in a food web (Schoener, 1989), this spatial variation demonstrates that there was much more variability in the predator community in the Peace compared to the Athabasca delta, and that there were strong interactions with season, as there was much more variation in August compared to June. Further, connectance, which was a marginally significant for the factor *Delta* (the Appendix), was generally lower in the Peace delta. Together with greater variation in web vulnerability in the Peace delta, this could mean that this delta is more vulnerable to future disturbances from encroaching oil sands work.

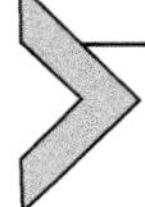

3. THE FUTURE OF TEXT MINING: NEW TOOLS FOR RAPID EXPANSION OF FOOD WEB DATABASES AND CHALLENGES TO THEIR WIDESPREAD ADOPTION FOR SEARCHING ECOLOGICAL LITERATURE

Results from the assessment of our text-mining pipeline indicate that more work needs to be done to effectively retrieve ecological trait information from the literature. Text mining is now widely used in the life sciences with major pharmaceutical companies and research institutes incorporating these techniques in advanced curation platforms supporting database creation. A primary challenge preventing text mining from being a fully realized tool for the ecological literature is the lack of a fully developed ontology for ecology (Baird et al., 2008, 2011; Rubach et al., 2011). Successful applications of text mining in other fields with more fully developed ontologies, such as ecotoxicology (Hardy et al., 2012) and biomedical research (Spasic et al., 2005), include gene name recognition, mutation impact extraction, pathway reconstruction, drug–drug, or protein–protein interaction. Within the ecological literature, by comparison, terminology is less precise and is particularly inconsistent across subdisciplines. In the context of our work, this was evident in the development of our gazetteers for body size, which often documented dozens of synonyms for the same term. For instance, "body size" can be assessed by dozens of methods, with units ranging from length values (e.g. total length, average width) to mass values (e.g. mass, size, weight) to even measurements that are proxies for body size (e.g. interoccular distance, carapace length, length of foreleg femora).

Despite these limitations, great strides can be made in ecological trait-based text mining once a robust ontology is constructed and made publicly available. Efforts on this front are already being made (e.g. Buttigieg et al., 2016; Frey and Cox, 2015; Madin et al., 2007), but more work is needed, particularly in getting ecologists to subscribe to a more precise ontology, which will improve the discoverability (sensu Dasgupta et al., 2007) of natural language ecological information. An additional, concurrent approach that could rapidly expedite advancement in text mining of ecological literature is machine learning (Witten et al., 2016), which can enhance rule-based text-mining applications, making them more robust. The power of machine learning is evident within our own pipeline, as its precision was extremely high in the retrieval of species names, which was facilitated by the **OrganismTagger** (OT) module (Naderi et al., 2011) that uses machine-learning algorithms. Specifically, OT is a hybrid rule-based and machine-learning system that extracts organism mentions from the biomedical literature, normalizes them to a scientific name, and provides grounding to entities in online data resources (e.g. NCBI Taxonomy database). OT has been benchmarked to a high standard (http://www.semanticsoftware.info/organismtagger) and performance has been evaluated on two manually annotated published corpora and achieves a precision and recall of >95% and >94% and a grounding accuracy of >97.4%. OT is published as open-source software described in detail in the published literature (Naderi et al., 2011). Tagging of organism mentions with OT in other pipelines will support the identification of traits for specific taxonomic groups, illustrating how incorporating modules that use machine learning can rapidly improve text-mining pipelines for ecological traits.

While our efforts to develop a text-mining pipeline for the retrieval of ecological trait information revealed many limitations, they should not detract from using text mining as an approach to focus and highlight relevant information that can subsequently be screened by humans. Particularly promising is our finding that our pipeline performed better in terms of precision than recall. Given our aim to expand existing trait databases by matching organism names with published traits, we are less concerned with obtaining all possible species-trait associations from the literature (i.e. recall) than we are providing information that is actually useful (i.e. precision) in the context of trait databases. Indeed, for many species (e.g. *Gammarus pulex*) there are likely thousands of estimates of traits like body size for particular taxa in the literature; obtaining all estimates of body size, however, is not the goal, as long as useful information is retrieved. In these cases, variance around body size estimates can be calculated, and weighted, evidence-based approaches can be used (e.g. eco-evidence approach, Norris et al., 2011; Webb et al., 2011, 2012; fuzzy coding, Chevene et al., 1994; Dolédec et al., 1999). Furthermore, even if we are unable to obtain estimates of all traits from all taxa in the literature, we can easily use our existing pipeline to greatly enhance the records available in current databases. Finally, the field of text mining will be further advanced as methods are developed to retrieve information from tables, figures, and non-ASCII text, approaches that will be particularly useful for extracting information from older papers (e.g. Fletcher and Kasturi, 1988; Kim et al., 2003; Yin et al., 2014). For example, Textfinder is an automated system to detect and recognize text in images (Wu et al., 1999), and Tesseract optical character recognition (Smith, 2007; Smith et al., 2009) can now be used to prepare text segments for further processing.

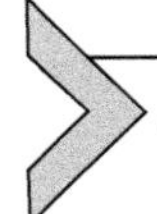

4. TOWARDS AN OPEN-SOURCE PIPELINE FOR THE RAPID CONSTRUCTION AND ASSESSMENT OF TRAIT-BASED FOOD WEBS FOR BIOMONITORING: THE PROMISE, CHALLENGES, AND NEXT STEPS

4.1 Advancing Network Approaches for Biomonitoring

As we face unprecedented global biodiversity loss (Dirzo et al., 2014), developing a more predictive food web ecology is necessary for understanding how ecosystems adapt to future change scenarios. Specifically, a major research challenge in food web ecology is predicting the emergence and reassembly of communities under global change scenarios (Albouy et al., 2014). We know that species invasions, biomass harvesting, range shifts, disturbances, and changes in land use strongly influence biodiversity (Murphy and Romanuk, 2014; Paillet et al., 2010; Vilà et al., 2011). Indeed, how these disturbances affect species composition is well described (e.g. Vilà et al., 2011; Warwick and Clarke, 1993), but forecasting resulting impacts on community structure and functioning requires a priori knowledge of potential interactions among species.

Predicting interactions among species that have never previously co-occurred is challenging, as traditional empirical methods of food web sampling, such as stomach content analysis, only provide limited inferences and are often laborious, yielding low-resolution taxonomic results. Predictive models of trophic interactions, leading to improved understanding of large-scale food web structure, can improve management of ecosystem functioning. Resolving food webs requires considerable knowledge about community members and their habits and associations. Without tools to refine text from journals and reports, accessing this knowledge is painstaking and expensive, and generating new knowledge incrementally using traditional sampling and observation is impractical for scaling up. Alternatively, DNA metabarcoding can reveal community composition in "high definition", but there is currently no complementary method to measure or otherwise generate the abundance and biomass data needed for food web analysis beyond quantified DNA sequence proxies (Gibson et al., 2015a). Moreover, inferring trophic interactions among species generated in DNA-based taxon lists would require a step change in our current knowledge of trophic links beyond the scope of current trophic link databases. Although species associations have been inferred from spatial and temporal co-occurrence (e.g. Vacher et al., 2016), not all co-occurrences represent true trophic interactions. In this study, we established the groundwork for future studies in wetlands and other biodiversity hotspots by (1) establishing an approach for synthesizing the vast taxonomic and functional information available to us through genomics information and traits databases, (2) providing a pipeline for the construction of heuristic food webs that can be used to evaluate predictions and hypotheses about actual food webs, and (3) demonstrating the utility of heuristic food webs for biodiversity and ecosystem function assessment.

Demonstrating the utility of heuristic food webs for wetland assessment is an important step in advancing ecological research because it allows us to wield and make sense of the rich taxonomic information that we are now afforded thanks to modern genomics approaches. Our goal here was to demonstrate the utility using heuristic food webs to visualize and assess the complexity in a species-rich wetland complex, and we were able to discern both spatial and temporal differences in food web properties. The next steps will be to assess what these changes could mean in other systems, including the many systems that harbour much less biodiversity than wetlands. Additionally, we need to do more work in systems associated with major perturbations, both natural and human-made, in order to assess the sensitivity of our heuristic food webs in detecting change along disturbance gradients. Ultimately, the goal will be to explore food web properties that are associated with resilience, such as vulnerability, trophic height, and connectance, in order to help us target systems of concern that are most subject to collapse.

4.2 The Need for Vastly Expanded Trait Databases

The prospect of developing a fully functional, open-source analytical pipeline that will allow for the rapid construction of heuristic food webs from DNA taxon lists and traits mined from the literature, however, will require several improvements concurrent with advancements in text mining. First, existing trait databases need to be greatly expanded with trophic interactions and other trait information generated from text-mined data sets, which will involve combing through thousands of published documents and require massive computing power along with improvements to our text-mining pipeline (see Section 4.1). Updated trait databases will enable annotation of co-occurrence information from DNA-generated taxon lists to support the development of heuristic food webs. Very few of the taxa identified from our DNA samples were found in existing trait databases; our text-mining pipeline, coupled with manual curation of traits associated with our taxa, allowed us to generate a trophic links database of over 20,000 pairwise interactions. As text-mining efforts are advanced, this and other trait databases can be vastly expanded, circumventing the tedious (and expensive) steps of manual curation, as many more taxa from future samples will be found in these databases. This will foster the development of a pipeline to gather and match existing trait information with DNA taxa lists. Script-based tools will allow for the execution of rapid taxon matching with associated traits from databases.

A concurrent challenge will be to fill in missing gaps in generated taxa-trait tables, as many taxa will not have published information on associated traits (apart from body size). Developing these approaches will allow for heuristic food web construction even before food web databases are complete. Building on prior knowledge of the rules of interactions (Fath et al., 2007) will allow for the refinement of the set of possible species interactions. One approach to filling in missing trophic interactions is using theoretical modelling approaches to identify remaining gaps in trophic interactions (sensu Gravel et al., 2016). The development of trait-matching models of interactions can be achieved by fitting theoretical models of network structure to observed data. Most ecological networks can be summarized with only a few axes, which can easily match to functional traits (Eklöf et al., 2013). For example, the niche model (Williams and Martinez, 2000) was fit directly with the assumption that the body size is the main niche axis and that the predator–prey body size relationship represents the optimal niche of a species (Gravel et al., 2013). This theory can be expanded by exploring network construction using niche models and phylogenetic relatedness. Another approach to filling in missing trophic information is phenomenological modelling, which is independent of theory. For example, hierarchical Bayesian statistics (Weinstein and Graham, 2017) and machine-learning techniques (Beauchesne et al., 2017; Desjardins-Proulx et al., 2017) can recreate pairwise interactions with a minimal amount of information. Both approaches—theoretical and phenomenological modelling—could be used to fill in gaps in our taxa-trait matrices, which are necessary for heuristic food web construction.

4.3 Assessing Uncertainty Within Ecological Networks and Heuristic Food Webs

A final challenge to the development of a heuristic food web pipeline is quantifying uncertainty within such models. Providing measures of uncertainty around targeted elements of network connection will elucidate how well heuristic food webs are expected to reflect actual food webs, allowing us to uncover emergent patterns (relating to, for example, diversity patterns, top-down or bottom-up trophic pressures, or diversity–complexity relationships) that might otherwise be masked (Cirtwill et al., 2018). Specifically, benchmark data sets will be needed to perform comparative analysis to assess the effectiveness of different models for inferring trophic interactions, and quantitative comparisons of methods will be necessary in order to identify their strengths and weaknesses. Additionally, probabilistic methods can be used to assess the uncertainty of these trophic inferences, from the pairwise interactions to network structure. Another way of improving heuristic food web pipelines is integrating techniques of stable isotope analysis with trait-based methods derived from food web theory. Model integration techniques originally proposed for species distribution models (Talluto et al., 2016) can be used to integrate both approaches. Specifically, Bayesian methods could be developed where prior distributions of parameters of mixing models are constrained by the feasibility of the interactions of a given heuristic model. Further, the energy flux between pairs of species in a mixing model could be constrained by a parameterized food web model based on body size and microhabitat use.

Another issue of assessing uncertainty of heuristic food webs generated from DNA information is error propagation across the three steps we describe to create these food webs (i.e. DNA metabarcoding, the text-mining pipeline, and the heuristic food web pipeline). This issue is an important one because error propagation could mean that while individual parts of a system function adequately, the system as a whole does not; alternatively, error propagation could mean that the system performs better than expected as a whole compared to its individual components (Arras, 1998; Oberkampf et al., 2002). Error propagation has been extensively studied in the context of GIS modelling pipelines (Heuvelink, 1998; Karssenberg and De Jong, 2005) and computer information systems (Avizienis et al., 2001; Ballou and Pazer, 1985). Because of the importance of understanding error propagation in machine learning, methods have been developed to extensively test these systems (e.g. Bottou, 2010; Bradley, 1997) that could be used to assess and improve current and future text-mining pipelines. Further, aspects of assessing error propagation in network speech recognition (Robinson and Fallside, 1991) could readily be adapted to assessing natural language text-mining approaches. Additionally, the importance of error propagation analysis, especially in the context of machine learning (sensu Robinson and Fallside, 1987; Rumelhart et al., 1985), presents a novel way of understanding the emergent properties of ecological networks. While individual components of a heuristic food web pipeline can readily be assessed, however, a detailed understanding of how error propagates across the entire pipeline (from DNA assessment and bioinformatics to the end point of heuristic food web construction) will require much more work in integrating information across very different disciplines (i.e. genomics, computer science, and ecology).

4.4 Assessing the Effectiveness of Heuristic Food Webs at Detecting Environmental Change

As pipelines are developed and refined for fully automated generation of heuristic food webs from DNA taxa lists, another important task will be testing the effectiveness of these networks in revealing the structural consequences of global change processes along natural disturbance gradients. We envision three approaches that can be used to validate heuristic food webs. First, natural abundance stable isotopes can be used to assess how the trophic height of a heuristic food webs relates to its actual measured height inferred by ^{15}N stable isotope values (sensu Jennings and van der Molen, 2015; Jennings et al., 2008). For example, prey-averaged trophic level measured from heuristic food webs could be compared to observed trophic height (from ^{15}N values) of individuals across samples, allowing for estimates of goodness of fit and inherent bias. This approach would allow the assessment of how trophic position within heuristic food webs compares to δ^{15}N values across different trophic groups (e.g. for filter feeders, shredders, predators), elucidating how well heuristic food webs estimate trophic height for organisms occupying different trophic levels.

A second approach that could be used to validate heuristic food webs is performing ^{15}N tracer experiments. Isotopic labelling approaches are powerful because they allow elements to be traced through the food web, providing incisive, quantifiable information on the fate of these elements to different food web compartments. Unlike natural abundance stable isotopes, which rely on fractionation assumptions and often suffer from overlapping end-members (Fry, 2006), ^{15}N tracers coupled with mass balance mixing models allow for the precise estimate the trophic height of local food webs. In particular, short-term ^{15}N tracer studies are useful because they eliminate confounding effects of nutrient enrichment seen in nutrient addition studies, such as abnormally long uptake lengths due to saturated N demand (Mulholland et al., 2002).

A third approach to validating heuristic food webs is comparing them to real food webs that have been measured in other systems around the world (i.e. the Broadstone Stream food web and others published in the Database of Trophic Interactions; Brose et al., 2005). Efforts to reconstruct food webs from the direct observation of feeding habits are rare as observations are laborious to obtain and involve a high degree of technical skill. Capitalizing on these existing efforts will enable the assessment of the generality and utility of heuristic food webs generated by our current and future pipelines (Fig. 2). Specifically, by deconstructing published food webs and running taxa lists through these pipelines (sensu Gray et al., 2015b), we can compare how well heuristic food webs reflect actual, measured food webs. While some metrics of food webs are not affected by network size, making them useful for comparison across systems (Schoener, 1989; Williams and Martinez, 2000), others are affected by the issue of diversity dependence, which is not always resolved by network standardization (reviewed in Pellissier et al., 2018). Consequently, novel approaches for comparing network structure of measured and heuristic food webs are required. For example, comparing empirical networks to null models can elucidate the factors shaping ecological networks coupled with functional traits (Pellissier et al., 2018). Additionally, analytical approaches that allow for the comparison of whole communities could be adapted for the comparison of heuristic and empirical food webs by integrating taxa, trait, and interaction matrices (e.g. Laigle et al., 2018), providing new insights into how these ecological networks differ across environmental gradients. All three of these approaches (i.e. natural abundance isotopes, isotopic tracers, and assessment of existing food webs) need to be done in conjunction with gradient studies to determine how heuristic food web approaches perform across diverse disturbance gradients in different bioregions.

5. PERSPECTIVES: THE FUTURE OF BIOMONITORING

We envision a future where scientists will be able to collect biological samples from almost any environment and rapidly process those samples through a series of standardized, open-source bioinformatics pipelines that will output detailed information about biological communities and their structure, allowing for powerful, predictive assessments of the health and function of an ecosystem (sensu Rooney and McCann, 2012). With these tools will come the possibility of acquiring near full-census community data for biological systems, moving us beyond the era of hypothesis testing and theoretical models (Anderson, 2008) and allowing us to finally get ahead of assessing the world's vanishing biodiversity (Dirzo et al., 2014) so that we can know what we have—and do something about it—before it is gone. Despite dramatic advances in the fields of genomics and computer science, this potential future will not be realized if these are not integrated in a meaningful way that will allow conservation practitioners to wield and make sense of the deluge of information afforded by modern genomics and text mining (Anderson, 2017; Bell et al., 2009; Hey and Trefethen, 2003). Others share our vision of the future of bioassessment, creating tools for the construction of food webs from existing databases (Gray et al., 2015b) and outlining how next-generation biomonitoring could be used to reconstruct large-scale ecological networks (Bohan et al., 2017). We have advanced this vision, demonstrating for the first time how DNA metabarcoding samples can be integrated into a food web construction pipeline that is enhanced by filters from text-mined information, and generating trait-based ecological networks, as envisioned by Bohan et al. (2017).

One of the key advantages of incorporating network-based tools, like heuristic food webs, into biomonitoring is that they provide taxon-free, diagnostic metrics of biodiversity and ecosystem function, allowing for comparison across the large, continental scales at which landscapes are being altered and global change is expected to occur (Melián et al., 2011). Additionally, the ecosystem models that are normally used to assess whole ecosystems typically ignore interactions among taxa, assuming that these connections are implicit and disconnected from the theory. This assumption, however, is one of the greatest sources of uncertainty in predicting how ecosystems will respond to global change (Poisot and Gravel, 2014). For example, while species may appear resilient to change individually, their persistence depends on how they respond to interactions with predators, prey, and competitors (Gilman et al., 2010). Coupled with the concept of "Biomonitoring 2.0" (Baird and Hajibabaei, 2012; Keck et al., 2017), trait-based food webs like the ones we have demonstrated will therefore allow us to assess global change in novel ways and at unprecedented scales. In order to anticipate how biodiversity and ecosystem stability will respond to global change, the next big challenge will be to use these heuristic food webs to understand how their structural properties mitigate or enhance different threats and to determine the consequences of structural change for ecosystem functioning.

ACKNOWLEDGEMENTS

We acknowledge the strong support of Parks Canada through granting of a research permit for this work, and in particular we would like to thank staff at Wood Buffalo National Park (Jeff Shatford, Queenie Gray, Ronnie and David Campbell, Sharon Irwin, Jason Straka, Rhona Kindopp, and Stu Macmillan) for help in accessing field sites in the Peace–Athabasca delta, as well as support for the collection of samples for DNA metabarcoding. A diverse set of funding sources made this work possible, including an LSARP grant (Ontario Genomics, Genome Canada) to M.H. and D.J.B. Support for the text-mining work was provided by Environment and Climate Change Canada (ECCC) program funds and by an NSERC Engage Grant (EG) (#486592-15), "Ecoinformatics data integration," awarded to IPSNP Computing Inc. and the University of New Brunswick (Scott Pavey, PI). Z.G.C. was supported by NSERC CRD funding to D.J.B. and also through a Visiting Fellowship from ECCC via the Canadian Federal Genomics Research & Development Initiative.

APPENDIX

Fixed effects results from linear mixed effects models, $y = Delta * Year * Season + (1 \mid Site/Sample)$. Separate models were conducted for eight different food web metrics: the number of nodes, the number of trophic links, directed connectance, degree of omnivory, unique trophic species, prey-averaged trophic height, mean vulnerability, and the standard deviation of normalized vulnerability. Bolded P-values indicate significant model terms ($\alpha = 0.05$).

Response	Term	Estimate	*t*-Value	*P*
Nodes	Intercept	75.57	53.13	**3.94 e-09**
	Delta	0.34	0.24	0.82
	Year	−1.95	−1.54	0.13
	Season	3.33	2.63	**0.01**
	Delta * Year	1.59	1.25	0.21
	Delta * Season	−0.81	−0.64	0.52
	Year * Season	−0.36	−0.29	0.78
	Delta * Year * Season	−0.027	−0.21	0.98
Trophic links	Intercept	710.09	20.87	**8.99 e-07**
	Delta	31.96	0.94	0.38
	Year	−36.85	−1.56	0.12
	Season	105.04	4.44	**4.04 e-05**
	Delta * Year	44.22	1.87	**0.067**
	Delta * Season	−15.46	−0.65	0.52
	Year * Season	5.69	0.24	0.81
	Delta * Year * Season	17.24	0.73	0.47
Connectance	Intercept	0.12	60.43	**1.31 e-09**
	Delta	0.0046	2.29	0.062
	Year	−0.00020	−0.16	0.87
	Season	0.0076	5.96	**1.24 e-07**
	Delta * Year	0.035	2.74	**0.0080**
	Delta * Season	0.00093	0.73	0.47
	Year * Season	0.025	1.99	0.051
	Delta * Year * Season	0.0012	0.84	0.40
Omnivory	Intercept	0.32	40.06	**1.77 e-08**
	Delta	0.012	1.43	0.20
	Year	−0.011	−2.87	**0.0053**
	Season	0.018	4.73	**9.76 e-06**

	Delta * Year	0.034	0.92	0.36
	Delta * Season	0.0031	0.82	0.41
	Year * Season	−0.0068	−1.84	0.069
	Delta * Year * Season	0.0072	1.95	0.055
Trophic species	Intercept	68.55	55.47	**9.03 e-09**
	Delta	0.046	0.037	0.97
	Year	−2.96	−2.81	**0.0063**
	Season	1.85	1.75	0.084
	Delta * Year	0.66	0.62	0.53
	Delta * Season	−0.61	−0.58	0.57
	Year * Season	−0.93	−0.89	0.38
	Delta * Year * Season	0.24	0.23	0.82
Trophic level	Intercept	3.47	150.14	**4.23 e-12**
	Delta	0.019	0.80	0.45
	Year	−0.020	−1.33	0.19
	Season	0.011	0.75	0.45
	Delta * Year	−0.012	−0.80	0.43
	Delta * Season	0.025	1.65	0.10
	Year * Season	−0.0025	−0.166	0.87
	Delta * Year * Season	0.015	0.98	0.33
Mean vulnerability	Intercept	9.14	33.17	**5.31 e-08**
	Delta	0.38	1.38	0.22
	Year	−0.23	−1.40	0.17
	Season	0.97	5.95	**7.28 e-08**
	Delta * Year	0.41	2.49	**0.015**
	Delta * Season	−0.039	−0.24	0.81
	Year * Season	0.14	0.89	0.38
	Delta * Year * Season	0.13	0.80	0.43

Normalized vulnerability	Intercept	9.45 e-01	81.40	**1.92 e-10**
	Delta	−2.65 e-02	−2.29	**0.062**
	Year	1.47 e-02	1.42	0.16
	Season	−3.12 e-02	−3.01	**0.0035**
	Delta * Year	−2.63 e-02	−2.53	**0.013**
	Delta * Season	2.83 e-03	0.27	0.79
	Year * Season	−1.75 e-02	−1.68	0.096
	Delta * Year * Season	7.51 e-05	0.0070	0.99

GLOSSARY

Black Listed Sentence Remover the final JAPE-rule-based processor which filters out blacklisted sentences

body size an important trophic trait that indicates the size, usually mass, of an organism

Body Size Black List Filter a JAPE-rule-based processor which removes sentences from the list of annotated sentence having blacklisted annotations from the step above in (9)

Body Size Value Determiner a collection of JAPE rules able to find a set of numerical patterns and determines to annotate them either as appropriate text or as blacklisted entities

Connectance the proportion of realized links to possible links

Co-occurrence Matcher a JAPE-rule-based processor which annotates sentences having co-occurrences of annotations for organisms, body size type, body size value, and measurement unit in windows of length 1, length 2, length 3, length 4, etc.

Corpus body of literature

Document Reset PR text-mining component that removes named annotation sets or reset the default annotation set, enabling a document to be reset to its original state by removing all the annotation sets and their contents

Gazeteer a list of abbreviations, symbols, and synonymous names for a term

Gazetteer-based Named Entity Extractor extracts body traits (e.g. dry mass, head length), measurement units (e.g. mg, mm), and blacklisted body size information (e.g. exuviae, eggs, growth)

General Architecture for Text Engineering (GATE) An open-source software architecture capable of creating robust, sustainable text processing pipeline. GATE is supported by an extensive community of developers, educators, and scientists (https://gate.ac.uk)

Heuristic food web a trait-based ecological network that integrates biological information with prior knowledge about taxon interactions, body size, and trophic habit

JAPE is the Java Annotation Patterns Engine, which is a key component of the GATE platform in computational linguistics

Large Number Remover a JAPE-rule-based number postprocessor which filters out numbers larger than 1000 because they mostly represent year which are irrelevant for this work

Natural language understanding (NLU) an aspect of natural language processing that deals with machine reading comprehension
Node a point in an ecological network connected to other nodes by some kind of association, including co-occurrence or trophic links
Numbers Tagger annotates numbers made up from numbers or numeric words
Omnivory the degree to which an organism has generalized feeding on other organisms, defined mathematically as the proportion of nodes that have two or more trophic links and have a non-integer trophic level
Organism Tagger open-source software for a hybrid rule-based/machine-learning system that extracts organism mentions from the biomedical literature, normalizes them to their scientific name, and provides grounding to the NCBI Taxonomy database. [http://www.semanticsoftware.info/organism-tagger]. It includes tools for automatically generating lexical and ontological resources from a copy of the NCBI Taxonomy database, thereby facilitating system updates by end users
Sentence Splitter text-mining component that splits text into sentences. The splitter uses a gazetteer list of abbreviations to help distinguish sentence-marking full stops from other kinds
Split Organism Sentence Combiner a JAPE-rule-based postprocessor which reassembles consecutive sentences which were split due to the abbreviated representation of an organism. For example, the presence of a period or full stop (.) in an abbreviated organism is often mistaken as the end of a sentence, therefore split into two sentences by the **Sentence Splitter** (3). After running the **Organism Tagger** (4), this step detects the split-point between two sentences and reassembles them into a complete sentence
Text mining an engineering discipline that offers a set of interoperable methods facilitating extraction of useful information, usually of very specific kinds, in the form of structured data, from documents written in natural languages, such as scientific publications, technical documentation, etc. Unlike research software attempting to implement general purpose NLU, text-mining programs are always specialized for some well-defined type of extracted information, which makes them highly tunable and allows to achieve high degrees of accuracy
Token minimal text segments, such as numbers, punctuations, and words of different types
Tokenizer text mining component that splits texts into token
Trophic level the position an organism occupies in a food chain
Trophic link an observed or inferred link between a consumer and a resource or prey item
Vulnerability the proportion of consumers to prey items

REFERENCES

Albouy, C., Velez, L., Coll, M., Colloca, F., Loc'h, F., Mouillot, D., et al., 2014. From projected species distribution to food-web structure under climate change. Glob. Chang. Biol. 20, 730–741.
Anderson, C., 2008. The end of theory: the data deluge makes the scientific method obsolete. Wired Magazine 16, 16-07.
Anderson, C., 2017. Data deluge: researchers turn to cloud computing as genomic sequencing data threatens to overwhelm traditional IT systems. Clin. OMICs 4, 26–29.
Arras, K.O., 1998. An introduction to error propagation: derivation, meaning and examples of equation CY = FX CX FXT. Technical report EPFL-ASL-TR-98-01 R3, ETH Zurich, 1–21. https://doi.org/10.3929/ethz-a-010113668.

Avizienis, A., Laprie, J.C., Randell, B., 2001. Fundamental concepts of computer system dependability. In: Workshop on Robot Dependability: Technological Challenge of Dependable Robots in Human Environments, pp. 1–16.
Baird, D.J., Hajibabaei, M., 2012. Biomonitoring 2.0: a new paradigm in ecosystem assessment made possible by next-generation DNA sequencing. Mol. Ecol. 21, 2039–2044.
Baird, D.J., Rubach, M.N., Van den Brinkt, P.J., 2008. Trait-based ecological risk assessment (TERA): the new frontier? Integr. Environ. Assess. Manag. 4, 2–3.
Baird, D.J., Baker, C.J., Brua, R.B., Hajibabaei, M., McNicol, K., et al., 2011. Toward a knowledge infrastructure for traits-based ecological risk assessment. Integr. Environ. Assess. Manag. 7, 209–215.
Ballou, D.P., Pazer, H.L., 1985. Modeling data and process quality in multi-input, multi-output information systems. Manag. Sci. 31, 150–162.
Barbier, E.B., 2011. Wetlands as natural assets. Hydrol. Sci. J. 56, 1360–1373.
Bartoldus, C.C., 1999. A Comprehensive Review of Wetland Assessment Procedures: A Guide for Wetland Practitioners. Environmental Concern, Inc., St. Michaels, MD.
Beauchesne, D., Desjardins-Proulx, P., Archambault, P., Gravel, D., 2017. Thinking outside the box—predicting biotic interactions in data-poor environments. Vie et Milieu Life Environ. 66 (3–4), 333–342.
Bell, G., Hey, T., Szalay, A., 2009. Beyond the data deluge. Science 323, 1297–1298.
Bijalwan, V., Kumar, V., Kumari, P., Pascual, J., 2014a. KNN based machine learning approach for text and document mining. Int. J. Database Theor. Appl. 7, 61–70.
Bijalwan, V., Kumari, P., Pascual, J., Semwal, V.B., 2014b. Machine learning approach for text and document mining. arXiv, 1–10, preprint arXiv:1406.1580.
Bik, H.M., Porazinska, D.L., Creer, S., Caporaso, J.G., Knight, R., et al., 2012. Sequencing our way towards understanding global eukaryotic biodiversity. Trends Ecol. Evol. 27, 233–243.
Bohan, D.A., Vacher, C., Tamaddoni-Nezhad, A., Raybould, A., Dumbrell, A.J., Woodward, G., 2017. Next-generation global biomonitoring: large-scale, automated reconstruction of ecological networks. Trends Ecol. Evol. 32, 477–487.
Bolón-Canedo, V., Sánchez-Maroño, N., & Alonso-Betanzos, A., 2015. Recent advances and emerging challenges of feature selection in the context of big data. Knowl. Based Syst. 86, 33–45.
Bottou, L., 2010. Large-scale machine learning with stochastic gradient descent. In: Proceedings of COMPSTAT 2010. Physica-Verlag HD, pp. 177–186.
Bradley, A.P., 1997. The use of the area under the ROC curve in the evaluation of machine learning algorithms. Pattern Recogn. 30, 1145–1159.
Brose, U., Blanchard, J.L., Eklöf, A., Galiana, N., Hartvig, M., Hirt, M.R., Kalinkat, G., Nordström, M.C., O'gorman, E.J., Rall, B.C., Schneider, F.D., 2017. Predicting the consequences of species loss using size-structured biodiversity approaches. Biol. Rev. 92 (2), 684–697.
Brose, U., Hillebrand, H., 2016. Biodiversity and ecosystem functioning in dynamic landscapes. Phil. Trans. R. Soc. B 371, 20150267.
Brose, U., Cushing, L., Berlow, E.L., Jonsson, T., Banasek-Richter, C., et al., 2005. Body sizes of consumers and their resources. Ecology 86, 2545.
Bunn, S.E., Arthington, A.H., 2002. Basic principles and ecological consequences of altered flow regimes for aquatic biodiversity. Environ. Manag. 30, 492–507.
Buttigieg, P.L., Pafilis, E., Lewis, S.E., Schildhauer, M.P., Walls, R.L., et al., 2016. The environment ontology in 2016: bridging domains with increased scope, semantic density, and interoperation. J. Biomed. Semant. 7, 57.
Carletti, A., De Leo, G.A., Ferrari, I., 2004. A critical review of representative wetland rapid assessment methods in North America. Aquat. Conserv. Mar. Freshwat. Ecosyst. 14, S103–S113.

Carvalho, D.A., Williner, V., Giri, F., Vaccari, C., Collins, P.A., 2017. Quantitative food webs and invertebrate assemblages of a large river: a spatiotemporal approach in floodplain shallow lakes. Mar. Freshw. Res. 68, 293–307.

Chepelev, L.L., Riazanov, A., Kouznetsov, A., Low, H.S., Dumontier, M., et al., 2011. Prototype semantic infrastructure for automated small molecule classification and annotation in lipidomics. BMC Bioinformatics 12, 303.

Chevene, F., Doléadec, S., Chessel, D., 1994. A fuzzy coding approach for the analysis of long-term ecological data. Freshw. Biol. 31, 295–309.

Cirtwill, A., Eklöf, A., Roslin, T., Wootton, K., Gravel, D., 2018. A quantitative framework for investigating the reliability of network construction, bioRxiv 332536. https://doi.org/10.1101/332536.

Closs, G.P., Lake, P.S., 1994. Spatial and temporal variation in the structure of an intermittent-stream food web. Ecol. Monogr. 64, 1–21.

Cohen, J.E., Jonsson, T., Carpenter, S.R., 2003. Ecological community description using the food web, species abundance, and body size. PNAS 100, 1781–1786.

Costello, M.J., May, R.M., Stork, N.E., 2013a. Response to comments on "Can we name Earth's species before they go extinct?" Science 341, 237.

Costello, M.J., May, R.M., Stork, N.E., 2013b. Can we name Earth's species before they go extinct? Science 339, 413–416.

Cunningham, H., Maynard, D., Bontcheva, K., 2011. Text Processing With Gate. Gateway Press, CA.

Curry, C.J., Gibson, J.F., Shokralla, S., Hajibabaei, M., Baird, D.J., 2018. Identifying North American freshwater invertebrates using DNA barcodes: are existing CO1 sequence libraries fit for purpose? Freshwat. Sci. 37, 178–189.

Dafforn, K.A., Baird, D.J., Chariton, A.A., Sun, M.Y., Brown, M.V., et al., 2014. Faster, higher and stronger? The pros and cons of molecular faunal data for assessing ecosystem condition. Adv. Ecol. Res. 51, 1–40.

Dasgupta, A., Ghosh, A., Kumar, R., Olston, C., Pandey, S., et al., 2007. The discoverability of the web. In: Proceedings of the 16th international conference on World Wide Web. ACM, pp. 421–430.

Davidson, N.C., 2014. How much wetland has the world lost? Long-term and recent trends in global wetland area. Mar. Freshw. Res. 65, 934–941.

Davis, J., Lake, P., Thompson, R., 2010. Freshwater biodiversity and climate change. In: -Jubb, I., Hopler, P., Cai, W. (Eds.), Greenhouse 09: Living With Climate Change. CSIRO Publishing, Melbourne, pp. 73–83.

Dee, L.E., Allesina, S., Bonn, A., Eklöf, A., Gaines, S.D., et al., 2017. Operationalizing network theory for ecosystem service assessments. Trends Ecol. Evol. 32, 118–130.

Delmas, E., Besson, M., Brice, M.H., Burkle, L.A., Dalla Riva, G.V., Fortin, M.J., Gravel, D., Guimarães Jr., P.R., Hembry, D.H., Newman, E.A., Olesen, J.M., Pires, M.M., Yeakel, J.D., Poisot, T., 2018. Analysing ecological networks of species interactions. Biol. Rev. https://doi.org/10.1111/brv.12433 (on-line).

Desjardins-Proulx, P., Laigle, I., Poisot, T., Gravel, D., 2017. Ecological interactions and the Netflix problem. PeerJ 5, e3644.

Dirzo, R., Young, H.S., Galetti, M., Ceballos, G., Isaac, N.J., et al., 2014. Defaunation in the anthropocene. Science 345, 401–406.

Dolédec, S., Statzner, B., Bournard, M., 1999. Species traits for future biomonitoring across ecoregions: patterns along a human-impacted river. Freshw. Biol. 42, 737–758.

Dudgeon, D., Arthington, A.H., Gessner, M.O., Kawabata, Z.-I., Knowler, D.J., et al., 2006. Freshwater biodiversity: importance, threats, status and conservation challenges. Biol. Rev. 81, 163–182.

Dunne, J.A., 2009. Food webs. In: Meyers, R.A. (Ed.), Encyclopedia of Complexity and Systems Science. Springer, New York, NY, pp. 3661–3682.

Eaton, W.D., McDonald, S., Roed, M., Vandecar, K.L., Hauge, J.B., Barry, D., 2011. A comparison of nutrient dynamics and microbial community characteristics across seasons and soil types in two different old growth forests in Costa Rica. Trop. Ecol. 52, 35–48.
Eaton, W.D., Anderson, C., Saunders, E.F., Hauge, J.B., Barry, D., 2012. The impact of Pentaclethra macroloba on soil microbial nitrogen fixing communities and nutrients within developing secondary forests in the Northern Zone of Costa Rica. Trop. Ecol. 53, 207–214.
Eklöf, A., Jacob, U., Kopp, J., Bosch, J., Castro-Urgal, R., et al., 2013. The dimensionality of ecological networks. Ecol. Lett. 16, 577–583.
Ercolini, D., 2013. High-throughput sequencing and metagenomics: moving forward in the culture-independent analysis of food microbial ecology. Appl. Environ. Microbiol. 79, 3148–3155.
Erwin, K.L., 2009. Wetlands and global climate change: the role of wetland restoration in a changing world. Wetl. Ecol. Manag. 17, 71–84.
Fath, B.D., Scharler, U.M., Ulanowicz, R.E., Hannon, B., 2007. Ecological network analysis: network construction. Econ. Model. 208, 49–55.
Finlayson, C.M., Davis, J.A., Gell, P.A., Kingsford, R.T., Parton, K.A., 2013. The status of wetlands and the predicated effects of global climate change: the situation in Australia. Aquat. Sci. 75, 73–93.
Fletcher, L.A., Kasturi, R., 1988. A robust algorithm for text string separation from mixed text/graphics images. IEEE Trans. Pattern Anal. Mach. Intell. 10, 910–918.
Fleuren, W.W., Alkema, W., 2015. Application of text mining in the biomedical domain. Methods 74, 97–106.
Frey, U., Cox, M., 2015. Building a diagnostic ontology of social-ecological systems. Int. J. Commons 9, 595–618.
Friberg, N., Bonada, N., Bradley, D.C., Dunbar, M.J., Edwards, F.K., et al., 2011. Biomonitoring of human impacts in freshwater ecosystems: the good, the bad and the ugly. Adv. Ecol. Res. 44, 1–68.
Fry, B., 2006. Stable Isotope Ecology. Springer, New York.
Gardham, S., Hose, G.C., Stephenson, S., Chariton, A.A., 2014. DNA metabarcoding meets experimental ecotoxicology: advancing knowledge on the ecological effects of copper in freshwater ecosystems. Adv. Ecol. Res. 51, 79–104.
Gibson, J.F., Shokralla, S., Curry, C., Baird, D.J., Monk, W.A., King, I., Hajibabaei, M., 2015a. Large-scale biomonitoring of remote and threatened ecosystems via high-throughput sequencing. PLoS One 10, e0138432.
Gibson, J.F., Stein, E.D., Baird, D.J., Finlayson, C.M., Zhang, X., Hajibabaei, M., 2015b. Wetland ecogenomics—the next generation of wetland biodiversity and functional assessment. Wetl. Sci. Pract. 32, 27–32.
Gilman, S.E., Urban, M.C., Tewksbury, J., Gilchrist, G.W., Holt, R.D., 2010. A framework for community interactions under climate change. Trends Ecol. Evol. 25, 325–331.
Gravel, D., Poisot, T., Albouy, C., Velez, L., Mouillot, D., 2013. Inferring food web structure from predator–prey body size relationships. Methods Ecol. Evol. 4, 1083–1090.
Gravel, D., Massol, F., Leibold, M.A., 2016. Stability and complexity in model meta-ecosystems. Nat Commun. 7, 12457.
Gray, C., Baird, D.J., Baumgartner, S., Jacob, U., Jenkins, G.B., O'Gorman, E.J., et al., 2014. Ecological networks: the missing links in biomonitoring science. J. Appl. Ecol. 51, 1444–1449.
Gray, C., Bista, I., Creer, S., Demars, B.O., Falciani, F., Monteith, D.T., Sun, X., Woodward, G., 2015a. Freshwater conservation and biomonitoring of structure and function: genes to ecosystems. In: Belgrano, A., Woodward, G., Jacob, U. (Eds.), Aquatic Functional Biodiversity: An Ecological and Evolutionary Perspective. Elsevier, London, UK, pp. 241–271.

Gray, C., Figueroa, D.H., Hudson, L.N., Ma, A., Perkins, D., Woodward, G., 2015b. Joining the dots: an automated method for constructing food webs from compendia of published interactions. Food Webs 5, 11–20.
Hajibabaei, M., Singer, G.A., Clare, E.L., Hebert, P.D., 2007a. Design and applicability of DNA arrays and DNA barcodes in biodiversity monitoring. BMC Biol. 5, 24.
Hajibabaei, M., Singer, G.A., Hebert, P.D., Hickey, D.A., 2007b. DNA barcoding: how it complements taxonomy, molecular phylogenetics and population genetics. Trends Genet. 23, 167–172.
Hardy, B., Apic, G., Carthew, P., Clark, D., Cook, D., 2012. Toxicology ontology perspectives. ALTEX 29, 139–156.
Heuvelink, G.B., 1998. Error Propagation in Environmental Modelling With GIS. CRC Press.
Hey, A.J., Trefethen, A.E., 2003. The data deluge: an e-science perspective. In: Berman, F., Fox, G., Hey, T. (Eds.), Grid Computing: Making the Global Infrastructure a Reality. John Wiley and Sons Ltd., West Sussex, UK, pp. 809–824.
Hudson, L.N., Emerson, R., Jenkins, G.B., Layer, K., Ledger, M.E., et al., 2013. Cheddar: analysis and visualisation of ecological communities in R. Methods Ecol. Evol. 4, 99–104.
Ings, T.C., Montoya, J.M., Bascompte, J., Blüthgen, N., Brown, L., et al., 2009. Ecological networks—beyond food webs. J. Anim. Ecol. 78, 253–269.
Inzalkar, S., Sharma, J., 2015. A survey on text mining-techniques and application. Int. J. Res. Sci. Eng. 24, 1–14.
Isbell, F., Gonzalez, A., Loreau, M., Cowles, J., Díaz, S., et al., 2017. Linking the influence and dependence of people on biodiversity across scales. Nature 546, 65.
Jennings, S., van der Molen, J., 2015. Trophic levels of marine consumers from nitrogen stable isotope analysis: estimation and uncertainty. ICES J. Mar. Sci. 72, 2289–2300.
Jennings, S., Barnes, C., Sweeting, C.J., Polunin, N.V.C., 2008. Application of nitrogen stable isotope analysis in size-based marine food web and macroecological research. Rapid Commun. Mass Spectrom. 22, 1673–1680.
Jonsson, T., Cohen, J.E., Carpenter, S.R., 2005. Food webs, body size, and species abundance in ecological community description. Adv. Ecol. Res. 36, 1–84.
Jordano, P., 2016. Chasing ecological interactions. PLoS Biol. 14, e1002559.
Junk, W.J., An, S., Finlayson, C.M., Gopal, B., Květ, J., Mitchell, S.A., et al., 2013. Current state of knowledge regarding the world's wetlands and their future under global climate change: a synthesis. Aquat. Sci. 75, 151–167.
Karssenberg, D., De Jong, K., 2005. Dynamic environmental modelling in GIS: 2. Modelling error propagation. Int. J. Geogr. Inf. Sci. 19, 623–637.
Keck, F., Vasselon, V., Tapolczai, K., Rimet, F., Bouchez, A., 2017. Freshwater biomonitoring in the information age. Front. Ecol. Environ. 15, 266–274.
Kim, S., Park, S., Kim, M., 2003. Central object extraction for object-based image retrieval. In: International Conference on Image and Video Retrieval. Springer, Berlin, Heidelberg, pp. 39–49.
Klein, A., Riazanov, A., Hindle, M.M., Baker, C.J., 2014. Benchmarking infrastructure for mutation text mining. J. Biomed. Semantics 5, 11.
Laigle, I., Aubin, I., Digel, C., Brose, U., Boulangeat, I., Gravel, D., 2018. Species traits as drivers of food web structure. Oikos 127, 316–326.
Layer, K., Hildrew, A.G., Jenkins, G.B., Riede, J.O., Rossiter, S.J., Townsend, C.R., Woodward, G., 2011. Long-term dynamics of a well-characterised food web: four decades of acidification and recovery in the broadstone stream model system. Adv. Ecol. Res. 44, 69–117.
Lemmens, P., Declerck, S.A., Tuytens, K., Vanderstukken, M., De Meester, L., 2018. Bottom-up effects on biomass versus top-down effects on identity: a multiple-lake fish community manipulation experiment. Ecosystems 21, 166–177.

Lu, X., Gray, C., Brown, L.E., Ledger, M.E., Milner, A.M., et al., 2016. Drought rewires the cores of food webs. Nat. Clim. Chang. 6, 875.
Madin, J., Bowers, S., Schildhauer, M., Krivov, S., Pennington, D., et al., 2007. An ontology for describing and synthesizing ecological observation data. Eco. Inform. 2, 279–296.
Mason, O.U., Hazen, T.C., Borglin, S., Chain, P.S., Dubinsky, E.A., Fortney, J.L., Mackelprang, R., 2012. Metagenome, metatranscriptome and single-cell sequencing reveal microbial response to Deepwater Horizon oil spill. ISME J. 6, 1715.
Massol, F., Gravel, D., Mouquet, N., Cadotte, M.W., Fukami, T., Leibold, M.A., 2011. Linking community and ecosystem dynamics through spatial ecology. Ecol. Lett. 14, 313–323.
McCune, B., Grace, J.B., Urban, D.L., 2002. Analysis of Ecological Communities. MjM Software Design, Gleneden Beach, OR.
McHugh, P.A., McIntosh, A.R., Jellyman, P.G., 2010. Dual influences of ecosystem size and disturbance on food chain length in streams. Ecol. Lett. 13, 881–890.
Melián, C.J., Vilas, C., Baldó, F., Gonzalez-Ortegon, E., Drake, P., Williams, R.J., 2011. Eco-evolutionary dynamics of individual-based food webs. Adv. Ecol. Res. 45, 225–268 (Academic Press).
Mitsch, W.J., Gosselink, J.G., 2007. Wetlands, fourth ed. Wiley, Hoboken.
Moor, H., Hylander, K., Norberg, J., 2015. Predicting climate change effects on wetland ecosystem services using species distribution modeling and plant functional traits. Ambio 44, S113–S126.
Mora, C., Rollo, A., Tittensor, D.P., 2013. Comment on "Can we name Earth's species before they go extinct?" Science 341, 237c.
Mulholland, P.J., Tank, J.L., Webster, J.R., Bowden, W.B., Dodds, W.K., et al., 2002. Can uptake length in streams be determined by nutrient addition experiments? Results from an interbiome comparison study. J. N. Am. Benthol. Soc. 21, 544–560.
Murphy, G.E., Romanuk, T.N., 2014. A meta-analysis of declines in local species richness from human disturbances. Ecol. Evol. 4, 91–103.
Naderi, N., Kappler, T., Baker, C.J., Witte, R., 2011. OrganismTagger: detection, normalization and grounding of organism entities in biomedical documents. Bioinformatics 27, 2721–2729.
Nelson, E., Mendoza, G., Regetz, J., Polasky, S., Tallis, H., et al., 2009. Modeling multiple ecosystem services, biodiversity conservation, commodity production, and tradeoffs at landscape scales. Front. Ecol. Environ. 7, 4–11.
Newsome, T.M., Greenville, A.C., Ćirović, D., Dickman, C.R., Johnson, C.N., et al., 2017. Top predators constrain mesopredator distributions. Nat. Commun. 8 ncomms15469.
Norris, R.H., Webb, J.A., Nichols, S.J., Stewardson, M.J., Harrison, E.T., 2011. Analyzing cause and effect in environmental assessments: using weighted evidence from the literature. Freshw. Sci. 31, 5–21.
Nyström, P., McIntosh, A.R., 2003. Are impacts of an exotic predator on a stream food web influenced by disturbance history? Oecologia 136, 279–288.
Oberkampf, W.L., DeLand, S.M., Rutherford, B.M., Diegert, K.V., Alvin, K.F., 2002. Error and uncertainty in modeling and simulation. Reliab. Eng. Syst. Saf. 75, 333–357.
Oliver, T.H., Heard, M.S., Isaac, N.J., Roy, D.B., Procter, D., et al., 2015. Biodiversity and resilience of ecosystem functions. Trends Ecol. Evol. 30, 673–684.
Orlofske, J.M., Baird, D.J., 2013. The tiny mayfly in the room: implications of size-dependent invertebrate taxonomic identification for biomonitoring data properties. Aquat. Ecol. 47, 481–494.
Paillet, Y., Bergès, L., Hjältén, J., Ódor, P., Avon, C., et al., 2010. Biodiversity differences between managed and unmanaged forests: meta-analysis of species richness in Europe. Conserv. Biol. 24, 101–112.
Pecl, G.T., Araújo, M.B., Bell, J.D., Blanchard, J., Bonebrake, T.C., et al., 2017. Biodiversity redistribution under climate change: impacts on ecosystems and human well-being. Science 355, eaai9214. https://doi.org/10.1126/science.aai9214.

Pellissier, L., Albouy, C., Bascompte, J., Farwig, N., Graham, C., et al., 2018. Comparing species interaction networks along environmental gradients. Biol. Rev. 93, 785–800.
Peralta, A.L., Matthews, J.W., Kent, A.D., 2010. Microbial community structure and denitrification in a wetland mitigation bank. Appl. Environ. Microbiol. 76, 4207–4215.
Peters, D.L., Prowse, T.D., Pietroniro, A., Leconte, R., 2006a. Flood hydrology of the Peace-Athabasca Delta, northern Canada. Hydrol. Process. 20, 4073–4096.
Peters, D.L., Prowse, T.D., Marsh, P., Lafleur, P.M., Buttle, J.M., 2006b. Persistence of water within perched basins of the Peace-Athabasca Delta, northern Canada. Wetl. Ecol. Manag. 14, 221–243.
Poelen, J.H., Simons, J.D., Mungall, C.J., 2014. Global biotic interactions: an open infrastructure to share and analyze species-interaction datasets. Eco. Inform. 24, 148–159.
Poisot, T., Gravel, D., 2014. When is an ecological network complex? Connectance drives degree distribution and emerging network properties. PeerJ 2, e251.
Poisot, T., Gravel, D., Leroux, S., Wood, S.A., Fortin, M.-J., et al., 2016. Synthetic datasets and community tools for the rapid testing of ecological hypotheses. Ecography 39, 402–408.
Pujolar, J.M., Marino, I.A., Milan, M., Coppe, A., Maes, G.E., et al., 2012. Surviving in a toxic world: transcriptomics and gene expression profiling in response to environmental pollution in the critically endangered European eel. BMC Genomics 13, 507.
Riazanov, A., Hindle, M.M., Goudreau, E.S., Martyniuk, C., Baker, C.J., 2012. Ecotoxicology Data Federation With SADI Semantic Web Services. SWAT4LS, pp. 1–18.
Robinson, A.J., Fallside, F., 1987. The Utility Driven Dynamic Error Propagation Network. University of Cambridge Department of Engineering.
Robinson, T., Fallside, F., 1991. A recurrent error propagation network speech recognition system. Comput. Speech Lang. 5, 259–274.
Rooney, N., McCann, K.S., 2012. Integrating food web diversity, structure and stability. Trends Ecol. Evol. 27, 40–46.
Rubach, M.N., Ashauer, R., Buchwalter, D.B., De Lange, H.J., Hamer, M., et al., 2011. Framework for traits-based assessment in ecotoxicology. Integr. Environ. Assess. Manag. 7, 172–186.
Rumelhart, D.E., Hinton, G.E., Williams, R.J., 1985. Learning Internal Representations by Error Propagation (No. ICS-8506). California University San Diego La Jolla Institute for Cognitive Science.
Scheffer, M., Carpenter, S.R., 2003. Catastrophic regime shifts in ecosystems: linking theory to observation. Trends Ecol. Evol. 18, 648–656.
Schoener, T.W., 1989. Food webs from the small to the large: the Robert H. MacArthur Award Lecture. Ecology 70, 1559–1589.
Sims, A., Zhang, Y., Gajaraj, S., Brown, P.B., Hu, Z., 2013. Toward the development of microbial indicators for wetland assessment. Water Res. 47, 1711–1725.
Smith, R., 2007. An overview of the Tesseract OCR engine. In: Ninth International Conference on Document Analysis and Recognition 2. IEEE, pp. 629–633.
Smith, R., Antonova, D., Lee, D.S., 2009. Adapting the Tesseract open source OCR engine for multilingual OCR. In: Proceedings of the International Workshop on Multilingual OCR. ACM, p. 1.
Spaak, J.W., Baert, J.M., Baird, D.J., Eisenhauer, N., Maltby, L., et al., 2017. Shifts of community composition and population density substantially affect ecosystem function despite invariant richness. Ecol. Lett. 20, 1315–1324.
Spangler, S., Wilkins, A.D., Bachman, B.J., Nagarajan, M., Dayaram, T., et al., 2014. Automated hypothesis generation based on mining scientific literature. In: Proceedings of the 20th ACM SIGKDD International Conference on Knowledge Discovery and Data Mining, pp. 1877–1886.
Spasic, I., Ananiadou, S., McNaught, J., Kumar, A., 2005. Text mining and ontologies in biomedicine: making sense of raw text. Brief. Bioinform. 6, 239–251.

Steffen, W., Broadgate, W., Deutsch, L., Gaffney, O., Ludwig, C., 2015. The trajectory of the Anthropocene: the great acceleration. Anthropocene Rev. 2, 81–98.

Talluto, M.V., Boulangeat, I., Ameztegui, A., Aubin, I., Berteaux, D., et al., 2016. Cross-scale integration of knowledge for predicting species ranges: a metamodelling framework. Glob. Ecol. Biogeogr. 25, 238–249.

Tan, M., Tsang, I.W., Wang, L., 2014. Towards ultrahigh dimensional feature selection for big data. J. Mach. Learn. Res. 15 (1), 1371–1429.

Thompson, R.M., Townsend, C.R., 1999. The effect of seasonal variation on the community structure and food-web attributes of two streams: implications for food-web science. Oikos 87, 75–88.

Thompson, R.M., Townsend, C.R., 2005. Energy availability, spatial heterogeneity and ecosystem size predict food-web structure in streams. Oikos 108, 137–148.

Thompson, R.M., Brose, U., Dunne, J.A., Hall, R.O., Hladyz, S., et al., 2012. Food webs: reconciling the structure and function of biodiversity. Trends Ecol. Evol. 27, 689–697.

Trebilco, R., Baum, J.K., Salomon, A.K., Dulvy, N.K., 2013. Ecosystem ecology: size-based constraints on the pyramids of life. Trends Ecol. Evol. 28, 423–431.

Trebilco, R., Dulvy, N.K., Anderson, S.C., Salomon, A.K., 2016. The paradox of inverted biomass pyramids in kelp forest fish communities. Proc. R. Soc. B 283 (1833). 20160816.

Tylianakis, J.M., Tscharntke, T., Lewis, O.T., 2007. Habitat modification alters the structure of tropical host–parasitoid food webs. Nature 445, 202–205.

Ushio, M., Murakami, H., Masuda, R., Sado, T., Miya, M., 2018. Quantitative monitoring of multispecies fish environmental DNA using high-throughput sequencing. Metabarcoding Metagenomics 2, e23297.

Vacher, C., Tamaddoni-Nezhad, A., Kamenova, S., Peyrard, N., Moalic, Y., et al., 2016. Learning ecological networks from next-generation sequencing data. Adv. Ecol. Res. 54, 1–39. Academic Press.

van Rijsbergen, C.J., 1979. Information Retrieval, second ed. Butterworth, London.

Vandervalk, B., McCarthy, E.L., Cruz-Toledo, J., Klein, A., Baker, C.J., et al., 2013. The SADI personal health lens: a web browser-based system for identifying personally relevant drug interactions. JMIR Res Protoc. 2, e14.

Vilà, M., Espinar, J.L., Hejda, M., Hulme, P.E., Jarošík, V., et al., 2011. Ecological impacts of invasive alien plants: a meta-analysis of their effects on species, communities and ecosystems. Ecol. Lett. 14, 702–708.

Wang, S., Loreau, M., 2016. Biodiversity and ecosystem stability across scales in metacommunities. Ecol. Lett. 19, 510–518.

Wang, H., Morrison, W., Singh, A., Weiss, H.H., 2009. Modeling inverted biomass pyramids and refuges in ecosystems. Ecol. Model. 220, 1376–1382.

Warren, P.H., 1989. Spatial and temporal variation in the structure of a freshwater food web. Oikos 55, 299–311.

Warwick, R.M., Clarke, K.R., 1993. Comparing the severity of disturbance: a metaanalysis of marine macrobenthic community data. Mar. Ecol. Prog. Ser. 92, 221–231.

Webb, J.A., Wealands, S.R., Lea, P., Nichols, S.J., de Little, S.C., Stewardson, M.J., et al., 2011. Eco Evidence: using the scientific literature to inform evidence-based decision making in environmental management. In: MODSIM2011 International Congress on Modelling and Simulation, pp. 2472–2478.

Webb, J.A., Nichols, S.J., Norris, R.H., Stewardson, M.J., Wealands, S.R., et al., 2012. Ecological responses to flow alteration: assessing causal relationships with Eco Evidence. Wetlands 32, 203–213.

Weinstein, B.G., Graham, C.H., 2017. On comparing traits and abundance for predicting species interactions with imperfect detection. Food Webs 11, 17–25.

Wilhelm, L., Besemer, K., Fasching, C., Urich, T., Singer, G.A., et al., 2014. Rare but active taxa contribute to community dynamics of benthic biofilms in glacier-fed streams. Environ. Microbiol. 16, 2514–2524.
Williams, R.J., Martinez, N.D., 2000. Simple rules yield complex food webs. Nature 404, 180.
Winemiller, K.O., 1990. Spatial and temporal variation in tropical fish trophic networks. Ecol. Monogr. 60, 331–367.
Witte, R., Baker, C.J., 2007. Towards a systematic evaluation of protein mutation extraction systems. J. Bioinform. Comput. Biol. 5, 1339–1359.
Witten, I.H., Frank, E., Hall, M.A., Pal, C.J., 2016. Data Mining: Practical Machine Learning Tools and Techniques. Morgan Kaufmann.
Woodward, G., Speirs, D.C., Hildrew, A.G., Hal, C., 2005. Quantification and resolution of a complex, size-structured food web. Adv. Ecol. Res. 36, 85–135.
Woodward, G., Gray, C., Baird, D.J., 2013. Biomonitoring for the 21st century: new perspectives in an age of globalisation and emerging environmental threats. Limnetica 32, 159–174.
Worm, B., Myers, R.A., 2003. Meta-analysis of cod–shrimp interactions reveals top-down control in oceanic food webs. Ecology 84, 162–173.
Wu, V., Manmatha, R., Riseman, E.M., 1999. Textfinder: an automatic system to detect and recognize text in images. IEEE Trans. Pattern Anal. Mach. Intell. 21, 1224–1229.
Yin, X.C., Yin, X., Huang, K., Hao, H.W., 2014. Robust text detection in natural scene images. IEEE Trans. Pattern Anal. Mach. Intell. 36, 970–983.
Zedler, J.B., Kercher, S., 2005. Wetland resources: status, trends, ecosystem services, and restorability. Annu. Rev. Env. Resour. 30, 39–74.
Zhan, A., Hulak, M., Sylvester, F., Huang, X., Adebayo, A.A., et al., 2013. High sensitivity of 454 pyrosequencing for detection of rare species in aquatic communities. Methods Ecol. Evol. 4, 558–565.

FURTHER READING

Gibson, J., Shokralla, S., Porter, T.M., King, I., van Konynenburg, S., Janzen, D.H., et al., 2014. Simultaneous assessment of the macrobiome and microbiome in a bulk sample of tropical arthropods through DNA metasystematics. Proc. Natl. Acad. Sci. U.S.A. 111, 8007–8012.
Hajibabaei, M., Shokralla, S., Zhou, X., Singer, G.A., Baird, D.J., 2011. Environmental barcoding: a next-generation sequencing approach for biomonitoring applications using river benthos. PLoS One 6, e17497.
Malaisé, V., Zweigenbaum, P., Bachimont, B., 2005. Mining defining contexts to help structuring differential ontologies. Terminology 11, 21–53.
Shokralla, S., Spall, J.L., Gibson, J.F., Hajibabaei, M., 2012. Next-generation sequencing technologies for environmental DNA research. Mol. Ecol. 21, 1794–1805.
Shokralla, S., Gibson, J.F., Nikbakht, H., Janzen, D.H., Hallwachs, W., Hajibabaei, M., 2014. Next-generation DNA barcoding: using next-generation sequencing to enhance and accelerate DNA barcode capture from single specimens. Mol. Ecol. Resour. 14, 892–901.

CHAPTER THREE

Volatile Biomarkers for Aquatic Ecological Research

Michael Steinke[1], Luli Randell, Alex J. Dumbrell, Mahasweta Saha[2]
University of Essex, School of Biological Sciences, Colchester, United Kingdom
[1]Corresponding author: e-mail address: msteinke@essex.ac.uk

Contents

1. Introduction 76
   1.1 Volatilomes: The Volatile Subset of Metabolomes 76
   1.2 Volatilomics: Terms and Techniques 77
2. Principal Techniques for Measuring Biogenic Volatiles 79
3. Medical Volatilomics Provides a Blueprint for Ecological Research 80
4. Role of Volatiles in Aquatic Ecological Interactions 81
   4.1 Case Study: Volatilomes of Freshwater Phytoplankton 82
5. Application of Volatilomics to Ecological Research: Using Volatilomics to 'Direct' Environmental Management 85
Acknowledgements 89
References 89

Abstract

All organisms and ecosystems emit and consume volatile organic compounds (VOCs). Traditionally, these have been qualitatively and quantitatively described in isolation without full consideration of the 'signatures' produced by the totality of all volatiles released. Here, we suggest that volatilomics, a research area applied to medical diagnostics, soil biology and pest control, can advance aquatic ecological research by providing a relatively fast diagnostic tool to investigate, for example, taxonomic and likely also functional diversity in aquatic systems—providing a novel technique for the biomonitoring of aquatic environments. Our case study demonstrates the utility of volatilomics to differentiate between four different algal genera using a principal component analysis. We highlight the utility of volatilomics to the monitoring of environmental processes and discuss its application to inform industrial mariculture procedures.

Keywords: Volatilomics, Volatile organic compounds (VOCs), Biomonitoring, Aquatic, Environmental management

[2] *Current address*: GEOMAR Helmholtz Centre for Ocean Research Kiel, Düsternbrooker Weg 20, 24105 Kiel, Germany.

Advances in Ecological Research, Volume 59
ISSN 0065-2504
https://doi.org/10.1016/bs.aecr.2018.09.002

1. INTRODUCTION

Terrestrial and aquatic plants and animals produce volatile organic compounds (VOCs) that provide signals acting as chemical cues, playing central roles in the recognition, attraction and deterrence of different species. Aquatic organisms emit volatiles when exposed to air or submerged underwater for intra- and interspecific communication (Mollo et al., 2017; Saha et al., 2018). 'Volatilomics' refers to the scientific study of volatile metabolites and, although relatively new in the 'omics' arena, is increasingly appreciated as a tool for the noninvasive presymptomatic diagnosis of human health (Amann and Smith, 2013). This has resulted in new methods for the early detection of various diseases in humans and can critically assist with identifying suitable strategies in the management of human health. Although technically feasible, similar applications have not yet been developed in aquatic research, yet these have the potential to be employed as a noninvasive, nondestructive, rapid and potentially highly cost-effective tool for biomonitoring, allowing users to readily survey the health status of marine and freshwater organisms and their environments. Given the embryonic stage of volatilomics research in being applied to aquatic biomonitoring, here we provide a brief overview of this developing technique and discuss the utility of aquatic volatilomics as a tool to advance freshwater and marine ecological research.

1.1 Volatilomes: The Volatile Subset of Metabolomes

Every organism produces metabolites that provide unique chemical footprints. The scientific study of these metabolites is referred to as metabolomics and aims to describe the diversity and abundance of metabolites in samples of differing complexities of their taxonomic composition, inter- and intraspecific interactions (e.g. sexual attraction, predator deterrence) or sampled environment (e.g. laboratory culture, microcosm or ecosystem). Hence, metabolomics offers a powerful approach to represent the link between genotype and molecular phenotype, and can be used to elucidate functions from the organism to the ecosystem scale (Goulitquer et al., 2012). For example, specific metabolomic signatures can be used to reveal both an organism's presence and its metabolic activity in the environment (Lee et al., 2012).

The 'volatilome' represents the volatile subset of the metabolome, and its study finds application for the development of noninvasive biomarkers in the clinical diagnosis and monitoring of human diseases (Amann and Smith, 2013). It is also used in pest control and yield improvement in agricultural systems (Pickett and Khan, 2016; Shrivastava et al., 2010) and sometimes focuses on the microbial production of volatiles in soils to characterise the function of plants, fungi and bacteria (Insam and Seewald, 2010; Kanchiswamy et al., 2015; Redeker et al., 2018).

1.2 Volatilomics: Terms and Techniques

Various terms are used to describe and classify volatiles (Achyuthan et al., 2017), and examples are presented in Table 1. Compounds in these classes are characterised by a relatively high vapour pressure at room temperature so that molecules of volatile compounds have a large tendency to escape from their liquid or solid form into the gas phase. Terms used to describe (components of) the volatilome include volatile inorganic compounds (VICs), biogenic volatile inorganic compounds (BVICs), volatile organic compounds (VOCs), nonmethane volatile organic compounds (NMVOCs), biogenic volatile organic compounds (BVOCs), microbial volatile organic compounds (MVOCs), oxygenated volatile organic compounds (OVOCs), hydrocarbons (HCs) and nonmethane hydrocarbons (NMHCs).

There is a confusing use of terms in the literature that describe volatiles released in biological processes. The three principle terms 'volatomics', 'volatolomics' and 'volatilomics' (and their derivatives 'volatome', 'volatolome' and 'volatilome') are currently used in the literature to describe the study of volatiles produced in biological processes. A search on the Web of Science for 'volatom*', 'volatolom*' and 'volatilom*' showed that the terms volatilomics and volatilome are most widely used in the published literature (559 citing articles, h-index of 12; 18 July 2018). However, here we briefly consider the meaning and definitions of these neologisms in an attempt to avoid possible confusion and encourage a coherent use of appropriate terms in the future.

The '*volatome*' is defined as '*... the sum of all released volatile organic compounds (VOCs) over a specific time and space...*' (D'Alessandro, 2006). By this definition, the term excludes inorganic volatiles (e.g. H_2S) that may add important diagnostic potential when studying the trace gas biology of aquatic environments. Furthermore, from a linguistic point of view, 'volatomics' may be confused with 'something to do with atoms' and we suggest that volatomics may be not the most appropriate term to use when describing the release of volatiles from biological systems.

Table 1 Classes of Volatiles With Typical Abbreviations, Their Definition and Example Compounds

Class	Abbreviation	Definition	Typical Compound(s)	References
Volatile inorganic compound	VIC	Carbon-free compound that has a high vapour pressure at room temperature	Argon, helium, nitrogen, hydrogen sulphide, ammonia	Kennes et al. (2016)
Biogenic volatile inorganic compound	BVIC	Volatile inorganic compound produced directly but not exclusively from biological processes	Nitrogen, hydrogen sulphide, ammonia	Kennes et al. (2016)
Volatile organic compounds	VOC	Carbon-containing compound that has a high vapour pressure at room temperature	Methane, acetone, dimethyl sulphide	Krupa and Fries (1971)
Nonmethane volatile organic compound	NMVOC	All VOCs but excluding methane	Acetone, dimethyl sulphide	Anastasi et al. (1991)
Biogenic volatile organic compounds	BVOC	Volatile organic compound produced directly but not exclusively from biological processes	Hydrogen sulphide, dimethyl sulphide, isoprene, ethene	Geron et al. (1994)
Microbial volatile organic compounds	MVOC	Volatile organic compound formed in the metabolism of fungi and bacteria (preferentially used in soil ecology literature)	Geosmin, dimethyl disulphide, hydrogen cyanide	Korpi et al. (2009)
Oxygenated volatile organic compounds	OVOC	Variety of volatile organic compound with oxygenated side groups	Methanol, acetone	Heikes et al. (2002)
Hydrocarbons	HC	Organic compound consisting entirely of hydrogen and carbon	Methane, pentane, ethene	
Nonmethane hydrocarbons	NMHC	All hydrocarbons excluding methane	Butane, isoprene	

As far as we are aware, a clear definition of '*volatolome*' is lacking in the literature. The term suggests a closer relationship with 'metabolome' and may be the most appropriate term when specifically addressing the volatile metabolome, i.e., the volatile subset of organic and inorganic metabolites produced by the collective metabolism(s) of organism(s), communities or entire ecosystems. It is important to note that a volatolome is the net result of volatile production and consumption processes.

The term '*volatilome*' includes the volatolome plus nonbiogenic, exogenously derived compounds that do not stem from metabolic processes (e.g. environmental contaminants; Insam and Seewald, 2010). For example, this term is often used in the medical study of human breath where the volatile composition is influenced by the combination of metabolically produced gases and volatiles exchanged with the atmosphere or ingested with food. Therefore, volatilome may be the best general term to use when the metabolic processes leading to the production of volatiles are unknown or poorly characterised.

Measurements of the volatilome can provide an *in vivo* metabolic footprint of the entirety of volatiles released without sample pretreatment or extraction (Insam and Seewald, 2010). In many settings, modern volatilomics may be superior to other biomarkers since it offers the possibility for an immediate, continuous (online) measurement which applies, for example, to quality control (e.g. food spoilage, mould detection; Mayr et al., 2003), biodiversity assessment (e.g. herbivore detection; Miresmailli et al., 2010) and medical diagnostics (see Section 3). It is also clear that individual components in environmental volatilomes are of value as information-conveying chemicals (infochemicals) in ecological research (see Section 4). This is because they are central to regulating (i) individuals' movement and behaviour, (ii) ecological interactions between and across populations and (iii) have the potential to affect the complexity of trophic structure in marine food webs (Nevitt et al., 1995; Steinke et al., 2002b, 2006).

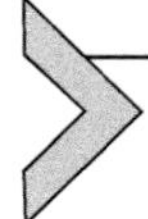

2. PRINCIPAL TECHNIQUES FOR MEASURING BIOGENIC VOLATILES

A number of techniques used in volatilomic analyses have recently been reviewed by Achyuthan et al. (2017). Most analyses use a type of gas chromatography (GC) coupled to various detectors including, for example, mass spectrometric detection (GC-MS; Hopkins et al., 2010), flame-ionisation detection (GC-FID; Steinke et al., 2018), flame-photometric detection (GC-FPD; Steinke et al., 2002a), photoionisation detection (GC-PID; Zhou et al., 2013) or Fourier transform infrared spectroscopy (GC-FTIR; Ketola et al., 2006). These techniques can be further combined with microextraction techniques including headspace solid-phase microextraction (HS-SPME; Vogt et al., 2008) or direct-immersion solid-phase microextraction (DI-SPME; Zhang et al., 2018) and may be extended to include two-dimensional separation techniques (e.g. GC × GC-TOF MS; Phillips et al., 2013). GC techniques use various column materials to separate individual volatiles in often complex mixtures, and the choice of column material and methods in general is selective for particular groups of volatiles. This suggests (i) that the choice of methodological approach strongly affects possible outcomes of a study and (ii) that GC analysis is typically incapable of identifying and quantifying the total volatilome. Other spectrometric methods may overcome such limitations: for example, proton-transfer reaction mass spectrometry (PTR-MS; Halsey et al., 2017; Mayr et al., 2003) and chemical ionisation mass spectrometry (CIMS; Hopkins et al., 2016) have sufficient sensitivity to allow continuous monitoring (online measurement) of selected gas samples.

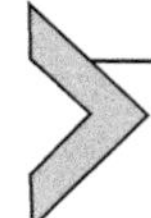

3. MEDICAL VOLATILOMICS PROVIDES A BLUEPRINT FOR ECOLOGICAL RESEARCH

Research in medical volatilomics has greatly advanced the application of biogenic volatiles for the rapid and noninvasive diagnosis of early bacterial infections and genetic disorders including cancer or Alzheimer's disease (Amann et al., 2014). Since metabolic reactions in cells, tissues or fluids result in the formation of volatiles, physiological abnormalities can be detected using methods that identify and quantify volatiles in urine, faeces or exhaled breath samples. For example, GC-MS analysis of urine samples demonstrates unique volatile compounds in transgenic mice with mutations on the amyloid precursor protein gene involved in the onset of Alzheimer's disease (Kimball et al., 2016). In patients suffering from cystic fibrosis (CF), secondary infections with *Pseudomonas aeruginosa* alter microbial diversity and produce a change in a patient's breath volatilome. This offers monitoring opportunities for airway infection as a responsive tool to reduce mortality rates in CF patients (Robroeks et al., 2010). HS-SPME coupled to GC-MS identified six compounds (2-pentanone, 2-heptanone, 3-methyl-3-buten-1-ol, ethyl acetate, ethyl propanoate and 2-methyl butanoate) that were only found in the headspace of cancerous cell lines (Silva et al., 2017). This knowledge can inform the development of novel technologies including GC-MS coupled to gold nanoparticle sensor arrays to distinguish between breath samples of cancer patients and healthy control groups (Peng et al., 2010).

We suggest that much of the existing technology used in medical research can readily inform the investigation of ecological processes. Volatile-mediated signalling in plants already finds application in agriculture (Pickett and Khan, 2016), whereas ecological research using volatiles in aquatic environments is in its infancy (Saha et al., 2018). For example, identifying substantial differences in the quantitative and/or qualitative differences in volatilomes of ecosystems before and during disturbance would accelerate developing robust sensor technology for ecosystem biomonitoring. This could provide novel fast, sensitive and relatively inexpensive tools to alert of an onset of disturbance including infestation with parasites (e.g. salmon lice infection in finfish mariculture) or the changes associated with the deterioration of symbiotic relationships (e.g. before and during coral bleaching). Such early warning would provide avenues for new management strategies that prevent or reduce the degradation of ecosystems. For example, early detection of salmon lice may minimise the geographic spread of infestations so that costly and environmentally damaging treatments can be optimised. Another example application could allow for combinations of coral and their symbionts to be manipulated and tested in coral restoration projects to identify holobionts with greater resistance to stress responses triggered by environmental change.

4. ROLE OF VOLATILES IN AQUATIC ECOLOGICAL INTERACTIONS

Volatiles shape a myriad of species interactions on land and critical ecological processes that depend on smell are commonplace (Ache and Young, 2005; Kessler and Baldwin, 2001). For example, volatiles often facilitate communication through their role in intra- and interspecific signalling in bacteria and fungi (e.g. hydrogen cyanide produced in bacteria; Piechulla et al., 2017), plants (e.g. the gaseous hormone ethylene; Wang et al., 2002), insects (e.g. trail-marking in ants, sex pheromones in moths; Wyatt, 2014) and vertebrates including humans (Nevitt, 2008; Wallraff, 2004; Wyatt, 2014).

In contrast to the terrestrial environment, our understanding of volatile-mediated processes in aquatic environments is limited, with previous studies restricted to a relatively small number of infochemicals (Giordano et al., 2017; Hay, 2009; Moelzner and Fink, 2015; Pohnert et al., 2007). This is surprising since aquatic environments are of particular relevance for volatile-mediated infochemistry with diffusion typically four orders of magnitude slower in water than in air. Furthermore, numerous microbial consumption processes, as exemplified for the abundant trace gases dimethyl sulphide (DMS; Schäfer et al., 2010) and isoprene (Alvarez et al., 2009), rapidly decrease background concentrations and substantially enhance the directional quality of chemical gradients providing the basis of efficient chemical communication. Hence, diffusive volatiles are among the most important parts of the 'chemical language' in the infochemistry of any aquatic organism as demonstrated by the identification of the volatile sexual pheromone ectocarpene in the seaweed *Ectocarpus siliculosus* about 30 years ago (Maier and Müller, 1986).

Aquatic volatiles can also transfer into the atmosphere to communicate food web interactions over relatively long distances (Nevitt et al., 1995; Savoca and Nevitt, 2014) and affect atmospheric processes and climate (Carpenter et al., 2012). For example, DMS, isoprene and numerous halocarbons are climatically important gases that have received global research attention and have been studied for their importance to ecological (Steinke et al., 2006), biogeochemical (Hopkins et al., 2010) and climate science (Vallina and Simó, 2007). Albatrosses and petrels rely on the volatile sulphur gas DMS to track highly productive areas (Nevitt, 2008), and brown seaweeds release volatile pheromones (Pohnert and Boland, 2002). Volatiles also react quickly in response to abiotic conditions and climate stressors (Peñuelas and Staudt, 2010). For example, the production of toxic cyanogen bromide (BrCN) by the microalgae *Nitzschia* cf. *pellucida* that kills surrounding biofilm organisms is light dependent with a short burst of BrCN immediately after sunrise (Vanelslander et al., 2012). Furthermore, different VOCs are emitted by healthy, senescent or stressed cells, apoptotic tissue or prey under predator attack (Achyuthan et al., 2017). These evidences provide us with a 'smoking gun' and make VOCs strong candidates for the interrogation of the state of health of aquatic organisms or communities, and thus the quantification of the volatilome as an ideal target for longer-term biomonitoring of aquatic ecosystems. To explore the basic functionality of volatilomics for biodiversity research, we reanalysed a dataset on freshwater volatiles and demonstrate proof of principle.

4.1 Case Study: Volatilomes of Freshwater Phytoplankton

We previously conducted measurements of isoprene and DMS in unialgal cultures of freshwater phytoplankton (Steinke et al., 2018). Using the raw gas chromatographic data, we explored the diversity of algal volatilomes in four species from different taxonomic classes: *Chlamydomonas reinhardtii* (Chlorophyceae), *Cryptomonas* sp. (Cryptophyceae), *Cyclotella meneghiniana* (Bacillariophyceae) and *Aphanizomenon flos-aquae* (Cyanophyceae). Method details are described in Steinke et al. (2018). Briefly, algal cultures were grown under constant conditions using culture-specific media, aliquoted into glass bottles that were closed with gas-tight glass stoppers (Winkler bottles) and incubated under growth conditions for approximately 4 h. At the end of the incubation, media were purged with N_2 gas at 80 mL/min and this sample gas stream was cryogenically enriched at −150°C before analysis using gas chromatography with flame ionisation detection (GC-FID). The diversity and abundance of peaks on the chromatograms indicated species-specific release of volatiles from the cultures (Fig. 1). The chromatogram traces were manually aligned for comparison, and the data deconstructed into a multivariate matrix on which principal component analysis (PCA) was applied to identify distinct data clusters associated with the four algal genera (Fig. 2).

With this limited dataset it may be premature to attempt an in-depth assessment of the suitability of volatilomics for algal research. For example, it remains to be tested whether the quantity of volatiles is directly proportional to algal biomass and how fluctuations in environmental conditions may affect the quantity and quality (plasticity) of volatilomes to identify robust volatile biomarkers for species of interest. Furthermore, it is also possible that composition and age of algal media may affect the outcome of volatilomic analyses. To counter this, algal media should be taken as controls to eliminate background noise. Nevertheless, the data suggest that the quantification of volatiles could be explored for quality control in algal biotechnology and may be useful in the assessment of biodiversity in complex mixtures and environmental samples.

It is important to note that environmental volatilomics should not be confused with an overly simplistic '*one signature compound per species or function*' approach but uses the entirety of the volatile footprint to inform how we describe biodiversity. The separation of thousands of volatile compounds is already feasible in human breath analysis (e.g. Phillips et al., 2013), and tens of thousands of molecular formulae can be identified in complex natural organic matter using Fourier transform ion cyclotron resonance mass spectra (FT-ICR-MS; Riedel and Dittmar, 2014). As already applied to other omics approaches, a volatilomic assessment of complex environmental samples would require the collection of large datasets in publicly accessible databases, preferentially using unified methodological approaches (see Section 5).

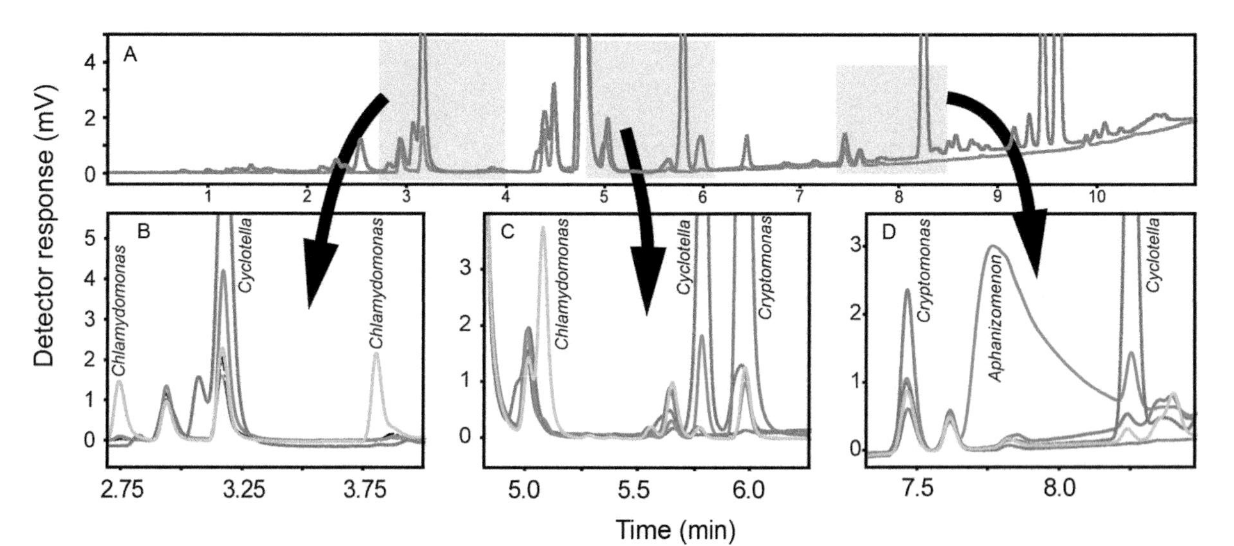

Fig. 1 Overlays of typical chromatograms for volatile organic compounds (VOC) produced from similar biomass in freshwater phytoplankton. (A) Overlay of data for *Cyclotella* sp. (*brown line*) and instrument blank (*grey line*) showing numerous signature peaks. *Grey rectangles* indicate position of timeshots in *lower panel*. (B–D) Overlays showing examples of selected signature volatiles in four species that illustrate the rich signature qualities for the identification of specific genera (*Chlamydomonas* (*green line*), *Cyclotella* (*brown line*), *Cryptomonas* (*red line*), *Aphanizomenon* (*blue-green line*)), based on their volatile profiles. *Grey line* shows instrument blank.

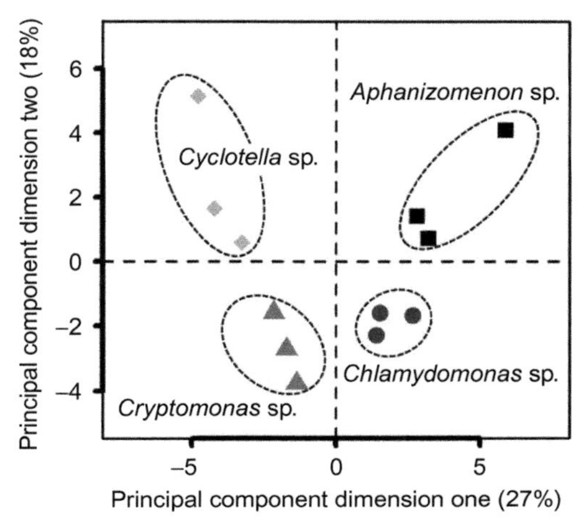

Fig. 2 Principal component analysis (PCA) of volatilomes from four specific genera (*Chlamydomonas, Cyclotella, Cryptomonas, Aphanizomenon*) demonstrates clear separation based on taxonomic identity of the unialgal cultures.

Once the volatilome patterns of major components to biodiversity are known, it will become possible to search for such patterns and identify key volatilome components for possible monitoring purposes. For example, if we demonstrate that harmful algal bloom species produce typical volatile footprints, a network of simple sensors that target a low number of volatiles may inform established but costly routine sampling that uses either microscopic enumeration or genetic identification of harmful species.

Our case study illustrates the utility of volatilomics to investigate taxonomic diversity; however, the production of volatiles is also affected by functional diversity. Microcosm experiments with the phytoplankton *Emiliania huxleyi* demonstrate a several-fold increase in the production of DMS in the presence of herbivores (Wolfe and Steinke, 1996) or during viral infection (Evans et al., 2007). As far as we are aware, a description of functional diversity with volatilomic data has not been attempted, but it is likely that volatilomic data can assist with explaining and predicting the impact of organisms on ecosystems.

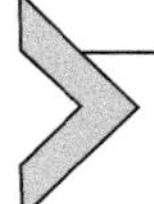

5. APPLICATION OF VOLATILOMICS TO ECOLOGICAL RESEARCH: USING VOLATILOMICS TO 'DIRECT' ENVIRONMENTAL MANAGEMENT

Volatilomes are currently used as biomarkers to detect food spoilage (Mayr et al., 2003) and human diseases (see Section 3). Early monitoring systems detecting stressors on aquatic ecosystems could provide signals or early warning prior to detrimental changes in the environment or changes to food web structures due to biodiversity loss or invasions. This could assist with identifying effective management strategies for degraded and/or fragile ecosystems and industrial processes. Comparing the patterns of BVOCs between healthy and unhealthy systems, for example in seaweed mariculture or salmon farming, would likely identify volatile biomarkers that are sensitive to changes in temperature, the infestation with parasites or infectious diseases (Saha et al., 2018). Our own pilot studies with seaweeds showed that BVOC profiles change during microbial infection that can harm and degrade seaweeds (data not shown). Many conventional tools to assess such changes including PAM-fluorometry or molecular characterisation of the genetic diversity are relatively slow and expensive, and require bench-top equipment and skilled personnel. This makes them unsuited for rapid field-testing and not amenable to automation so that real-time monitoring tools for predicting the state of health of marine organisms or systems are lacking. We suggest that volatilomes should be assessed for such monitoring tasks with the aim of developing simple miniaturised sensor systems to monitor selected biomarkers specific to the organism, community or environment under investigation. This is analogous to many proposed methods for deploying biomonitors for rapid biodiversity assessment based on next-generation sequencing (NGS) technologies (e.g. Bohan et al., 2017), which provide data that can be used to reconstruct entire ecological networks to assess ecosystem health (Derocles et al., 2018). The advantage of volatilomics over these NGS-based methods is that it has the potential to be deployed in real time, continuously tracking the volatilomic footprint, whereas NGS methods currently (and probably for the near future) rely on (an often time consuming) extraction/isolation of nucleic acids from environmental samples. This may make volatilomic biomonitoring more suitable than NGS approaches for certain environments; for example, highly dynamic ecosystems or those where rapid detection of changes (albeit at a potentially lower resolution of information) to the system health are required (e.g. aquaculture/mariculture). However, a potential disadvantage to volatilomic approaches is the associated informatics required to process data are far less well developed than the bioinformatics methods available for NGS analysis, although with very similar analytical needs and a huge potential for sharing knowledge and adapting other 'omics' informatics approaches. Any comparison of volatilome samples based on chromatography must denoise data and provide a base-line correction, account for retention drift and correctly align the chromatogram traces from different samples so that statistical comparisons are comparing equivalent data, pick peaks from the chromatograms and merge the resulting processed data across samples. Many methods already exist for all or part of this (e.g. Hoffmann et al., 2012; Zheng et al., 2017), but generally larger-scale pipelines for processing massive amounts of data are less well developed (for a review of current approaches, see Smolinska et al., 2014). However, dedicated analytical packages are emerging within the research community (e.g. Ottensmann et al., 2018) and as more follow, a rapid increase in the depth of volatilome research will not be far behind. These approaches allow comparisons of data without the need to specifically associate any single VOC signals to individual organisms, and once chromatograms are aligned and the data have been extracted, differences in sample profiles can be statistically evaluated using a standard suite of multivariate analyses and ordinations. However, volatilome analysis will always be strengthened if specific organisms (or metabolic processes) can be identified within the data. For this to happen, volatilome signatures need to be matched to those housed in appropriately curated databases, with identifying tags associated with each specific volatilome signature. This is analogous to the bioinformatic methods used to assign taxonomy or function to NGS data and suffers from the same main limitation—a lack of information, poor curation or incorrect information available in the databases. Arguably the biggest limitation currently is a lack of information on resolved (i.e. links between species and volatilome signatures are established) volatilome signatures from aquatic ecosystems and their organisms. However, this is beginning to change and various platform-specific databases exist for the identification of chromatographic and/or mass-spectrometric data including the National Institute of Standards and Technology (NIST) databases for mass-spectral data, a GC retention index collection and various freely available data analysis tools, or the METLIN metabolite database that provides different metabolite-searching tools. But as aforementioned, identification of metabolites is not strictly required to interpret volatilomes as demonstrated in Figs. 1 and 2—the key, as with all methods, is to provide the appropriate data for the particulare biomonitoring context being considered. In this instance, we believe the suggested volatilomic approach, in combination with established tools and techniques highlighted in Fig. 3, could be employed as a future method for the biomonitoring of aquatic ecosystems and provide an exceptional early warning system for rapid changes in response to environmental perturbations within these ecosystems.

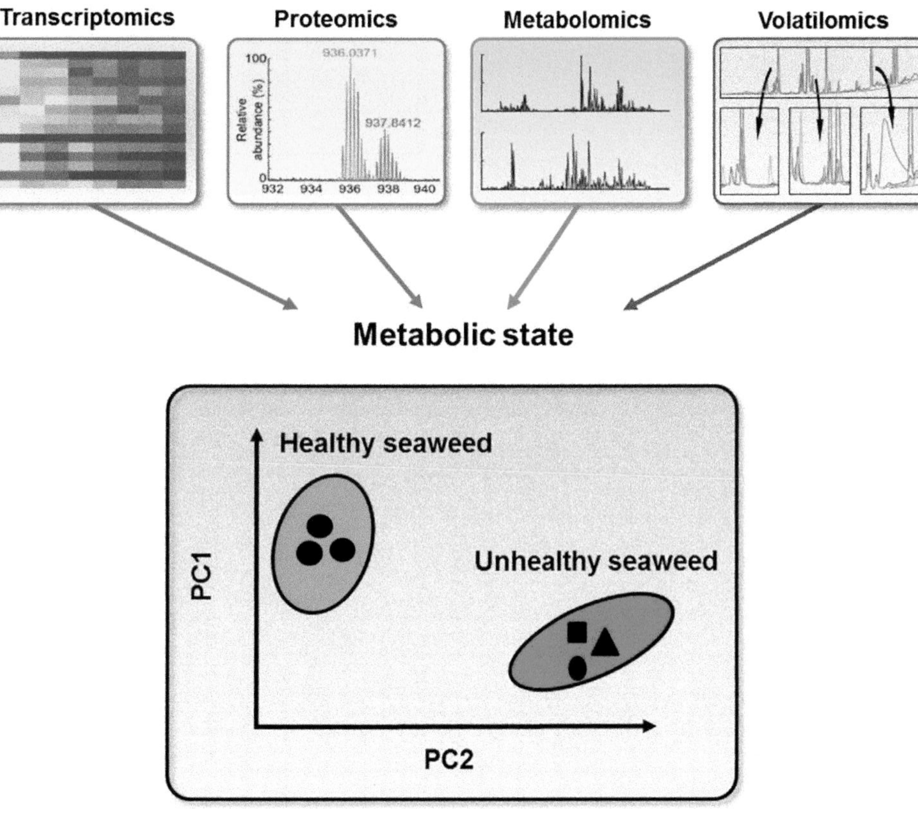

Fig. 3 Can we use volatilomics as a stand-alone tool or in combination with other 'omics' techniques to characterise metabolic finger- and footprints in the assessment of the state of health of aquatic organisms (e.g. healthy vs. unhealthy seaweeds)? *Figure adapted from Brodie, J., Ball, S.G., Bouget, F.-Y., Chan, C.X., De Clerck, O., Cock, J.M., Gachon, C., Grossman, A.R., Mock, T., Raven, J.A., Saha, M., Smith, A.G., Vardi, A., Yoon, H.S., Bhattacharya, D., 2017. Biotic interactions as drivers of algal origin and evolution. New Phytol. 216, 670–681.*

ACKNOWLEDGEMENTS

Constructive comments from two anonymous reviewers improved an earlier version of the manuscript. Bettina Hodapp, Rameez Subhan and Dominik Martin-Creuzburg provided assistance with generating data for Fig. 1. The Konstanz-Essex Development Fund provided financial support to Michael Steinke and Dominik Martin-Creuzburg. Mahasweta Saha acknowledges financial support from the German Research Foundation (DFG) under Grant number SA 2571/2-1.

REFERENCES

Ache, B.W., Young, J.M., 2005. Olfaction: diverse species, conserved principles. Neuron 48, 417–430.

Achyuthan, K., Harper, J., Manginell, R., Moorman, M., 2017. Volatile metabolites emission by in vivo microalgae—an overlooked opportunity? Metabolites 7, 39.

Alvarez, L.A., Exton, D.A., Timmis, K.N., Suggett, D.J., McGenity, T.J., 2009. Characterization of marine isoprene-degrading communities. Environ. Microbiol. 11, 3280–3291.

Amann, A., Smith, D., 2013. Volatile Biomarkers—Non-Invasive Diagnosis in Physiology and Medicine. Elsevier.

Amann, A., Costello, B., Miekisch, W., Schubert, J., Buszewski, B., Pleil, J., Ratcliffe, N., Risby, T., 2014. The human volatilome: volatile organic compounds (VOCs) in exhaled breath, skin emanations, urine, feces and saliva. J. Breath Res. 8, 17.

Anastasi, C., Hopkinson, L., Simpson, V.J., 1991. Natural hydrocarbon emissions in the United Kingdom. Atmos. Environ. A. Gen. Top. 25, 1403–1408.

Bohan, D.A., Vacher, C., Tamaddoni-Nezhad, A., Raybould, A., Dumbrell, A.J., Woodward, G., 2017. Next-generation global biomonitoring—large-scale, automated reconstruction of ecological networks. Trends Ecol. Evol. 32, 477–487.

Carpenter, L.J., Archer, S.D., Beale, R., 2012. Ocean-atmosphere trace gas exchange. Chem. Soc. Rev. 41, 6473–6506.

D'Alessandro, M., 2006. Assessing the Importance of Specific Volatile Organic Compounds in Multitrophic Interactions. PhD thesis, University of Neuchâtel.

Derocles, S.A.P., Bohan, D.A., Dumbrell, A.J., Kitson, J.J.N., Massol, F., Pauvert, C., Plantegenest, M., Vacher, C., Evans, D.M., 2018. Biomonitoring for the 21st century: integrating next generation sequencing into ecological network analysis. Adv. Ecol. Res. 58, 1–62.

Evans, C., Kadner, S.V., Darroch, L.J., Wilson, W.H., Liss, P.S., Malin, G., 2007. The relative significance of viral lysis and microzooplankton grazing as pathways of dimethylsulfoniopropionate (DMSP) cleavage: an *Emiliania huxleyi* culture study. Limnol. Oceanogr. 52, 1036–1045.

Geron, C.D., Guenther, A.B., Pierce, T.E., 1994. An improved model for estimating emissions of volatile organic-compounds from forests in the Eastern United States. J. Geophys. Res. Atmos. 99, 12773–12791.

Giordano, G., Carbone, M., Ciavatta, M.L., Silvano, E., Gavagnin, M., Garson, M.J., Cheney, K.L., Mudianta, I.W., Russo, G.F., Villani, G., Magliozzi, L., Polese, G., Zidorn, C., Cutignano, A., Fontana, A., Ghiselin, M.T., Mollo, E., 2017. Volatile secondary metabolites as aposematic olfactory signals and defensive weapons in aquatic environments. Proc. Natl. Acad. Sci. U.S.A. 114, 3451–3456.

Goulitquer, S., Potin, P., Tonon, T., 2012. Mass spectrometry-based metabolomics to elucidate functions in marine organisms and ecosystems. Mar. Drugs 10, 849–880.

Halsey, K.H., Giovannoni, S.J., Graus, M., Zhao, Y., Landry, Z., Thrash, J.C., Vergin, K.L., de Gouw, J., 2017. Biological cycling of volatile organic carbon by phytoplankton and bacterioplankton. Limnol. Oceanogr. 62, 2650–2661.

Hay, M.E., 2009. Marine chemical ecology: chemical signals and cues structure marine populations, communities, and ecosystems. Ann. Rev. Mar. Sci. 1, 193–212.
Heikes, B.G., Chang, W.N., Pilson, M.E.Q., Swift, E., Singh, H.B., Guenther, A., Jacob, D.J., Field, B.D., Fall, R., Riemer, D., Brand, L., 2002. Atmospheric methanol budget and ocean implication. Global Biogeochem. Cycles 16, 13.
Hoffmann, N., Keck, M., Neuweger, H., Wilhelm, M., Högy, P., Niehaus, K., Stoye, J., 2012. Combining peak- and chromatogram-based retention time alignment algorithms for multiple chromatography-mass spectrometry datasets. BMC Bioinf. 13, 214.
Hopkins, F.E., Turner, S.M., Nightingale, P.D., Steinke, M., Bakker, D., Liss, P.S., 2010. Ocean acidification and marine trace gas emissions. Proc. Natl. Acad. Sci. U.S.A. 107, 760–765.
Hopkins, F.E., Bell, T.G., Yang, M., Suggett, D.J., Steinke, M., 2016. Air exposure of coral is a significant source of dimethylsulfide (DMS) to the atmosphere. Sci. Rep. 6, 36031.
Insam, H., Seewald, M.S.A., 2010. Volatile organic compounds (VOCs) in soils. Biol. Fertil. Soils 46, 199–213.
Kanchiswamy, C.N., Malnoy, M., Maffei, M.E., 2015. Chemical diversity of microbial volatiles and their potential for plant growth and productivity. Front. Plant Sci. 6, 151.
Kennes, C., Abubackar, H.N., Chen, J., Veriga, M.C., 2016. Microorganisms application for volatile compounds degradation. In: Długoński, J. (Ed.), Microbial Biodegradation: From Omics to Function and Application. Caister Academic Press, pp. 183–196.
Kessler, A., Baldwin, I.T., 2001. Defensive function of herbivore-induced plant volatile emissions in nature. Science 291, 2141–2144.
Ketola, R.A., Kiuru, J.T., Tarkiainen, V., Kokkonen, J.T., Rasanen, J., Kotiaho, T., 2006. Detection of volatile organic compounds by temperature-programmed desorption combined with mass spectrometry and Fourier transform infrared spectroscopy. Anal. Chim. Acta 562, 245–251.
Kimball, B.A., Wilson, D.A., Wesson, D.W., 2016. Alterations of the volatile metabolome in mouse models of Alzheimer's disease. Sci. Rep. 6, 19495.
Korpi, A., Jarnberg, J., Pasanen, A.L., 2009. Microbial volatile organic compounds. Crit. Rev. Toxicol. 39, 139–193.
Krupa, S., Fries, N., 1971. Studies on ectomycorrhizae of pine. I. Production of volatile organic compounds. Can. J. Bot. 49, 1425–1431.
Lee, D.Y., Park, J.-J., Barupal, D.K., Fiehn, O., 2012. System response of metabolic networks in *Chlamydomonas reinhardtii* to total available ammonium. Mol. Cell. Proteomics 11, 973–988.
Maier, I., Müller, D.G., 1986. Sexual pheromones in algae. Biol. Bull. 170, 145–175.
Mayr, D., Margesin, R., Klingsbichel, E., Hartungen, E., Jenewein, D., Schinner, F., Märk, T.D., 2003. Rapid detection of meat spoilage by measuring volatile organic compounds by using proton transfer reaction mass spectrometry. Appl. Environ. Microbiol. 69, 4697–4705.
Miresmailli, S., Gries, R., Gries, G., Zamar, R.H., Isman, M.B., 2010. Herbivore-induced plant volatiles allow detection of *Trichoplusia ni* (Lepidoptera: Noctuidae) infestation on greenhouse tomato plants. Pest Manag. Sci. 66, 916–924.
Moelzner, J., Fink, P., 2015. Gastropod grazing on a benthic alga leads to liberation of food-finding infochemicals. Oikos 124, 1603–1608.
Mollo, E., Garson, M.J., Polese, G., Amodeo, P., Ghiselin, M.T., 2017. Taste and smell in aquatic and terrestrial environments. Nat. Prod. Rep. 34, 496–513.
Nevitt, G.A., 2008. Sensory ecology on the high seas: the odor world of the procellariiform seabirds. J. Exp. Biol. 211, 1706–1713.
Nevitt, G.A., Veit, R.R., Kareiva, P., 1995. Dimethyl sulphide as a foraging cue for Antarctic Procellariiform seabirds. Nature 376, 680–682.

Ottensmann, M., Stoffel, M.A., Nichols, H.J., Hoffman, J.I., 2018. GCalignR: an R package for aligning gas-chromatography data for ecological and evolutionary studies. PLoS One 13, e0198311.
Peng, G., Hakim, M., Broza, Y.Y., Billan, S., Abdah-Bortnyak, R., Kuten, A., Tisch, U., Haick, H., 2010. Detection of lung, breast, colorectal, and prostate cancers from exhaled breath using a single array of nanosensors. Br. J. Cancer 103, 542.
Penuelas, J., Staudt, M., 2010. BVOCs and global change. Trends Plant Sci. 15, 133–144.
Phillips, M., Cataneo, R.N., Chaturvedi, A., Kaplan, P.D., Libardoni, M., Mundada, M., Patel, U., Zhang, X., 2013. Detection of an extended human volatome with comprehensive two-dimensional gas chromatography time-of-flight mass spectrometry. PLoS One 8, 8.
Pickett, J.A., Khan, Z.R., 2016. Plant volatile-mediated signalling and its application in agriculture: successes and challenges. New Phytol. 212, 856–870.
Piechulla, B., Lemfack, M.C., Kai, M., 2017. Effects of discrete bioactive microbial volatiles on plants and fungi. Plant Cell Environ. 40, 2042–2067.
Pohnert, G., Boland, W., 2002. The oxylipin chemistry of attraction and defense in brown algae and diatoms. Nat. Prod. Rep. 19, 108–122.
Pohnert, G., Steinke, M., Tollrian, R., 2007. Chemical cues, defence metabolites and the shaping of pelagic interspecific interactions. Trends Ecol. Evol. 22, 198–204.
Redeker, K.R., Cai, L.L., Dumbrell, A.J., Bardill, A., Chong, J.P.J., Helgason, T., 2018. Noninvasive analysis of the soil microbiome: biomonitoring strategies using the volatilome, community analysis, and environmental data. Adv. Ecol. Res. 59, 93–132.
Riedel, T., Dittmar, T., 2014. A method detection limit for the analysis of natural organic matter via Fourier transform ion cyclotron resonance mass spectrometry. Anal. Chem. 86, 8376–8382.
Robroeks, C.M.H.H.T., van Berkel, J.J.B.N., Dallinga, J.W., Jöbsis, Q., Zimmermann, L.J.I., Hendriks, H.J.E., Wouters, M.F.M., van der Grinten, C.P.M., van de Kant, K.D.G., van Schooten, F.-J., Dompeling, E., 2010. Metabolomics of volatile organic compounds in cystic fibrosis patients and controls. Pediatr. Res. 68, 75.
Saha, M., Berdalet, E., Carotenuto, Y., Fink, P., Harder, T., John, U., Not, F., Pohnert, G., Potin, P., Selander, E., Vyverman, W., Wichard, T., Zupo, V., Steinke, M., 2018. Babylonian towers in a blue world—using chemical language to shape future marine health. Front. Ecol. Environ. under review.
Savoca, M.S., Nevitt, G.A., 2014. Evidence that dimethyl sulfide facilitates a tritrophic mutualism between marine primary producers and top predators. Proc. Natl. Acad. Sci. U.S.A. 111, 4157–4161.
Schäfer, H., Myronova, N., Boden, R., 2010. Microbial degradation of dimethylsulphide and related C-1-sulphur compounds: organisms and pathways controlling fluxes of sulphur in the biosphere. J. Exp. Bot. 61, 315–334.
Shrivastava, G., Rogers, M., Wszelaki, A., Panthee, D.R., Chen, F., 2010. Plant volatiles-based insect pest management in organic farming. Crit. Rev. Plant Sci. 29, 123–133.
Silva, C.L., Perestrelo, R., Silva, P., Tomás, H., Câmara, J.S., 2017. Volatile metabolomic signature of human breast cancer cell lines. Sci. Rep. 7, 43969.
Smolinska, A., Hauschild, A.C., Fijten, R.R., Dallinga, J.W., Baumbach, J., van Schooten, F.J., 2014. Current breathomics—a review on data pre-processing techniques and machine learning in metabolomics breath analysis. J. Breath Res. 8, 027105.
Steinke, M., Malin, G., Archer, S.D., Burkill, P.H., Liss, P.S., 2002a. DMS production in a coccolithophorid bloom: evidence for the importance of dinoflagellate DMSP lyases. Aquat. Microb. Ecol. 26, 259–270.
Steinke, M., Malin, G., Liss, P.S., 2002b. Trophic interactions in the sea: an ecological role for climate relevant volatiles? J. Phycol. 38, 630–638.

Steinke, M., Stefels, J., Stamhuis, E., 2006. Dimethyl sulfide triggers search behavior in copepods. Limnol. Oceanogr. 51, 1925–1930.
Steinke, M., Hodapp, B., Subhan, R., Bell, T.G., Martin-Creuzburg, D., 2018. Flux of the biogenic volatiles isoprene and dimethyl sulfide from an oligotrophic lake. Sci. Rep. 8, 630.
Vallina, S.M., Simó, R., 2007. Strong relationship between DMS and the solar radiation dose over the global surface ocean. Science 315, 506–508.
Vanelslander, B., Paul, C., Grueneberg, J., Prince, E.K., Gillard, J., Sabbe, K., Pohnert, G., Vyverman, W., 2012. Daily bursts of biogenic cyanogen bromide (BrCN) control biofilm formation around a marine benthic diatom. Proc. Natl. Acad. Sci. U.S.A. 109, 2412–2417.
Vogt, M., Turner, S., Yassaa, N., Steinke, M., Williams, J., Liss, P., 2008. Laboratory inter-comparison of dissolved dimethyl sulphide (DMS) measurements using purge-and-trap and solid-phase microextraction techniques during a mesocosm experiment. Mar. Chem. 108, 32–39.
Wallraff, H.G., 2004. Avian olfactory navigation: its empirical foundation and conceptual state. Anim. Behav. 67, 189–204.
Wang, K.L.C., Li, H., Ecker, J.R., 2002. Ethylene biosynthesis and signaling networks. Plant Cell 14, S131–S151.
Wolfe, G.V., Steinke, M., 1996. Grazing-activated production of dimethyl sulfide (DMS) by two clones of *Emiliania huxleyi*. Limnol. Oceanogr. 41, 1151–1160.
Wyatt, T.D., 2014. Pheromones and Animal Behavior: Chemical Signals and Signatures. Cambridge University Press.
Zhang, L., Gionfriddo, E., Acquaro, V., Pawliszyn, J., 2018. Direct immersion solid-phase microextraction analysis of multi-class contaminants in edible seaweeds by gas chromatography-mass spectrometry. Anal. Chim. Acta 1031, 83–97.
Zheng, Q.-X., Fu, H.-Y., Li, H.-D., Wang, B., Peng, C.-H., Wang, S., Cai, J.-L., Liu, S.-F., Zhang, X.-B., Yu, Y.-J., 2017. Automatic time-shift alignment method for chromatographic data analysis. Sci. Rep. 7, 256.
Zhou, Y.Y., Yu, J.F., Yan, Z.G., Zhang, C.Y., Xie, Y.B., Ma, L.Q., Gu, Q.B., Li, F.S., 2013. Application of portable gas chromatography-photo ionization detector combined with headspace sampling for field analysis of benzene, toluene, ethylbenzene, and xylene in soils. Environ. Monit. Assess. 185, 3037–3048.

CHAPTER FOUR

Noninvasive Analysis of the Soil Microbiome: Biomonitoring Strategies Using the Volatilome, Community Analysis, and Environmental Data

Kelly R. Redeker*,[1], Leda L. Cai*, Alex J. Dumbrell†, Alex Bardill*, James P.J. Chong*, Thorunn Helgason*

*Department of Biology, University of York, York, United Kingdom
†School of Biological Sciences, University of Essex, Colchester, United Kingdom
[1]Corresponding author: e-mail address: kelly.redeker@york.ac.uk

Contents

1. Introduction 94
2. An Overview of the Soil Volatilome 101
   2.1 Microbes 102
   2.2 Plants 104
   2.3 Soils 105
3. Understanding Essex UK Salt Marsh Sediments Through the Volatilome 108
   3.1 Methods 109
   3.2 Relationships Between Environment, Microbial Community, and the Volatilome 113
4. Conclusion 125
Acknowledgements 127
References 127
Further Reading 132

Abstract

Within soils there are microorganisms that act to break down complex substrates (saprophytes), microorganisms that actively aid nutrient delivery (mycorrhizal fungi and nitrogen-fixing bacteria), and others that hijack the system to their own benefit (parasitic bacteria and fungi). The complex interaction between plants, these microbes, and the soil determines how effectively nutrients will be recycled, with a significant impact on regional productivity and biodiversity. Each microbe plays a role in overall soil function but, despite the critical role they play, soil microbial communities and their functions remain challenging to accurately quantify.

Advances in Ecological Research, Volume 59
ISSN 0065-2504
https://doi.org/10.1016/bs.aecr.2018.07.001

The functional behaviour of soils is difficult to quantify, in part due to the effects of disturbance when sampling. This suggests that noninvasive analytical tools are necessary to diagnose current soil function and to predict changes in soil behaviour with changing climate or land use. Microbial communities, the drivers of soil function, are diverse, and their individual metabolisms are often tightly coupled, such that the microbial community in aggregate may be considered to have a "net" metabolism. This net metabolism can be described by the volatile signatures that propagate from the soil into the atmosphere and, by proxy, allowing a noninvasive analysis of the microbial community active in the subsurface.

Here, we detail the complexities of the soil volatile metabolism, propose a "fingerprint" strategy to describe this complex community that uses trace gas fluxes combined with environmental data, and describe the promising outcomes from an initial foray using this method.

1. INTRODUCTION

Delivery of sustainable agriculture and effective conservation approaches rely on resilient soil functionality under a range of soil conditions (Kibblewhite et al., 2008). Understanding whether a soil is resilient and functional, often termed "soil health", requires a broad understanding of physical, chemical, and biological soil properties (Allen et al., 2011). Many of the basic chemical and physical properties of soil are changing (driven by climate, land use, and agricultural pressures) with poorly understood impacts on soil biological communities and processes. Efficient and appropriate biomonitoring methods are needed to identify the ecological impacts of these changes and to manage ecosystems effectively before we lose functionality and critical ecoservice delivery.

Nearly all current methods used to explore these important subsurface aspects/processes require that the soil be collected and treated; thus modifying an ecosystem that is inherently sensitive to disturbance and producing potentially misleading conclusions (Choi et al., 2017). Furthermore, soil microbial communities are often monitored using DNA-based methods, which (i) generally require substantial modification of the soil environment; (ii) conflate the presence of a microbe with its activity, or functional impact; and (iii) tend to focus on community members with greatest abundance. In order to support sustainable management options and deliver better agricultural yields while maintaining soil capacity and function, it is important to establish a noninvasive, biomonitoring methodology that allows characterization of soil microbial community and function without soil disturbance.

We propose that the use of the volatile fraction of microbial metabolisms (the volatilome) allows us to bypass many of these concerns. This noninvasive, surface-sampling technique allows multiple analyses over time at the same location, reflects on-going microbial community function and activity, and therefore allows a more accurate assessment of the state of soils.

Soil volatile fluxes originate from microbial metabolic processes. Microbes in soil environments require constant metabolic activity; for cell maintenance, to grow and replicate, and to maintain functional capacity during periods of stasis. This metabolism either generates chemical outputs (as a metabolic endpoint, like CO_2, or as a signalling compound, or for use in defence) or takes in chemicals for purpose (loose amino acids can be harvested for nitrogen or direct use, or sugars can be harvested for carbon/energy). These resources are obtained from, or are released into, the soil matrix (Fig. 1).

The soil matrix consists of solid soil particles, soil particle aggregates, and open air- and/or water-filled volumes, called pore space. The soil pore space is important for root growth, water movement, water storage, and gas movement within the soil. Soils interact with the air above them through (primarily) diffusive transport, exchanging topsoil pore space air with the air lying over the soil surface. The topsoil pore spaces interact with the pore spaces in the layers below, leading to overall material movement throughout the soil column (Redeker et al., 2015). This air exchange allows organisms within the soil to interact with the overlying air, either through delivery of excess metabolic by-products from soil organisms to the atmosphere (e.g. carbon dioxide from plant and microbial respiration) or through consumption of delivered chemical compounds via atmospheric air (e.g. methane consumption in aerobic forest soils) (Conrad, 1996). This chemically complex combination of efflux (soil-to-atmosphere) and influx (atmosphere-to-soil) has been termed the "soil volatilome" and has been proposed as a tool to noninvasively study soil health and soil microbial communities (Insam and Seewald, 2010; Muller et al., 2013; Schmidt et al., 2015) (Fig. 1).

The volatilome has already been used for some single-species diagnostics under controlled conditions. In particular, pathogens have been identified through headspace analysis of single organism cultures (headspace analysis is the practice of containing organisms in trapped air volumes and sampling that air volume at the beginning and end of a set period of time) (Fig. 2). These analyses have led to the creation of profiles of specific volatiles and/or blends of volatiles by which contamination by these pathogenic organisms may be confirmed. Medicine, food, air, and water have all been

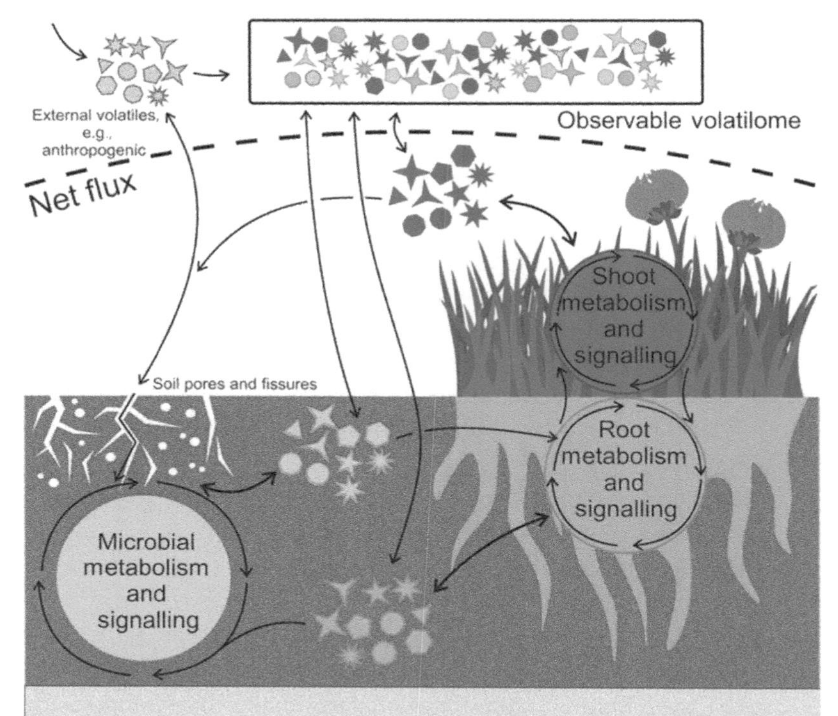

Glossary

Headspace—Sampled, entrapped air in contact with an organism or soil or ecosystem over a period of time

Boundary layer—The part of the atmosphere directly influenced by surface features, leading to turbulent mixing at higher elevations (>1 m) and primarily diffusive flow at the surface interface (0 m).

Pore space—Volumes within soils and sediments that can either be filled with air or water. Not considered to be directly connected to the boundary layer.

Efflux—Movement of volatiles from the sampled organism/volume to the headspace. Associated with metabolic and abiotic production processes.

Influx—Movement of volatiles from the headspace into the sample volume/organism. Associated with consumption by organisms or degradation by chemical means.

Diffusion—Volatile movement driven by random chemical movement that tends, over time, to move along chemical gradients delivering material from regions of high concentrations to regions of low concentration. Very slow relative to other movement processes.

Ebullition—Bubble-driven movement of gases. Rapid relative to diffusion.

Signalling—Volatiles generated by living organisms to provoke a response in other individuals. Examples of signalling include: (a) an alarm compound to provoke a defense response in others of the same species and (b) an attractant to draw another organism towards the signalling individual.

Inhibition—Volatiles generated by organisms to reduce efficacy of another organism. Examples of inhibition include: (a) antibiotic production where plants, bacteria of fungi generate compounds which act to inhibit or kill pathogenic bacteria growth.

Gross metabolism—Any given organism will generate certain compounds (e.g. carbon dioxide through respiration) and will consume others (e.g. sugars and oxygen). We characterize this individual behaviour and outcome as a "gross metabolism".

Net metabolism—In many cases volatiles generated by one organism for one purpose may be consumed by another organism. For instance, methanogens produce methane which is eaten by methanotrophs. Another, more complex example is methyl bromide which is possibly produced as a defense compound (toxic at certain concentrations) by plants and fungi but is consumed by a range of methyltrophic bacteria. We identify "Net metabolism" as the final, additive outcome of these various "gross metabolic" processes.

Cometabolism—In some cases it is required that multiple metabolites are processed at the same time. An example of this is cometabolism of halogenated hydrocarbons with other substrates, where the halogenated material can not be metabolised alone (Wahman et al., 2005).

Fig. 1 A schematic of the volatilome within the soil and boundary layer, including a glossary of terms.

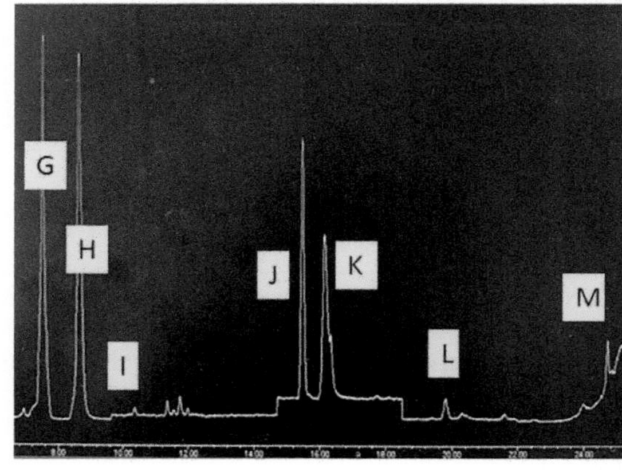

A: PAR transparent headspace sampling chamber over PVC chamber base

B: Glass-coated stainless steel transfer line, including ~20 cm Ascarite trap

C: Electropolished stainless steel sampling canister

D: Liquid nitrogen condensation trap

E: Standard cylinders attached to the condensation trap rig

F: Gas chromatograph mass spectrometer, equipped with a 30 m PoraBondQ column

Output from the mass spectrometer in selected ion mode

G: methyl chloride

H: CFC-12

I: methyl bromide

J: dichloromethane and methyl iodide

K: CFC-11 and dimethyl sulfide

L: isoprene

M: chloroform

Fig. 2 Volatilome sampling. Volatiles are sampled within a trapped airspace (A), first at an initial time point and then 9 min later. This allows the analyst to ascribe changes in

(Continued)

tested for contamination through analyses of these signature volatiles (Garcia-Alcega et al., 2016; Insam and Seewald, 2010 and references therein). Similarly, soil analyses have been performed to detect agricultural plant pathogens and moulds in soils through their identified volatile signatures (Peñuelas et al., 2014 and references therein). It has been proposed that bacteria transformed with a plant-derived methyltransferase enzyme can be used to report on subsurface microbial gene expression dynamics through volatilized methyl halides (Cheng et al., 2016).

Moving from single-species identification to determination of soil microbial communities through the volatilome has been challenging, although some environmental conditions allow simple functional assignments. Anoxic, saturated subsurface environments may support methanogens, which can be confirmed through methane release from soils (LeMer and Roger, 2001), whereas methane, methanol, or methyl halide (CH_3X, where X=Cl, Br, or I) uptake by soils indicates methylotrophs active within a (most likely) aerobic subsurface (Chistoserdova et al., 2009). In plant-free soils a substantial efflux of methyl halides appears to be indicative of aerobic conditions and fungal activity (Redeker and Kalin, 2012; Redeker et al., 2004; Watling and Harper, 1998).

The signal received at the soil surface is the net combination of all volatile metabolisms from all subsurface organisms, often acting counter to each other (e.g. metabolites produced by one organism affect or are consumed by another) (Fig. 1). The identification of soil microbial communities through the volatilome has been hampered by this complexity and requires advanced statistical/mathematical methods to disentangle the disparate signals received at the surface. To use these tools we must have a greater understanding of the emissions/consumption profiles for many of the organisms expected within the soil column.

Fig. 2—Cont'd the sampled air mass to processes occurring in the trapped airspace over the sampling period while avoiding excessive stress on the living system. Samples are drawn from the chamber into the sampling canister (C) via an Ascarite trap (B), which reduces water and carbon dioxide content of the sampled air. Once returned to the lab the canisters are attached to the liquid nitrogen condensation trap (D) which cools the air under vacuum, removing nitrogen, oxygen, and argon from the sample and leaving only compounds normally present in the part-per-billion range. This condensed, concentrated sample is delivered to the GC–MS (F), which separates the sample (gas chromatography) and analyses each individual compound (G, H, I, J, K, L, and M) based on their mass (mass spectrometry).

Comprehensive databases that link microbes to defined metabolic uptake and outputs are crucial for effective biomonitoring, since they provide the metabolic profiles that allow specific identification of community (functional) members. Comprehensive volatile databases combined with noninvasive, surface-based sampling may be of more use in determining subsurface microbial communities than DNA-based methods, analysed through extant 16S databases since noninvasive methods are linked to on-going microbial activity as opposed to DNA methods which quantify the presence of organisms in the soil. Therefore, even if we accept the problems of disruption caused by sampling soils for DNA-based outcomes, DNA-based methods are still unlikely to provide data on which organisms remain active within the soil profile, which is often the specific information required. Unlike genomics (DNA based), both transcriptomics (based on RNA) and proteomics (based on expressed proteins) approaches can probe microbial function. However, both still require soil disturbance for analysis, limiting their efficacy for understanding in situ soil community structure and function. They may be of greater use in creating effective databases linking microbial community members to specific metabolic outcomes. At least one database for the volatilome has been initiated. It incorporates published volatile emissions from over 10,000 microbial specimens (Lemfack et al., 2014).

The available databases for volatile profiles are still in the formative stage. Within the Lemfack et al. (2014) database the organisms reported are those organisms that are culturable, and the reported volatiles are those of sufficient concentration that they may be sampled, separable, and identifiable through the methods chosen. This tends to bias the collected VOC profiles towards subsets of chemical compounds observable with commonly used methods (hydrocarbons, terpenes, alcohols, aldehydes, and ketones) and the organisms to those that may be cultured within the lab, known to be less than 30% of the total microbial communities found within soils (Alessi et al., 2018). Microbes and plants have been shown to differ strongly in the total amount and the profile of sulphur-bearing, halogenated, and metallic compounds volatilized (Bentley and Chasteen, 2004; Gribble, 2003; Parks et al., 2013; Redeker et al., 2004), suggesting that these types of compounds may have substantial predictive benefit, yet these are not fully represented in current databases since the relevant analytical methods are not as commonly used. It is also worth noting that the database focusses on produced volatiles and does not currently include consumed volatiles, a critical gap that needs to be addressed in order to understand complex metabolite interactions within soils. Despite current limitations, the database has confirmed that some species can be predictably identified from VOC profiles (Lemfack et al., 2014; Peñuelas et al., 2014). The challenge remains, however, to identify microbial communities within the subsurface based entirely on the net sampled volatilome.

The challenge has been aided by the development of new soil community databases and paired isotopic analyses. These approaches are inappropriate for long-term profiling of soil communities (either due to their high specificity or to their disruptive nature) but they allow greater scope and confidence in assignments made through our proposed noninvasive volatile profiling. For instance, greater specificity of identification within the volatilome can be obtained from isotopic analyses of emissions, either singly (Redeker and Kalin, 2012) or paired (Oduro et al., 2013). The most common example of function-specific isotopic information comes from the methane cycle, where methanogens and methylotrophs generate very different isotopic profiles (von Fischer and Hedin, 2007); however, species-specific isotopic signatures have been observed for methyl halide emissions from plants and fungi (Harper et al., 2001; Saito and Yokouchi, 2008). Paired isotopic analyses provide greater specificity than individual isotopic analyses, having been used successfully to characterize cycling of bacterial sulphur volatiles (Oduro et al., 2013) and to analyse nitrous oxide emissions from nitrification and denitrification pathways (Zhang et al., 2016).

Furthermore, recent studies using soil-disruptive 'omics technologies and databases derived from these data allow us to identify which members of the soil microbial community are more likely to coexist, and under which environmental conditions (Fierer et al., 2009; Griffiths et al., 2011; Serna-Chavez et al., 2013). These data, in combination with local soil environmental data, may be used either (i) to generate more precise microbial community predictions from noninvasive, nondestructive volatilome analyses or (ii) to ground truth volatilome-derived community assignments.

Even with these additional technologies, successful characterization of soil microbial communities using the soil volatilome requires a fingerprint strategy. Criminal forensic fingerprinting (used to profile human culprits) relies on several dozen distinct minutiae, or patterns within a fingerprint, to accurately identify specific individuals and to avoid false positives. We expect that soil communities will also require multiple identification points to reduce the possible range of community assignments and provide greater accuracy in identification. At the moment only broad strokes descriptors of microbial communities can be depicted, with basic environment, functionality, and family level organisms identified (e.g. anoxic environments leading to methane generation by methanogenic archaea), but as databases incorporating functionality, favourable environments of specific organisms, likely coexistance of community members and species-specific volatilomes are developed more identification points may be added and genera or species level identification may be possible.

This chapter will explore the utility of a volatile "fingerprint" approach to describing microbial communities. We begin by outlining the current state of knowledge regarding how major biological communities influence soil and sediment volatiles. We include a discussion of interactions driven purely through soil chemical and physical properties as well as an examination of complex environment/community interactions. We will then describe the methods and approaches used in a study that characterized UK salt marsh sediment microbial communities through analysis of environmental parameters, microbial community, and a subset of volatile fluxes. The results of this study are presented to demonstrate the potential of the volatilome to coherently and self-consistently predict members of the soil community and the sediment environment.

2. AN OVERVIEW OF THE SOIL VOLATILOME

Components of the volatilome can be loosely classed based on their function within the soil. The volatilome includes many basic metabolites and they interconnect with all major biogeochemical cycles (Falkowski et al., 2008). These metabolites include end products (e.g. CO_2, Yuste et al., 2007; CH_4, LeMer and Roger, 2001), intermediates (e.g. N_2O, Stein and Yung, 2003; acetic acid; Krzycki and Zeikus, 1984 and others; Peñuelas et al., 2014), and consumed metabolites (e.g. CH_4, LeMer and Roger, 2001; and other methyl donors; McAnulla et al., 2001). Metabolites can also fall into multiple categories, for instance, methane and nitrous oxide are produced and consumed by different organisms within the same soil profile. The consumption of volatiles is often overlooked in discussions regarding the volatilome, despite the important cues that this provides for soil pH, anoxia, redox potential, and nutrient status.

VOCs produced by one group of organisms can promote or inhibit other organisms and their behaviours. For instance, some bacterial volatiles have been shown to promote growth in *Arabidopsis thaliana* (Ryu et al., 2003; Zhang et al., 2007), while volatiles from an Ascomycete (*Fusarium culmorum*) induce greater motility in some strains of bacteria (Schmidt et al., 2016). A number of plausible antifungal and antibacterial compounds have been identified from plant, fungal, and bacterial emissions (Schmidt et al., 2015). These compounds do not consistently generate significant reduction in bacterial or fungal load, perhaps (i) due to differences in soil chemistry/reactivity between study sites, (ii) due to spatiotemporal variation between regional soil microbial communities, or (iii) due to greater efficacy of the antimicrobial compound when combined with other volatiles emitted from other members of the microbial community (Gallucci et al., 2009; Hemaiswarya and Doble, 2010). Although acute impacts of these produced antimicrobial compounds appear to be limited, long-term, chronic effects from these compounds on microbial activity and composition may still represent an important component of ecosystem function by modifying local plant–microbe interactions.

Volatiles act to communicate information within soils, generating defence and/or stress tolerance responses which have been reported for plant–plant (Pierik et al., 2014), bacteria–plant (Toljander et al., 2007), bacteria–bacteria (Schmidt et al., 2015; Schulz-Bohm et al., 2015), and fungi–bacteria (Schmidt et al., 2015; Schulz-Bohm et al., 2015) interactions. More complex, multitrophic signalling has also been observed, as between plants signalling predatory nematodes to consume herbivorous insects (Ali et al., 2011), or between virally infected plants signalling nematodes to provide transportation for the virus (Turlings et al., 2012). These signalling volatiles are targeted towards specific organismal interactions and are therefore some of the most likely components of the volatilome to provide species-, or genera-, specific identification when they have been characterized.

While the primary actors within soils that influence the volatilome are microbes and plants, other biological actors can affect these primary actors through predation (Griffiths et al., 1999) and viral attack (Pan et al., 2014), and these may have significant impact on consumed and produced metabolites.

2.1 Microbes

The microbial community in soils consists of four separate (large in themselves) subgroups, including bacteria, fungi, archaea, and protists. These organisms fulfil five primary functional roles as autotrophs (not fungi), decomposers, mutualists, predators/pathogens (not archaea), or lithotrophs (not fungi or protists) which tie them intimately to the biogeochemical cycling of chemical elements in soils. Many reviews and studies exist for bacterial volatiles (e.g. Insam and Seewald, 2010 and references therein) but far fewer for fungi (and those are limited to pathogenic, ECM or saprophytic fungi; very little or no data exists for arbuscular mycorrhizal fungi despite their critical importance), and importantly there are no substantial reviews regarding non-greenhouse gas VOC emissions from archaea (but several reviews regarding their capacity to live in extreme environments, leading to potentially unique metabolisms with unique volatile signatures).

Attempts to classify functional group or microbial type by their volatile profile have recently been published (Muller et al., 2013; Peñuelas et al., 2014). These studies have either performed analyses or collected reported fluxes for comparison across a range of organisms. Both studies demonstrate the challenge; most volatile metabolites are common across all microbial life. However, it is clear that fungal groups (ECM, pathogenic and saprophytic; Muller et al., 2013) or different classes of bacteria and fungi (Actinobacteria, Ascomycetes, Bacteroides, Cyanobacteria, Firmicutes, Alpha-, and Gammaproteobacterial; Peñuelas et al., 2014) may be separable through their distinct volatile profiles. There exist differences in the reported signature volatiles between studies; for instance sesquiterpene compounds were common across all fungal groups tested in the study performed by Muller et al. (2013), whereas in Peñuelas et al. (2014) the fungal group was identified by their oxygenated metabolites (including alcohols, ketones, and furans) and terpenes were not indicated. Also, Peñuelas et al. (2014; and associated database) do not appear to include the broad range of halogenated compounds generated by fungi and bacteria as reported by Gribble (2012) and Field (2016).

It is probable that the differences in reported identifying compounds are due to one or a combination of (i) context-specific responses, (ii) different organisms studied within the same broad classification, and (iii) differing methods of analysing the volatiles released from these organisms. For example, the method used by Muller et al. (2013) preferentially selects for less volatile, more hydrophobic compounds leading to a probable underestimate of low-molecular weight, more polar, halogenated volatiles and emphasizing the less polar, less volatile, higher molecular weight sesquiterpenes. This highlights one of the primary challenges facing the volatilome community. Multiple methods of analyses targeting different chemical families and functions are necessary in order to identify the most useful compounds for identification, which will then allow us to characterize the soil microbial community in the most versatile and robust manner.

2.2 Plants

On the whole, volatile metabolites generated by aboveground plant tissues have been intensively studied (e.g. Loreto et al., 2006; Matsui et al., 2012; Niinemets et al., 2013; Ul Hassan et al., 2015). Subsurface plant volatile interactions are less intensively studied than aboveground emissions but some aspects have been documented, including plant–microbe, plant–pest, and multitrophic interactions (Massalha et al., 2017).

Root emissions are quite different from volatiles emitted by aboveground tissues and the environment in which they exist affects the manner in which these volatiles move or are metabolized (Read et al., 2003). Root exudates influence their local environment, producing acids and surfactants to aid nutrient acquisition (Jones, 1998; Mukherjee and Lal, 2013), toxins to defend against herbivory (Jassbi et al., 2010), and compounds that attract predators that consume herbivorous pests (Ali et al., 2011). Roots also influence their local environment by removing water through transpiration (Nobrega et al., 2017; Verstraeten et al., 2008) and mobilization/translocation of water through root networks (Prieto et al., 2012), leading to modification of available pore air space within the rhizosphere which may affect volatile compound diffusivity prior to soil microbial consumption, chemical degradation, or emission to the atmosphere (Redeker et al., 2015).

Aerenchymous plant roots (aerenchymous plants have large air spaces in leaf, stem, and root tissues allowing easier air passage between plant components) influence the local volatile environment through provision of a transit route into/out of the soil that is more rapid than diffusion through waterlogged soils (Laanbroek, 2010). The role of aerenchymous roots is to provide ready access of atmospheric oxygen to the subsurface plant components, and this allows subsurface volatiles to escape using the same transit route. Direct and indirect influences on soil hydrology/diffusivity from physical and chemical root impacts are varied and have been the subject of multiple studies and reviews (e.g. Bodner et al., 2014 and references therein).

While root-based emissions are not as well profiled or quantified as aboveground emissions there are several families of plants that are known for specific trace gas emissions. The brassicas (family Brassicaceae, the cabbages) in particular are known for their production of volatile, sulphurous compounds that are characteristic of the family (Virtanen, 1965). These sulphurous compounds are also produced by Brassicaceae root tissues and specifically in the form of an herbivory response that generates isothiocyanates (through enzymatic degradation of glucosinolates) (Doheny-Adams et al., 2017). Sulphur metabolisms have been linked to halogen-compound metabolisms through halide/thiol methyltransferases, and the Brassicaeae have been reported to be the plants that produce the most methyl halides per gram plant tissue, although this has yet to be tested substantially on root tissues (Saini et al., 1995).

2.3 Soils

Soils and sediments should be considered multilayered compartments. In each compartment, biogenic and abiotic production may be occurring, along with metabolic consumption and abiotic degradation (although biological processes tend to dominate; Redeker and Kalin, 2012). Simultaneously each compound will be diffusing along a concentration gradient between its current location and the soil surface (Table 1; Fig. 1). This diffusive rate determines the overall degradation and consumption of the compound of interest since it regulates the contact period with active sites and organisms within the soil matrix. It is estimated that as much as 90% of methane generated in the subsurface is consumed before it can be emitted at the surface whereas ebullition and/or transport through aerenchymous plant tissues, which lowers transit time, substantially reduces this metabolic loss (Zhu et al., 2012). The overall flux of compounds from the soil surface, the volatilome, will be determined by this balance between production, consumption, and rate of movement (Fig. 1).

The soil and sediment matrix through which volatiles pass determines the residence time of the compound within the soil (Redeker and Kalin, 2012). Greater porosity leads to more rapid diffusion while greater pore water content reduces diffusive rates (Redeker et al., 2015). Surface winds may reduce residence time through advection or rotational pumping of pore air spaces (Redeker et al., 2015). Soils produce many volatiles through abiotic reactions and most biogenic compounds will react in the soil environment given sufficient time (Insam and Seewald, 2010).

The soil environment may directly influence the volatilome by affecting one or more of the above processes. Indirect effects from the soil environment are also possible. Abiotic stress tends to lead to changes in emissions, for instance, drought and flooding have been shown to modify the type and extent of volatiles generated by bacteria (Asensio et al., 2007). Soil saturation and drought fill and empty soil pore spaces, leading to reduced and enhanced diffusivity, respectively. Extreme drought can also lead to cracking at the surface, providing deep advective access to the subsurface, and creates open channels within the soil where root–soil connections are lost as the soil shrinks. Soil salinity, pH, nutrient status, and temperature have been shown

Table 1 Biological and Chemical Information for Analysed Volatiles

Volatile	Abbreviation	Chemical formula	Soil Interactions						Purported Biological Role
			Physical	Chemical	Plant	Fungi	Bacteria	Archaea	
Methane	CH_4	CH_4	(i) Diffuses rapidly (ii) Strongly prefers gas to liquid phase	(i) Very limited chemical decomposition (ii) Very limited chemical production	(i) Direct transit between soil and atmosphere provided by aerenchymous plants	(i) Very limited fungal production (ii) Very limited fungal consumption	(i) Not reported as a producer (ii) Known consumers	(i) Known producers (ii) Known consumers	(a) Energy by synthesis (b) Energy by oxidation
Methyl chloride	MeCl	CH_3Cl	(i) Diffuses $\sim 2\times$ slower than methane (ii) $75\times$ more soluble in water than methane	(i) Decomposition reaction with OH^- in water (ii) Limited production in soil	(i) Production in roots and aboveground biomass (ii) No reported consumption	(i) Production via sapprotrophs and ectomycorrhizal fungi. Arbuscular and ericoid fungi uncertain (ii) No reported consumption	(i) Not reported as a producer (ii) Known consumers	(i) ??? (ii) ???	(a) Energy by oxidation (b) Oxidative stress tolerance (c) Targeted methylation, specialized metabolism (d) Remains uncertain
Methyl bromide	MeBr	CH_3Br	(i) Diffuses $\sim 2.5\times$ slower than methane (ii) $100\times$ more soluble in water than methane	(i) Decomposition reaction with Cl^- or OH^- in water (ii) No reported production in soil	(i) Production in roots and aboveground biomass (ii) Reported consumption in plant biomass at high ambient concentrations	(i) Production via sapprotrophs and ectomycorrhizal fungi. Arbuscular and ericoid fungi uncertain (ii) No reported consumption	(i) Not reported as a producer (ii) Known consumers	(i) ??? (ii) ???	(a) Energy by oxidation (b) Biocidal, potential defence compound (c) Targeted methylation, specialized metabolism (d) Remains uncertain

Methyl iodide	MeI	CH_3I	(i) Diffuses ~3 × slower than methane (ii) 150 × more soluble in water than methane	(i) Decomposition reaction with Cl^-, Br^-, or OH^- in water (ii) No reported production	(i) Production in roots and aboveground biomass (ii) No reported consumption	(i) Production via sapprotrophs and ectomycorrhizal fungi. Arbuscular and ericoid fungi uncertain (ii) No reported consumption	(i) Known producers (ii) Known consumers	(i) ??? (ii) ???	(a) Biocidal, potential defence compound (b) Targeted methylation, specialized metabolism (c) Remains uncertain
Dimethyl sulphide	DMS	CH_3SCH_3	(i) Diffuses ~2 × slower than methane (ii) 400 × more soluble in water than methane	(i) Decomposition reaction with OH^- in water (ii) No reported production	(i) Known producers (ii) Certain plant pigments react with DMS	(i) No reported production (ii) No reported consumption	(i) Known producers (ii) Known consumers	(i) Known producers (ii) Known consumers	(a) Signalling compound (b) Bioactive, potential defence compound (c) Remains uncertain
Isoprene	Isoprene	C_5H_8	(i) Diffuses ~2 × slower than methane (ii) 10 × more soluble in water than methane	(i) Very limited chemical decomposition (ii) Very limited chemical production	(i) Known producers (ii) No reported consumption	(i) No reported production (ii) ???	(i) Known producers (ii) Known consumers	(i) ??? (ii) ???	(a) Abiotic stress tolerance (b) Constituent metabolite of other signalling compounds (c) Other roles remain unclear
Chloroform	$CHCl_3$	$CHCl_3$	(i) Diffuses ~3 × slower than methane (ii) 200 × more soluble in water than methane	(i) Very limited chemical decomposition (ii) Very limited chemical production	(i) Known producers (ii) No reported consumption	(i) Known producers (ii) No reported consumption	(i) Not reported as a producer (ii) Known consumers	(i) ??? (ii) ???	(a) Bioactive, potential defence compound (b) Remains uncertain

to be determining factors for soil microbial community composition (Lozupone and Knight, 2007) and therefore function (Jones et al., 2014; Nicol et al., 2008). Redox status also substantially influences the volatiles produced and released from the soil/sediment system (Devai and DeLaune, 1995). Greater surface exposure leads to enhanced winds over the soil surface, acting to remove volatiles from the soil rapidly and reducing the soil concentration gradient (Redeker et al., 2015).

3. UNDERSTANDING ESSEX UK SALT MARSH SEDIMENTS THROUGH THE VOLATILOME

Multiple challenges remain to the use of the volatilome as an effective, noninvasive tool for diagnosing soil microbial communities. The databases that are currently established need to be maintained, effectively curated, and expanded to include the volatilome of more individual organisms, across a broader range of volatiles, and to include consumption of volatiles as well as emission.

This database approach is currently limited by our capacity to culture soil microorganisms in the lab. Most soil organisms are not currently culturable and are unlikely to become so. It is therefore important to develop 'omics techniques sufficiently that a full range of functional behaviour is ascribable, even when encountering unusual and/or rare organisms. During this development period, we will need to target functional behaviours of specific, poorly understood organisms which lab-based culturing and 'omics approaches are capable of resolving in order to provide maximum database coverage.

To develop an effective, volatilome-centred, noninvasive approach to monitoring microbial communities, in which specific volatile fluxes are linked to individual functions or organisms, it is important to place these microorganisms in environmental and ecological context. Some organisms (e.g. aerobes) will rarely cohabit in large numbers with other organisms (e.g. anaerobes) dependent upon both competitive and environmental pressures. By understanding organisms within a community and environmental context, we can constrain other probable community inhabitants once a detected volatile signature is ascribed to a specific organism. Therefore, databases should also begin to incorporate environmental parameters and microbial community members with which the organisms (and their volatile fluxes) are associated. It remains an open question as to whether the fingerprint strategy can be applied across a range of soil types and conditions, or whether it will be ecosystem specific. To address this question, it is important to begin to apply this solution to a range of soils and sediments to observe the consistency of outcomes across a range of ecotypes.

We describe later the first effort to integrate many of these necessary components using data from Essex, UK salt marsh sediments. We have placed community members in context, both in terms of their preferred local environment and in terms of their common microbial (and plant) associates through genomics analysis of sediment paired with aboveground biomass and several important sediment characteristics. We have measured the net flux of a number of informative volatile compounds, taking advantage of both emission and consumption profiles. We discuss our preliminary findings in terms of volatile predictions of community members and environmental parameters.

3.1 Methods

3.1.1 The Study Sites

Three paired (natural and managed) salt marsh sites were selected along the Essex coastline within the Colne and Blackwater estuaries (Fig. 3). The Colne Estuary provides a particularly rich resource for understanding how microbial communities have adjusted with historic environmental change (Nedwell et al., 2016). Sites included a location within the Fingringhoe Wick Nature Reserve (Fingringhoe Wick Range: 51° 49′ 53.95″ N 0° 58′ 11.54″ E), one near Mersea Island (51° 47′ 50.29″ N 0° 55′ 16.68″ E) and the third was located at the Essex Wildlife Trust Abbotts Hall Farm (51° 47′ 11.95″ N 0° 51′ 40.78″ E). Managed saltmarshes were paired with adjacent natural salt marshes and elevations of sampling sites were matched within 20 cm among marshes.

3.1.2 Sampling the Volatilome and Sediment in the Field

Microbial communities, environmental parameters, and volatile fluxes were based on sequences and samples obtained from paired natural and managed salt marsh sediments (5–10 and 10–15 cm depths) sampled once per season between July 2014 and June 2015. Four different vegetation-dominant communities were studied, including *Atriplex* spp., *Limonium* spp., *Puccinellia* spp., and unvegetated mud. Comparative volatile fluxes between natural and managed salt marshes were taken on the same date while sediment cores to describe local sediment characteristics were taken from within the volatile sampling "footprint" (Fig. 2). Aboveground biomass and sediment

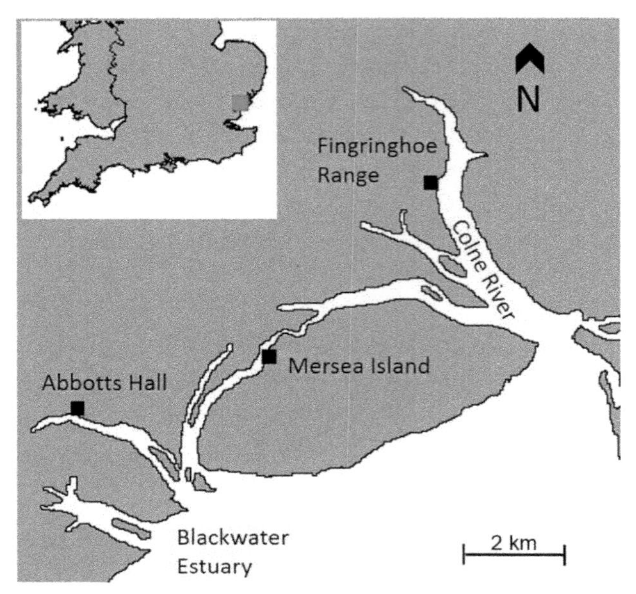

Fig. 3 Essex UK salt marsh sampling sites. The furthest site westward is the Abbotts Hall (AH) site, while the furthest site eastward is the Fingringhoe Wick Range (FWR) site. The Mersea Island (MI) sampling site is centrally located, but sheltered, relative to the other sites. The sites were paired, both natural and managed marshes were adjacent to each other and are represented by single demarcations.

microbial community cores were also taken from within the volatile sampling footprint directly after volatile sampling.

Seasonal samples (Summer 2014–Springer 2015) for a limited suite of volatilome fluxes (CH_4, CH_3Cl, CH_3Br, CH_3I, dimethyl sulphide, isoprene, and $CHCl_3$; Table 1) were taken using standard static chamber enclosures in each marsh for each of the four dominant vegetation covers with three replicates for each. Chamber bases were pinned to the sediment to avoid cutting effects on roots, but otherwise methodology followed the same protocol as described in Redeker and Kalin (2012) (Fig. 2). Isoprene, sulphur, and halogenated compound samples were stored in electropolished stainless steel canisters until analysis, within a 2-week period in which the canisters have been demonstrated to be stable for reactive compounds (Low et al., 2003). Methane samples were stored in 12 mL exetainers until analysis.

Vegetation was harvested from within the volatilome chamber base after trace gas sampling. Once the chamber footprint was clear two sediment cores were taken from the central region of the chamber, one for environmental parameters (e.g. pH, granulometry, nutrients, bulk density) and the other for microbial community analysis.

Sediment cores were stored on ice until arrival at the University of York where they were stored at 4°C and processed within 48h. Sections of the sediment cores between 5–10 cm and 10–15 cm were selected to avoid root contamination and subsamples from within the interior of the cores at these depths were collected with aseptic technique, then frozen in liquid nitrogen immediately and stored at −80°C until DNA extraction. Sediment parameter samples were collected from the same depths as the microbial community samples but were processed for granulometry, pH, bulk density, and nutrient measurements according to standard methodologies.

3.1.3 Analysing Volatilome and Sediment Samples

The reactive volatilome samples (isoprene, halogen, and sulphur compounds) were condensed onto a liquid nitrogen condensation trap and transferred to a Restek© PoraBond Q column (30m, 0.32mm ID, 0.5-μm thickness) within a HP 5972 MSD running in selective ion mode. Greenhouse gas samples were run on a PerkinElmer Autosystem XL GC ECD/FID using a hand-packed Restek© PoraPak Q column (2 m, 1/8″ OD, 50–80 mesh). Analyses and calibration of trace gases followed the methodology described in Redeker et al. (2007) (Fig. 2).

Bulk density, nutrient, and pH processing were performed on samples immediately after removal from the sediment cores. Samples were dried at 70°C for 7 days and dry weight divided by known sample volume (corrected for compression during coring) was used to calculate bulk density. To normalize the ionic strength of the salt marsh sediment samples each was placed in a 1.0 M KCl solution prior to analysis on a Jenway Ltd. 3310 pH meter. Sediment samples were treated with hydrogen peroxide (30% H_2O_2 concentration) to remove organic matter prior to grain size analysis on "Malvern Mastersizer 2000" (UK manufacturer, Malvern). Nutrient samples were prepared in accordance with Houba et al. (1995) with 1.0 M KCl, and stored at −20°C until analysis. Analysis for NH_4^+ and NO_2^-/NO_3^- was performed using a Seal Analytical AutoAnalyzer3.

3.1.4 DNA Extraction and Sequencing

NGS-based analysis of soils and sediments provides the current gold standard of microbial community (based on presence) and functional potential but does not allow repeat sampling for nondestructive applications (Derocles et al., 2018; Jackson et al., 2016). In this early stage of volatilome assessment, we have aligned the observed volatilomics with this excellent data source to overcome incomplete volatile database issues. Total DNA metagenome sequencing was applied to avoid primer-based artefacts in the data and to gain additional information on gene functions. Total DNA was extracted using the MoBio Powersoil kit (now Qiagen DNeasy Powersoil kit), using 250 mg of soil. Samples were quantified by Qubit fluorometric quantitation (Thermo Fisher), diluted down to 0.2 $ng\,\mu L^{-1}$ and requantified (again by Qubit). A total 1 ng DNA was used in library preparation using the Nextera XT DNA library prep kit v2, with set A indices (dual 8 bp indices). The library preparation size selected the DNA for 300–600 bp in size. Final libraries were quantified by Qubit on HS DNA bioanalyzer chips prior to pooling, at approximately equimolar ratios for sequencing. All samples were run on a single lane on an Illumina HiSeq 3000 using 150 bp paired end reads.

3.1.5 Microbial Community Sequence Analysis

Sequencing produced a total of 328 million pairs of reads, ranging from 3.9 to 13.2 million read pairs per sample (mean 8.65 million). Reads were adaptor-trimmed using cutadapt (Martin, 2011) and uploaded to the OneCodex analysis platform (Minot et al., 2015). Taxa abundances were estimated from the number of matches to the OneCodex database. Taxa abundance tables provided by OneCodex were converted to BIOM format and correlations with environmental data were calculated using the observation_metadata_correlation.py script in QIIME (Caporaso et al., 2010).

3.1.6 16S rRNA Analysis

Read files were separated by site and season before being assembled into contigs using Megahit (v 0.3.3-a, Li et al., 2015). Contigs greater than 30 kb were removed using a custom Python script (see Supplementary Data in the online version at https://doi.org/10.1016/bs.aecr.2018.07.001) (<0.02% of contigs from each sample) before contigs containing 16S/23S and 18S/28S sequences were identified using SortMeRNA-2.1 (Kopylova et al., 2012) then classified using SSuMMo (Leach et al., 2012) and visualized using the Interactive Tree of Life (Letunic and Bork, 2016) (Fig. 4, see Supplementary Material in the online version at https://doi.org/10.1016/bs.aecr.2018.07.001 for a full scale image of Fig. 4).

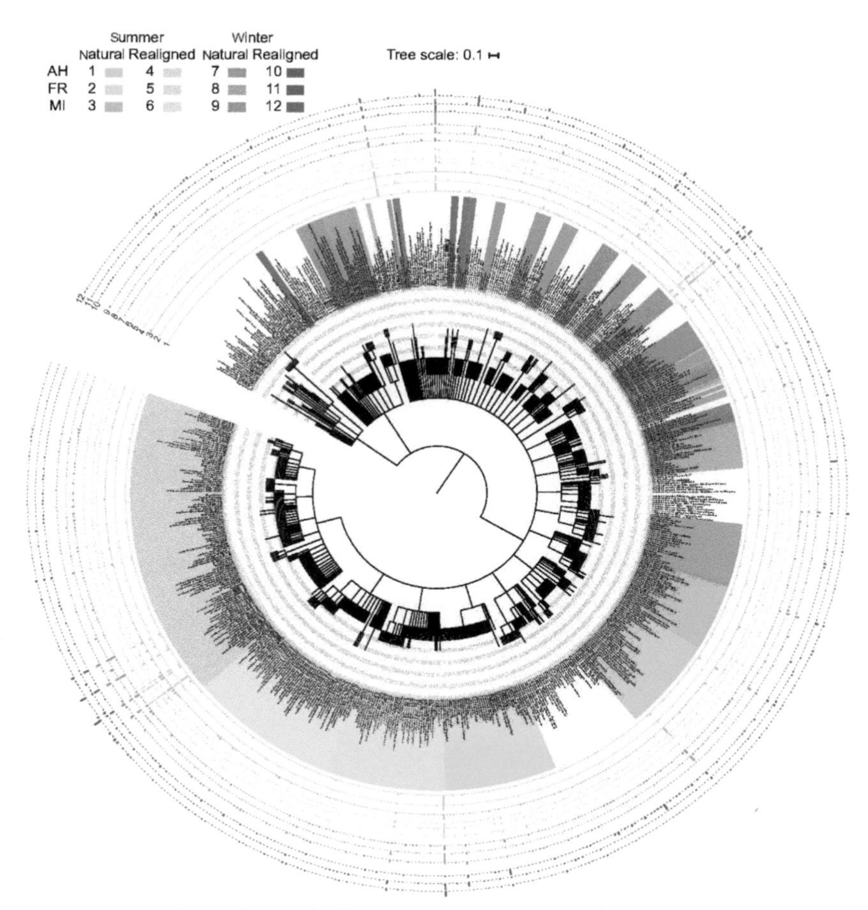

Fig. 4 SSuMMo-derived microbial community comparison between sites and seasons assigned by Class. *Bars* indicate positive presence of the closest microbial match and size of the bar is indicative of the abundance of the organism within the sample. Samples are as indicated, from the inside ring out: ordered as Abbots Hall, Fingringhoe Wick, and Mersea Island, summer natural and realigned marshes then winter natural and realigned marshes. Archaea were found to be between 2% and 12% of the identified community and generally more abundant in winter at realigned sites.

3.2 Relationships Between Environment, Microbial Community, and the Volatilome

Communities, environmental parameters, and volatile fluxes described later are based on sequences and samples obtained from natural and managed salt marsh sediments (5–10 cm depth) during July/August 2014 (Summer) and February/March 2015 (Winter). To simplify we will not consider the

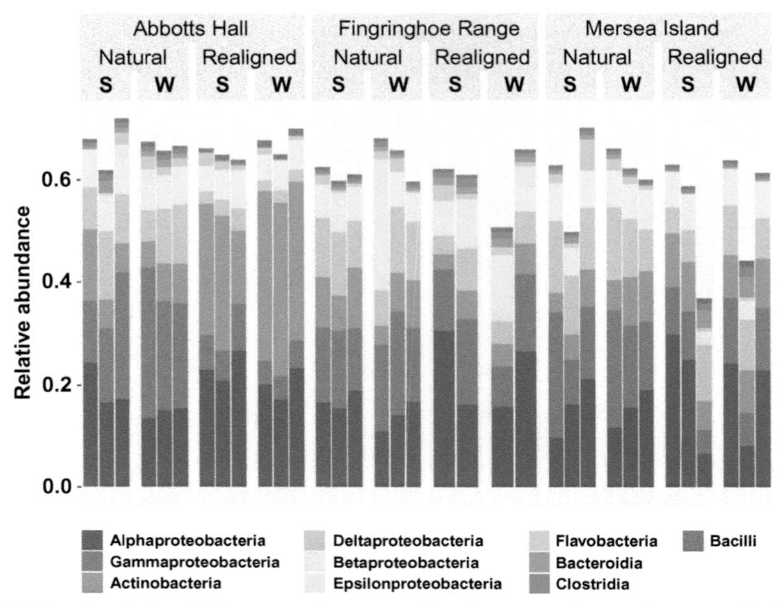

Fig. 5 Class level microbial community comparison between sites and seasons. Only the 10 most abundant classes are shown.

sediment under the influence of plants and vegetation within the salt marshes and have focussed entirely upon mud pan samples (see Fig. 2). Figs 4 and 5 and Tables 2 and 3 demonstrate how the microbial communities, volatilome, and environmental parameters behave across space and time.

3.2.1 Mud Pan Microbial Communities Across Essex Salt Marshes

Even after aggregating sequences derived from three replicate mud pan samples per location and sampling date, the sediment microbial communities show differences in space and time (Figs 4 and 5). While individual species changes are too numerous to explicitly detail (Fig. 4) there are several broad trends that can be easily observed. The Actinobacteria are much more represented at the managed Abbotts Hall site than in either managed Fingringhoe Wick or Mersea Island samples or in any of the natural site samples (Fig. 5). There are greater numbers of representative species from within the Rhodobacteraceae, Gammaproteobacteria, and Epsilon/Deltaproteobacteria in the Fingringhoe Wick mud samples, particularly when compared to the samples from Abbotts Hall (Fig. 5). Conversely, Fingringhoe Wick mud samples have poor representation in members of the Bacteroidetes relative to Abbotts Hall and Mersea Island (Fig. 4).

Table 2 Averaged Selected Environmental Parameters

Season	Site	Natural or Managed	pH	Stderr	Bulk Density ($g\,cm^{-3}$)	Stderr	Pore Water (%)	Stderr	NH_4 (mg/kg sediment)	Stderr	NO_x (mg/kg Sediment)	Stderr
S	AH	N	5.2	1.1	0.33	0.09	61.7	4.7	24.3	4.7	4.9	3.5
W	AH	N	3.5	0.4	0.38	0.09	65.2	4.6	112.6	76.7	10.8	4.3
S	MI	N	6.4	0.8	0.38	0.03	61.3	0.1	4.8	3.1	0.0	0.0
W	MI	N	6.6	0.8	0.42	0.05	63.0	2.8	11.5	2.2	0.5	0.5
S	FWR	N	6.7	0.1	0.50	0.05	51.1	0.6	13.7	1.0	18.7	12.3
W	FWR	N	5.0	1.9	0.41	0.06	57.2	4.0	66.0	22.7	11.2	11.2
S	AH	M	6.6	0.1	1.29	0.11	23.8	3.6	15.0	6.5	3.2	2.3
W	AH	M	6.7	0.1	1.33	0.05	22.9	0.9	3.1	1.4	3.9	1.8
S	MI	M	7.7	0.2	0.38	0.01	59.3	2.4	9.8	4.2	2.7	1.8
W	MI	M	6.4	1.3	0.38	0.02	63.3	1.8	16.8	3.7	0.3	0.3
S	FWR	M	7.1	0.6	0.20	0.02	76.1	1.6	100.6	46.6	0.0	0.0
W	FWR	M	4.5	1.7	0.17	0.03	76.9	2.2	66.1	25.5	0.0	0.0

Table 3 Averaged Measured Fluxes of Selected Components of the Volatilome

Season	Site	Natural or Managed	CH_4	Stderr	CH_3Cl	stderr	CH_3Br	stderr	CH_3I	stderr	DMS	stderr	Isoprene	stderr	$CHCl_3$	stderr
S	AH	N	−6.8E−03	5.6E−03	1.8E−05	2.6E−05	3.4E−06	1.5E−06	2.0E−06	4.0E−07	2.1E−03	2.3E−03	−1.1E−05	7.4E−06	−2.4E−07	6.6E−06
W	AH	N	−2.5E−02	3.3E−03	−2.4E−04	2.3E−04	−8.4E−06	1.3E−05	6.9E−06	8.6E−06	1.7E−03	1.2E−03	−4.3E−06	1.5E−06	−4.9E−06	6.3E−06
S	MI	N	−2.4E−02	4.7E−02	−2.4E−06	5.7E−06	−1.9E−06	2.4E−06	1.6E−06	9.1E−07	1.8E−04	7.5E−05	−6.1E−06	8.1E−06	−1.3E−06	1.5E−06
W	MI	N	1.8E−02	9.1E−03	4.0E−06	5.5E−06	2.2E−07	1.3E−07	2.7E−07	9.2E−08	1.5E−05	2.6E−05	−9.4E−08	1.1E−07	3.1E−06	1.6E−06
S	FWR	N	−1.2E−02	4.5E−03	3.7E−05	1.8E−05	2.8E−07	5.9E−07	1.6E−08	6.3E−07	1.1E−05	3.6E−06	−1.6E−06	1.3E−05	7.3E−07	7.4E−07
W	FWR	N	4.1E−02	3.1E−02	1.5E−05	1.4E−05	3.5E−07	9.4E−08	1.0E−06	3.5E−07	2.7E−05	1.6E−05	−1.3E−07	1.6E−07	−4.6E−07	5.1E−07
S	AH	M	−1.8E−02	4.6E−03	8.0E−06	4.5E−05	2.3E−06	2.0E−06	8.5E−08	5.8E−07	5.9E−04	3.4E−04	−2.7E−06	3.5E−07	−4.5E−07	1.0E−06
W	AH	M	1.4E−02	4.5E−02	−4.6E−05	7.2E−05	3.6E−06	2.7E−06	2.2E−05	9.0E−06	3.1E−03	3.0E−04	−8.8E−07	2.1E−06	4.3E−05	9.6E−06
S	MI	M	−2.2E−02	3.7E−02	−2.7E−05	1.3E−05	2.1E−06	7.5E−07	6.1E−06	7.0E−07	6.5E−04	1.5E−04	7.2E−06	6.2E−06	1.3E−06	7.6E−07
W	MI	M	−2.5E−02	8.4E−03	−3.3E−05	6.8E−06	−1.5E−07	1.8E−07	9.3E−07	4.8E−07	2.7E−05	3.1E−05	−1.4E−07	1.3E−07	−1.2E−07	1.1E−06
S	FWR	M	−1.7E−03	1.8E−03	8.2E−06	8.1E−06	2.2E−07	4.0E−07	6.0E−07	2.8E−07	4.9E−04	3.8E−04	4.3E−06	3.5E−06	−4.7E−07	1.0E−06
W	FWR	M	6.9E−02	6.8E−02	−1.3E−06	6.1E−06	2.3E−07	2.7E−07	1.1E−06	2.1E−07	1.6E−03	1.6E−03	1.6E−06	1.6E−06	5.4E−07	2.0E−07

Negative fluxes indicate that sediments degraded or consumed the compound from the chamber headspace, while positive fluxes denote movement of the material from the soil to the air. All fluxes are in $g\,m^{-2}\,day^{-1}$.

The NGS sequence data were dominated by bacterial sequences, and more sediment samples will need to be analysed to obtain functional information regarding the rarer species. A greater number of samples is also needed to provide statistically useful volatilome/rare microbiome comparisons. For instance, the only identified, moderately abundant fungal community member was an uncultured Ascomycete, which was only present in realigned sites (Fig. 4; AH Summer and FRW Winter). Archaeal community members were more commonly distributed, with notable differences including greater Halobacteriaceae presence in winter sampling periods and a reduced abundance of the Chrysiogenaceae in Fingringhoe Range Wick relative to their abundance in Abbotts Hall marshes (Fig. 4).

Two of the microbial community organisms with greatest sequence representation are obligate anaerobes that lie within the Phylum Chloroflexi in the family Anaerolineaceae (Fig. 4). However, in Mersea Island and Fingringhoe Wick the dominant Anaerolineaceae genera are *Levilinea* while in Abbotts Hall the dominant genera is *Longilinea*. The primary description of these organisms suggests that they have similar optimal temperatures (37°C) and pH (7.0) for growth but that *Levilinea* has a greater temperature range (25–50°C) but a smaller pH range (6.0–7.2) relative to *Longilinea* (30–40°C and 5.0–8.5 pH, respectively) (Yamada et al., 2006, 2007).

3.2.2 Environmental Conditions in Sediments Within Essex Salt Marshes

Other broad trends reveal themselves when comparing environmental parameters obtained from the replicate mud pan samples (Table 2). Abbotts Hall samples from natural salt marshes tend to be more acidic than those from Mersea Island or Fingringhoe Wick Range. Managed sites tended to be more pH neutral than the natural marsh samples and overall there was a bimodal distribution of pH, with the majority of samples falling within a pH range of 6.5–8.0 ($n=20$), however, a substantial minority of sediment samples were acidic, falling in the pH range of 2.5–4.5 ($n=10$) (Fig. 6). The differences in pH within Essex sediments may have some explanatory capacity for the observed variance in Anaerolineaceae genera (phylum: Chloroflexi) since the lower pH sediments tend to harbour the *Longilinea*, which have been shown to tolerate a broader range of pH in lab cultures (Yamada et al., 2007).

Differences in bulk density within sediments also appear to have explanatory power (Table 2; Fig. 5). Abbotts Hall sediments in managed salt marshes were consistently the densest of all sediments, with Fingringhoe

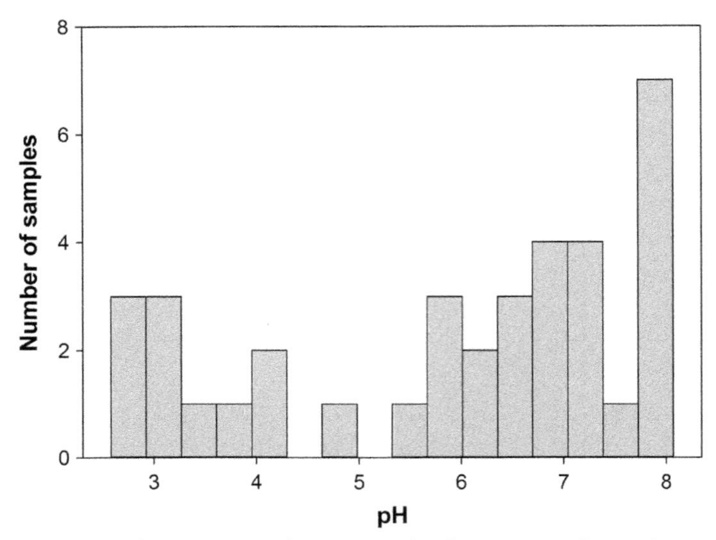

Fig. 6 Distribution of pH across sediment samples from Essex salt marshes.

Wick samples the least dense. This is opposite the trend observed in the natural marshes, where Abbotts Hall samples tended to be least dense and Fingringhoe Wick samples most dense. While the observed changes in sediment density in natural sediments does not appear to have substantial impacts, the high density managed sediments from Abbotts Hall clearly have greater Actinobacterial presence, suggesting that Actinobacteria outcompete other classes in high-density sediments.

Nitrogen availability and form varies between locations and season but is not obviously connected to microbial community (Table 2, Figs 4 and 5). While there is great variation between samples taken from different locations in the same salt marsh on the same day there appears to be a trend in ammonia concentrations within natural salt marshes such that Abbotts Hall sediments have the greatest concentrations and Mersea Island sediments the least. Natural sediments demonstrate a strong seasonality, with greater concentrations of sediment-bound ammonia in the winter relative to the summer. In contrast there is no obvious seasonal trend in ammonia within managed salt marsh sediments.

There was significantly less NO_x species when compared to ammonia in all sediments with the exception of the natural Fingringhoe Wick site. In all others oxidized nitrogen is less prevalent than reduced and there is no obvious seasonal trend in either natural or managed sediments.

3.2.3 The Volatilome Across Essex Salt Marshes

We measured fluxes of seven biologically informative trace gases (Table 1) from Essex salt marsh sediments, quantifying the net metabolism of these sediments over four seasons, three sites, and two management conditions. Fluxes of volatilome constituents from sediments are as variable as subsurface microbial communities and environmental parameters. There were significant differences in the fluxes between compounds, sites, and seasons (Fig. 7; Table 3). Both net production of compounds and net consumption of compounds were observed. DMS and methyl iodide were both consistently emitted from salt marsh sediments across all sites and seasons, while methane tended towards efflux during winter and towards influx during the summer sampling period. Unsurprisingly the greatest effluxes and influxes are observed for methane (>10 mg CH_4 m^{-2} day^{-1}), but dimethyl sulphide follows directly with average and consistent efflux across all sites, at some locations exceeding 2 mg DMS m^{-2} day^{-1}.

Methane behaves consistently between the natural and managed salt marshes; in both cases there is uptake of methane during the summer (-14 ± 10 mg CH_4 m^{-2} day^{-1}) and efflux of methane during the winter (15 ± 40 mg CH_4 m^{-2} day^{-1}). Methyl bromide is emitted by managed salt marshes (1.4 ± 1.5 μg CH_3Br m^{-2} day^{-1}), while fluxes from natural salt marshes are, on average, not different from zero (0.5 ± 3.3 μg CH_3Br m^{-2} day^{-1}).

While most other compounds were too variable between sites to generalize results from management style or seasonality there were some site-specific fluxes of interest. Abbotts Hall produced the greatest effluxes of methyl iodide across the study (managed: 11 ± 9 μg CH_3I m^{-2} day^{-1}; natural: 5 ± 9 μg CH_3I m^{-2} day^{-1}) as well as the most dimethyl sulphide (1.9 ± 1.5 mg DMS m^{-2} day^{-1} both natural and managed), and the Abbotts Hall managed site was the only site to consistently generate and export chloroform from the sediment surface (43 ± 10 μg $CHCl_3$ m^{-2} day^{-1} during winter). Mersea Island sites generated the least DMS (0.2 ± 0.1 mg DMS m^{-2} day^{-1}) and Fingringhoe Wick sites produced the least methyl iodide (0.7 ± 0.5 μg CH_3I m^{-2} day^{-1}).

At this gross, regional scale of analysis the links between volatile emissions and microbial community members are quite weak. Coastal sites rich in Bacteroidetes, known isoprene producers, do not generate more isoprene than other locations and sulphur-volatile generating bacteria from the Gammaproteobacteria, the Firmicutes, and Actinobacteria are not strongly linked to lesser or greater dimethyl sulphide fluxes.

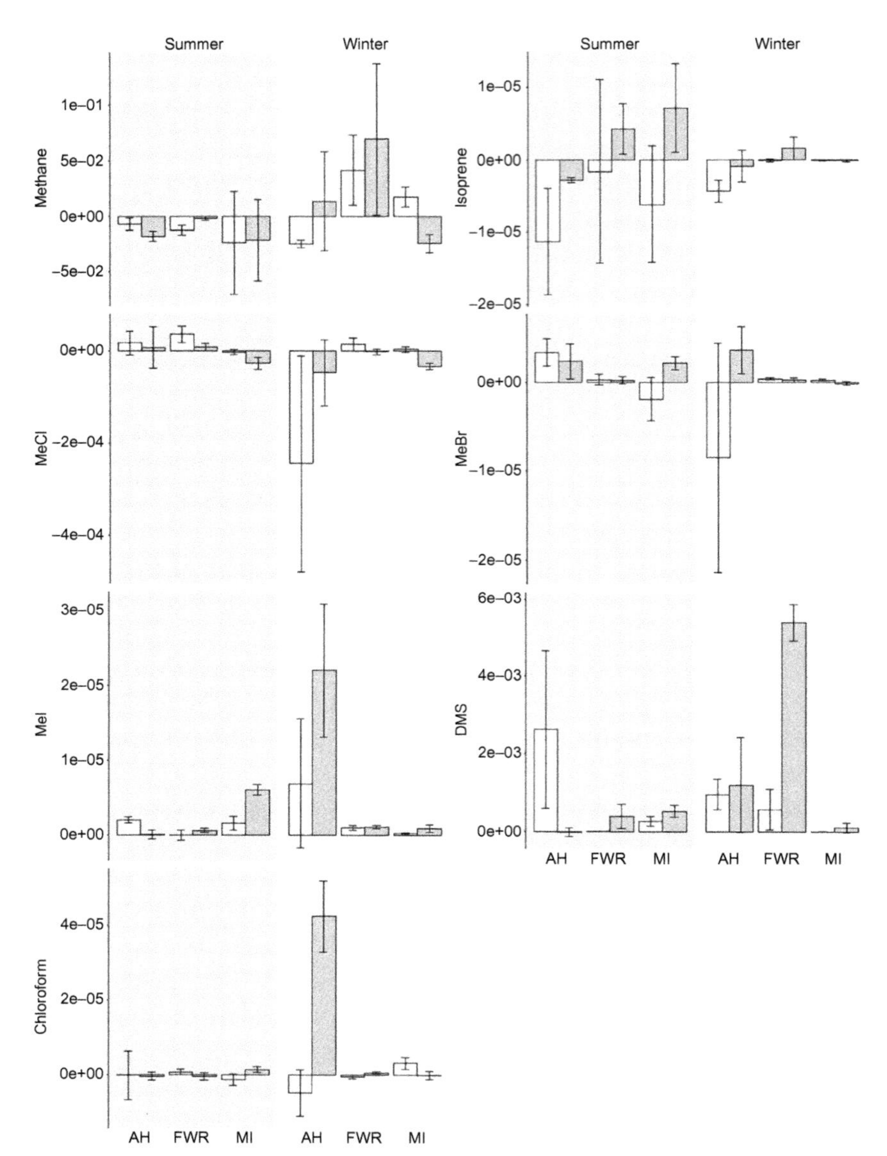

Fig. 7 Averaged trace gas fluxes for managed (*grey columns*) and natural (*white columns*) salt marsh sediments across summer (stippled, *left*) and winter (clear, *right*) seasons. Note that the fluxes are shown at a range of scales and include both efflux from (positive) and influx to (negative) the sediment. Fluxes shown are all $g\,m^{-2}\,day^{-1} \pm$ stderr.

3.2.4 Tying Together the Environment, Volatilome, and Microbial Communities

Comparison of microbial communities, environmental parameters, and volatile fluxes is hampered by the substantial variability exhibited by natural and managed sediments. This variability across space and time does, however, provide a continuum of environmental conditions and volatile fluxes that can be correlated against microbial sequence abundance. When all sediment data are combined and analysed through Pearson correlations, we can explore and confirm interactions suggested by the aggregate analysis. We focus here on the strongest correlations but, due to the often nonparametric nature of the data and limited replicate numbers, we were unable to completely correct for variable (environment, volatile, microbe) interactions. Therefore, we may underestimate some of the listed probabilities due to type I statistical errors.

It is important to note in the following discussion that we are discussing greater or lesser abundance of sequenced DNA as opposed to presence vs absence and correlations as described here are based on these relative abundances. As is clear from Figs 4 and 5 all major classes are present in all sediment samples (bacterially dominated, but inclusive of archaea and fungi), if not inclusive of every individual species observed. Therefore, positive correlations suggest greater abundance rather than sudden appearance and negative correlations suggest that a greater volatile flux or environmental condition is linked to a smaller representation of the microbial organism within the overall community. It should be noted that, negative correlations with volatile flux may indicate either (i) fewer organisms exist where greater emissions occur or (ii) that larger microbial sequence abundances occur where greater influx occurs.

Environmental parameters appear to predict abundance of Actinobacteria (bulk density) and most abundant genera within the Chloroflexi (pH). In aggregate samples this may be true but when examined individually (environmental and microbial samples from sediment below 0.1 m^2 volatile "footprints") the relationship with pH is not as apparent. Overall the Phylum Chloroflexi is poorly correlated with pH (Pearson's R^2 [hereafter Pearson's correlation coefficient will be annotated as PR^2] $= \sim 0.3$, $P = 0.15$), with individual genera/species showing both significant negative correlations (*Chloroflexus* spp. $PR^2 = -0.56$, $P < 0.001$, *Dehalococcoides mccartyi* spp. $PR^2 = -0.41$, $P < 0.001$) and significantly positive correlations (SAR 202 cluster $PR^2 = 0.48$, $P < 0.001$, *Herpetosiphon aurantiacus* $PR^2 = 0.48$, $P < 0.001$). While the observed preference of *Levilinea* for more neutral pH sediments is not directly observed, there are strong correlations between the abundances of sequences derived from *Levilinea* and *Dehalococcoides mccartyi* ($PR^2=0.83$, $P<1e^{-9}$) and *Levilinea* and *Chloroflexus* ($PR^2=0.83$, $P<1e^{-9}$), both of which were the most significantly, negatively correlated genera/species with pH. This suggests that we may be witnessing a loose cohort effect, in which organisms that tend to associate may be driving some of the observed correlations between community and environmental parameters.

The relationship between Actinobacteria and bulk density is much more evident. The phylum Actinobacteria is strongly correlated with bulk density of salt marsh sediments ($PR^2=0.89$, $P<1e^{-15}$) as is the class Actinobacteria ($PR^2=0.88$, $P<1e^{-15}$). As well as the positively associated Actinobacteria there are several other microbial groupings that strongly negatively correlate with bulk density, including Chrysiogenetes ($PR^2=-0.71$, $P<1e^{-6}$), Deferribacteres ($PR^2=-0.60$, $P<1e^{-4}$), and Aquificae ($PR^2=-0.55$, $P<0.001$) as well as groupings that are insensitive, including Gemmatimonadetes and Elusimicrobia ($P>0.4$).

A selection of other reliable interactions between microbes and environmental parameters includes pH (Deferribacteres $PR^2=-0.50$, $P<0.005$), ammonia content (Gammaproteobacteria $PR^2=0.78$, $P=1e^{-8}$), and NO_x concentration (Halobacteria $PR^2=0.56$, $P<0.001$). Unsurprisingly, a number of fungal groupings held strong preferences for lower sediment water content (Hypocreomycetidae, Pezizomycotina, Leotiomyceta, Saccharomyceta; $PR^2\leq 0.80$, $P<5e^{-11}$), while none showed a strong preference for saturated sediments. Other fungal preferences observed include a strong positive correlation with ammonia and bulk density for several Ascomycetes (ammonia—Cryphonectriaceae = 0.80, $P=2e^{-9}$; Diaporthales = 0.7, $P=1e^{-8}$) (bulk density—Sordariomycetes = 0.87, $P=1e^{-14}$; Leotiomyceta = 0.87, $P=3e^{-14}$).

The volatile fluxes from sediment surface are indicative of a different suite of microbial community members. Strong interactions between microbes and measured volatile compounds include DMS (Defluviimonas $PR^2=0.94$, $P<1e^{-15}$), chloroform (Actinobacteria $PR^2=0.61$, $P<1e^{-4}$ and Chrysiogenetes $PR^2=-0.39$, $P=0.01$), methyl iodide (Chrysiogenetes $PR^2=-0.40$, $P=0.01$), isoprene (the Euryarchaeota Methanomicrobia $PR^2=0.53$, $P<0.001$ and the Cyanobacteria Microchaete and Mastigocoleus $PR^2=0.5$, $P<0.001$), and methane (Gemmatimonadetes $PR^2=-0.49$, $P<0.005$). Fungal associations with trace gas fluxes include a strong positive correlation between the Ascomycetes Fusarium,

Nectriaceae, Glomerellales, and chloroform ($PR^2=0.74$, $P<5e^{-8}$; $PR^2=0.70$, $P<5e^{-7}$; $PR^2=0.58$, $P=1e^{-4}$, respectively) and the Sordariaceae with methyl iodide ($PR^2=0.54$, $P=5e^{-4}$).

These relationships/interactions are crucial for community assembly since these organisms can be reliably associated with certain sediment characteristics or functional outputs. Once the relative abundance of a number of organisms has been identified, their relationships with other microbial populations can fill in the broad picture of microbial community. In this way, if we explore the community associated with the Euryarchaeota we find that they associate with themselves consistently with the Chloroflexi ($PR^2=0.75$, $P<1e^{-8}$), the Thermotogae ($PR^2=0.70$, $P<1e^{-6}$), the Omnitrophica and the Latescibacteria ($PR^2=0.68$, $P<1e^{-6}$). In contrast the Euryarchaeota are consistently negatively correlated with the members of the deltaproteobacteria order Burkholderiales ($PR^2=-0.54$, $P<5e^{-4}$).

Using this approach, a preliminary analysis of two important players in element cycling in marine sediments, the Desulfobulbaceae (cable bacteria involved in sulphur oxidation) and the Thaumarchaeota (ammonia-oxidizing archaea), suggests that a coherent community picture can be generated (Fig. 8). The Desulfobulbaceae tend to be negatively correlated with pH in sediments ($PR^2\sim-0.3$, $P<0.1$), and are strongly negatively correlated with bulk density ($PR^2=-0.7$, $P<1e^{-5}$). The Thaumarchaeota on the other hand show a moderate negative correlation with ammonia in sediments ($PR^2\sim-0.2$, $P<0.1$) and a strong positive correlation with pH ($PR^2=0.5$, $P<0.005$) and bulk density ($PR^2=0.7$, $P<1e^{-12}$). This suggests in dense, more alkaline sediments we should expect to find a greater abundance of Thaumarchaeota and a reduced presence of Desulfobulbaceae and in less compact, more acidic sediments the reverse should be true.

The volatiles associated with these organisms are also complementary, where the Desulfobulbaceae are negatively correlated with chloroform fluxes ($PR^2=-0.4$, $P<0.05$) and methyl iodide ($PR^2=-0.3$, $P<0.05$) but are weakly positively correlated with isoprene ($PR^2=0.4$, $P<0.05$). In contrast, the Thaumarchaeota are positively correlated with chloroform ($PR^2=0.7$, $P<1e^{-5}$) and methyl iodide ($PR^2=0.7$, $P<0.0005$) and show negative correlations with methyl bromide ($PR^2=-0.4$, $P<0.05$) and methane ($PR^2=-0.4$, $P<0.05$). The volatile signals agree again, that where there are higher chloroform and methyl iodide fluxes we should expect greater numbers of Thaumarchaeota and reduced abundance for Desulfobulbaceae (Fig. 8).

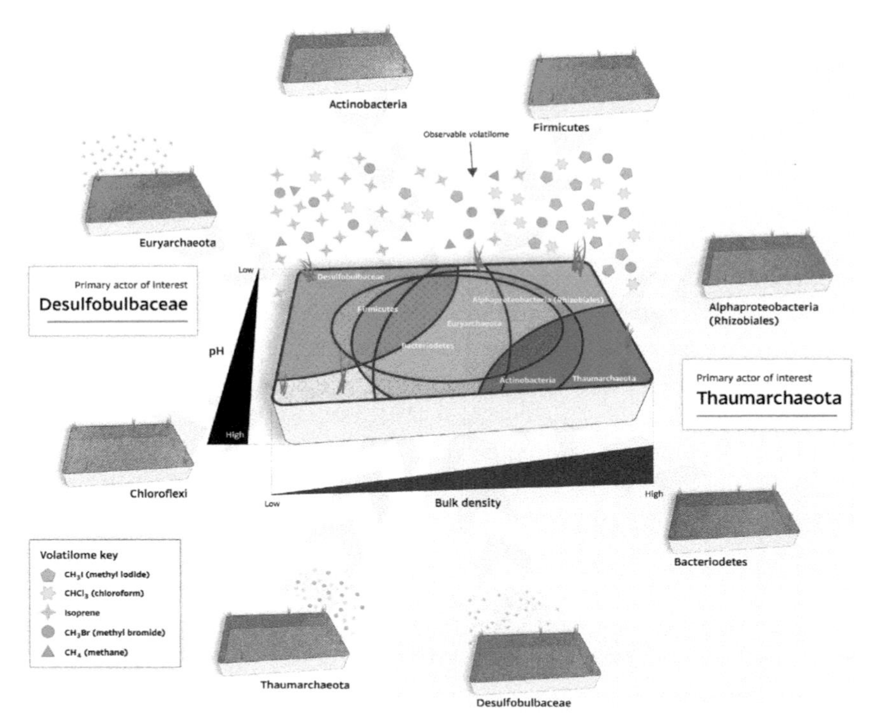

Fig. 8 The combined interactions between microbes, their environment, and the net volatilome observed. Each microbe is shown individually (satellite images) in its preferred environmental state (high or low pH vs greater or lesser bulk density) and, where appropriate, with associated volatile fluxes (*filled symbols* indicate greater flux, *open symbols* indicate lesser-to-negative flux). The central image shows the combined, coherent community with microbial assemblages within environmental context and with net volatilome.

The microbial communities associated with these specific groups are also in agreement (Fig. 8). Desulfobulbaceae are commonly found in association with Chloroflexi spp. ($PR^2=0.5$, $P<0.001$), Euryarchaeota spp. ($PR^2=0.4$, $P<0.001$), and Firmicutes spp. ($PR^2=0.6$, $P<1e^{-5}$) and do not tend to associate with Thaumarchaeota spp. ($PR^2=-0.4$, $P<0.005$) or Actinobacteria spp. ($PR^2=-0.7$, $P<1e^{-7}$). The Thaumarchaeota are found in regions with high population representation of Alphaproteobacteria (Rhizobiales) spp. ($PR^2=0.8$, $P<1e^{-10}$) and reduced presence of Bacteriodetes spp. ($PR^2=-0.5$, $P<0.0005$) and Firmicutes spp. ($PR^2=-0.5$, $P<0.0005$). The microbial associations agree that where larger numbers of Firmicutes are found we can expect fewer Thaumarchaeota and greater numbers of Desulfobulbaceae.

4. CONCLUSION

Within the context of variable and complex salt marsh sediments, we have demonstrated that there are a number of organisms whose relative abundance can be predicted by volatile emissions, and these inhabit a limited range of environmental conditions and associate with a limited subset of other organisms. These correlations and interactions lead to basic predictions of individual phyla abundance derived from measured volatilome fluxes. Importantly, these interactions, based on separate measurements of sediment characteristics, volatile trace gas fluxes, and sediment microbial communities generate self-consistent outcomes. This preliminary success, based here on limited numbers of samples and volatiles, suggests that analysis of a small number of volatiles will allow us to constrain the subsurface microbial community to the phylum level once community builder algorithms have been trained in common relationships. That this has been possible despite the limited number of sediment samples and the reduced range of observed volatiles suggests that this method has promise for more accurate noninvasive diagnosis of soil microbial communities and function.

It is important to note that we do not claim that the volatiles chosen in this analysis are the most useful for all future approaches. The compounds chosen in this analysis are not often used in current diagnostic approaches and may provide an insight into a different set of biological functions than more commonly reported compounds. Indeed, the process of selecting the most informative compounds for use in a "fingerprint" volatile analysis of subsurface microbial communities is one of the greater challenges facing the application of this approach. Future work will need to explore single-species emissions and metabolic uptake (to identify "known unknowns") and these studies will need to be combined with scanning of net volatilomes from soils and sediments (to identify "unknown unknowns"). Once a suite of informative compounds has been identified they may be individually targeted for maximum sensitivity in analysis (scanning methods tend to be less sensitive than targeted methods) and the selected list of compounds may be refined after comparison to soil and sediment communities.

We focus in our analysis on community structure as described through individual species abundance. However, our sampling and analytical approach, particularly WGS sequencing of sediments paired with trace gas fluxes, also allow us to study the community structure through the lens of functionality. This is often done in a targeted manner, by looking for the total abundance of specific genetic sequences that allow for individual metabolisms or reactions to be carried out. For instance, conserved genetic sequences that allow for nitrification or denitrification are carried across a range of microorganisms and these sequences can be targeted in sediment samples to identify the overall nitrification/denitrification (functional) potential of the microbial community, regardless of which specific organisms carry the genetic sequence. A caveat to this approach; specific, genetically encoded enzymes may play different or opposing roles in different organisms and under different environmental pressures. For example, the enzyme Dsr catalyses the reduction of sulphur in some sulphur-cycling bacteria and the opposite, oxidation reaction in others (Wasmund et al., 2017). Likewise, nitrogen reductase has been shown to oxidize as well as reduce nitrogen compounds, while the ammonia-oxidizing enzyme, ammonia monooxygenase, also oxidizes methane and other organics (Prosser, 2015). Despite these concerns, there is a substantial value in exploring the metagenome using this functional approach, although clearly better definition and characterization of multifunctional genes are required for optimal outcomes.

We propose that analysis of the net volatilome should become an integral part of future biomonitoring strategies. We have shown here that this technique avoids many of the common issues surrounding other soil sampling and community and analysis methods, it (i) provides a nondestructive sampling route, allowing us to understand subsurface environment and community without disruption; (ii) specifically focusses on net functional outcomes (the net volatilome) rather than the present microbial community; and (iii) this allows us to focus on community members with greatest impact rather than abundance.

A targeted volatilomics approach is already in use to detect and deter specific parasites, pests, and diseases across a range of applications. However, the net volatilome fingerprint can also be used to observe changes in the subsurface that are driven by changing climate (often linked to enhanced pest and disease incidence), providing a prior warning diagnostic rather than an after-the-fact signal. The volatilome fingerprint can be used to better understand local soil cycling of nutrients and within agroecosystems may provide information on maximally efficient routes for nutrient delivery to crops. Transformed and managed landscapes can be monitored through this noninvasive technology at the same locations over time to observe the impacts of land use on soil function.

For this method to reach its full potential there remain a number of important further steps, however, including (i) maintenance and expansion of the current volatilome databases (to include both efflux and influx, and to grow to include AMF, archaea, and unculturable bacteria), (ii) expansion of analytical methodologies to provide a more comprehensive range of volatiles per organism, (iii) a greater number of replicates across a range of soil and sediment environments, (iv) attention to spatial and temporal variability, and (v) robust statistical approaches to link complex volatile fluxes to individual species within complex microbial communities. Assuming that the listed challenges are met, assembling subsurface microbial communities through analysis of the volatilome will require attention to the chemical and physical properties of the soil, as well as an appreciation for the impact that nonmicrobial organisms may have on microbial community behaviour and function.

ACKNOWLEDGEMENTS

J.P.J.C. is a Royal Society Industry Fellow. The authors thank the staff of the Essex Wildlife Trust, particularly those at Abbotts Hall and Fingringhoe Wick, for access, support, and advice. We would also like to thank the Colchester Wildfowling and Conservation Club for access to the Mersea Island site. The authors thank Phil Brailey and Emma Rand for figure creation and assistance. This work used the York Biosciences Technology Facility and the authors thank Peter Ashton and Sally James for their assistance. K.R., A.D., and T.H. acknowledge Natural Environment Research Council award no. NE/K01546X/1.

REFERENCES

Alessi, A., Redeker, K.R., Chong, J.P.J., 2018. A practical introduction to microbial molecular ecology through the use of iChips. Ecol. Evol. (in review).

Ali, J.G., Alborn, H.T., Stelinski, L.L., 2011. Constitutive and induced subterranean plant volatiles attract both entomopathogenic and plant parasitic nematodes. J. Ecol. 99 (1), 26–35. http://www.jstor.org/stable/41058834.

Allen, D.E., Singh, B.P., Dalal, R.C., 2011. Soil health indicators under climate change: a review of current knowledge. In: Singh, B., Cowie, A., Chan, K. (Eds.), Soil Health and Climate Change. In: Soil Biology, vol. 29. Springer, Berlin, Heidelberg.

Asensio, D., Penuelas, J., Filella, I., Llusià, J., 2007. On-line screening of soil VOCs exchange responses to moisture, temperature and root presence. Plant Soil 291, 249–261.

Bentley, R., Chasteen, T.G., 2004. Environmental VOSCs—formation and degradation of dimethyl sulfide, methanethiol and related materials. Chemosphere 55, 291–317.

Bodner, G., Leitner, D., Kaul, H.P., 2014. Coarse and fine root plants affect pore size distributions differently. Plant Soil 380, 133–151. https://doi.org/10.1007/s11104-014-2079-8.

Caporaso, J.G., Kuczynski, J., Stombaugh, J., Bittinger, K., Bushman, F.D., Costello, E.K., Fierer, N., Peña, A.G., Goodrich, J.K., Gordon, J.I., Huttley, G.A., Kelley, S.T., Knights, D., Koenig, J.E., Ley, R.E., Lozupone, C.A., McDonald, D., Muegge, B.D., Pirrung, M., Reeder, J., Sevinsky, J.R., Turnbaugh, P.J., Walters, W.A., Widmann, J., Yatsunenko, T., Zaneveld, J., Knight, R., 2010. QIIME allows analysis of high-throughput community sequencing data. Nat. Methods 7, 335–336. https://doi.org/10.1038/nmeth.f.303.

Cheng, H.Y., Masiello, C.M., Bennett, G.N., Silberg, J.J., 2016. Volatile gas production by methyl halide transferase: an in-situ reporter of microbial gene expression in soil. Environ. Sci. Technol. Lett. 50, 8750–8759.

Chistoserdova, L., Kalyuzhnaya, M.G., Lidstrom, M.E., 2009. The expanding world of methylotrophic metabolism. Annu. Rev. Microbiol. 63, 477–499. https://doi.org/10.1146/annurev.micro.091208.073600.

Choi, S., Song, H., Tripathi, B.M., Kerfahi, D., Kim, H., Adams, J.M., 2017. Effect of experimental soil disturbance and recovery on structure and function of soil community: a metagenomic and metagenetic approach. Sci. Rep. 7. https://doi.org/10.1038/s41598-017-02262-6. Article number: 2260.

Conrad, R., 1996. Soil microorganisms as controllers of atmospheric trace gases (H_2, CO, CH_4, OCS, N_2O, and NO). Microbiol. Rev. 60, 609–620.

Derocles, S.A.P., Bohan, D.A., Dumbrell, A.J., Kitson, J.J.N., Massol, F., Pauvert, C., Plantegenest, M., Vacher, C., Evans, D.M., 2018. Biomonitoring for the 21st century: integrating next generation sequencing into ecological network analysis. Adv. Ecol. Res. 58, (in press).

Devai, I., DeLaune, R.D., 1995. Formation of volatile sulfur compounds in salt marsh sediment as influenced by soil redox condition. Org. Geochem. 23 (4), 283–287.

Doheny-Adams, T., Redeker, K., Kittipol, V., Bancroft, I., Hartley, S.E., 2017. Development of an efficient glucosinolate extraction method. Plant Methods 13, 17. https://doi.org/10.1186/s13007-017-0164-8.

Falkowski, P.G., Fenchel, T., Delong, E.F., 2008. The microbial engines that drive earth's biogeochemical cycles. Science 320, 1034. https://doi.org/10.1126/science.1153213.

Field, J.A., 2016. Natural production of organohalide compounds in the environment. In: Adrian, L., Löffler, F.E. (Eds.), Organohalide-Respiring Bacteria. Springer-Verlag, Berlin Heidelberg. https://doi.org/10.1007/978-3-662-49875-0_2.

Fierer, N., Strickland, M.S., Liptzin, D., Bradford, M.A., Cleveland, C.C., 2009. Global patterns in belowground communities. Ecol. Lett. 12, 1238–1249. https://doi.org/10.1111/j.1461-0248.2009.01360.x.

Gallucci, M.N., Oliva, M., Casero, C., Dambolena, J., Luna, A., Zygadlo, J., et al., 2009. Antimicrobial combined action of terpenes against the food-borne microorganisms *Escherichia coli*, *Staphylococcus aureus* and *Bacillus cereus*. Flavour Fragr. J. 24, 348–354.

Garcia-Alcega, S., Nasir, Z.A., Ferguson, R., Withby, C., Dumbrell, A.J., Colbeck, I., Gomes, D., Tyrrel, S., Coulon, F., 2016. Fingerprinting outdoor air environments using microbial volatile organic compounds (MVOCs). Trends Anal. Chem. 86, 75–83.

Gribble, G.W., 2003. The diversity of naturally produced organohalogens. Chemosphere 52, 289–297.

Gribble, G.W., 2012. Recently discovered naturally occurring heterocyclic organohalogen compounds. Heterocycles 84 (1), 157–207. https://doi.org/10.3987/rev-11-sr(p)5.

Griffiths, B.S., Bonkowski, M., Dobson, G., Caul, S., 1999. Changes in soil microbial community structure in the presence of microbial-feeding nematodes and protozoa. Pedobiologia 43, 297–304.

Griffiths, R.I., Thomson, B.C., James, P., Bell, T., Bailey, M., Whiteley, A.S., 2011. The bacterial biogeography of British soils. Environ. Microbiol. 13, 1642–1654. https://doi.org/10.1111/j.1462-2920.2011.02480.x.

Harper, D.B., Kalin, R.M., Hamilton, J.T.G., Lamb, C., 2001. Carbon isotope ratios for chloromethane of biological origin: potential tool in determining biological emissions. Environ. Sci. Technol. Lett. 35 (18), 3616–3619.

Hemaiswarya, S., Doble, M., 2010. Synergistic interaction of phenylpropanoids with antibiotics against bacteria. J. Med. Microbiol. 59, 1469–1476.

Houba, V.J.G., Van der Lee, J.J., Novozinsky, I., 1995. Soil Analysis Procedures, Other Procedures. Vol. 5B. Wageningen Agricultural University, Wageningen.

Insam, H., Seewald, M.S.A., 2010. Volatile organic compounds (VOCs) in soils. Biol. Fertil. Soils 46, 199. https://doi.org/10.1007/s00374-010-0442-3.

Jackson, M.C., Weyl, O.L.F., Altermatt, F., Durance, I., Friberg, N., Dumbrell, A.J., Piggott, J.J., Tiegs, S.D., Tockner, K., Krug, C.B., Leadley, P.W., Woodward, G., 2016. Recommendations for the next generation of global freshwater biological monitoring tools. Adv. Ecol. Res. 55, 615–636.

Jassbi, A.R., Zamanizadehnajari, S., Baldwin, I.T., 2010. Phytotoxic volatiles in the roots and shoots of *Artemisia tridentata* as detected by headspace solid-phase microextraction and gas chromatographic-mass spectrometry analysis. J. Chem. Ecol. 36, 1398–1407. https://doi.org/10.1007/s10886-010-9885-0.

Jones, D., 1998. Organic acids in the rhizosphere—a critical review. Plant Soil 205 (1), 25–44. http://www.jstor.org/stable/42949904.

Jones, C.M., Spor, A., Brennan, F.P., Breuil, M.-C., Bru, D., Lemanceau, P., Griffiths, B., Hallin, S., Philippot, L., 2014. Recently identified microbial guild mediates soil N_2O sink capacity. Nature Climate Change 4, 801–805. https://doi.org/10.1038/NCLIMATE2301.

Kibblewhite, M.G., Ritz, K., Swift, M.J., 2008. Soil health in agricultural systems. Philos. Trans. R. Soc. Lond. B Biol. Sci. 363, 685–701. https://doi.org/10.1098/rstb.2007.2178.

Kopylova, E., Noé, L., Touzet, H., 2012. SortMeRNA: fast and accurate filtering of ribosomal RNAs in metatranscriptomic data. Bioinformatics 28, 3211–3217.

Krzycki, J.A., Zeikus, J.G., 1984. Acetate catabolism by *Methanosarcina barkeri*: hydrogen-dependent methane production from acetate by a soluble cell protein fraction. FEMS Microbiol. Lett. 25 (1), 27–32. https://doi.org/10.1111/j.1574-6968.1984.tb01369.x.

Laanbroek, H.J., 2010. Methane emission from natural wetlands: interplay between emergent macrophytes and soil microbial processes. A mini-review. Ann. Bot. 105, 141–153. https://doi.org/10.1093/aob/mcp201.

Leach, A.L., Chong, J.P., Redeker, K.R., 2012. SSuMMo: rapid analysis, comparison and visualization of microbial communities. Bioinformatics 28, 679–686.

LeMer, J., Roger, P., 2001. Production, oxidation, emission and consumption of methane by soils: a review. Eur. J. Soil Biol. 37, 25–50.

Lemfack, M.C., Nickel, J., Dunkel, M., Preissner, R., Piechulla, B., 2014. mVOC: a database of microbial volatiles. Nucleic Acids Res. 42 (Database issue), D744–D748. https://doi.org/10.1093/nar/gkt1250.

Letunic, I., Bork, P., 2016. Interactive tree of life (iTOL) v3: an online tool for the display and annotation of phylogenetic and other trees. Nucleic Acids Res. 44, W242–W245.

Li, D., Liu, C.M., Luo, R., Sadakane, K., Lam, T.W., 2015. MEGAHIT: an ultra-fast single-node solution for large and complex metagenomics assembly via succinct de Bruijn graph. Bioinformatics 31, 1674–1676.

Loreto, F., Barta, C., Brilli, F., Nogues, I., 2006. On the induction of volatile organic compound emissions by plants as consequence of wounding or fluctuations of light and temperature. Plant Cell Environ. 29, 1820–1828. https://doi.org/10.1111/j.1365-3040.2006.01561.x.

Low, J.C., Wang, N.-Y., Williams, J., Cicerone, R.J., 2003. Measurements of ambient atmospheric C_2H_5Cl and other ethyl and methyl halides at coastal California sites and over the Pacific Ocean. J. Geophys. Res. 108 (D19), 4608. https://doi.org/10.1029/2003JD003620.

Lozupone, C.A., Knight, R., 2007. Global patterns of bacterial diversity. Proc. Natl. Acad. Sci. U. S. A. 104 (27), 11436–11440.

Martin, M., 2011. Cutadapt removes adapter sequences from high-throughput sequencing reads. EMBnet. J. 17, 10–12.

Massalha, H., Korenblum, E., Tholl, D., Aharoni, A., 2017. Small molecules below-ground: the role of specialized metabolites in the rhizosphere. Plant J. 90, 788–807. https://doi.org/10.1111/tpj.13543.

Matsui, K., Sugimoto, K., Mano, J., Ozawa, R., Takabayashi, J., 2012. Differential metabolisms of green leaf volatiles in injured and intact parts of a wounded leaf meet distinct ecophysiological requirements. PLoS One 7 (4), e36433. https://doi.org/10.1371/journal.pone.0036433.

McAnulla, C., McDonald, I.R., Murrell, J.C., 2001. Methyl chloride utilising bacteria are ubiquitous in the natural environment. FEMS Microbiol. Lett. 201, 151–155. https://doi.org/10.1111/j.1574-6968.2001.tb10749.x.

Minot, S.S., Krumm, N., Greenfield, N.B., 2015. One codex: a sensitive and accurate data platform for genomic microbial identification. bioRxiv, 027607. https://doi.org/10.1101/027607.

Mukherjee, A., Lal, R., 2013. Biochar impacts on soil physical properties and greenhouse gas emissions. Agronomy 3, 313–339. https://doi.org/10.3390/agronomy3020313.

Muller, A., Faubert, P., Hagen, M., Zu Castell, W., Polle, A., Schnitzler, J.P., et al., 2013. Volatile profiles of fungi–chemotyping of species and ecological functions. Fungal Genet. Biol. 54, 25–33. https://doi.org/10.1016/j.fgb.2013.02.005.

Nedwell, D.B., Underwood, G.J.C., McGenity, T.J., Whitby, C., Dumbrell, A.J., 2016. The Colne estuary: a long-term microbial ecology observatory. Adv. Ecol. Res. 55, 227–281.

Nicol, G.W., Leininger, S., Schleper, C., Prosser, J.I., 2008. The influence of soil pH on the diversity, abundance and transcriptional activity of ammonia oxidizing archaea and bacteria. Environ. Microbiol. 10 (11), 2966–2978. https://doi.org/10.1111/j.1462-2920.2008.01701.x.

Niinemets, Ü., Kännaste, A., Copolovici, L., 2013. Quantitative patterns between plant volatile emissions induced by biotic stresses and the degree of damage. Front. Plant Sci. 4, 262. https://doi.org/10.3389/fpls.2013.00262.

Nobrega, R.L.B., Guzha, A.C., Torres, G.N., Kovacs, K., Lamparter, G., Amorim, R.S.S., et al., 2017. Effects of conversion of native cerrado vegetation to pasture on soil hydro-physical properties, evapotranspiration and streamflow on the Amazonian agricultural frontier. PLoS One 12 (6), e0179414. https://doi.org/10.1371/journal.pone.0179414.

Oduro, H., Kamyshny, A., Zerkle, A., Li, Y., Farquhar, J., 2013. Quadruple sulfur isotope constraints on the origin and cycling of volatile organic sulfur compounds in a stratified sulfidic lake. Geochim. Cosmochim. Acta 120, 251–262. https://doi.org/10.1016/j.gca.2013.06.039.

Pan, D., Watson, R., Wang, D., Tan, Z.H., Snow, D.D., Weber, K.A., 2014. Correlation between viral production and carbon mineralization under nitrate-reducing conditions in aquifer sediment. ISME J. 8, 1691–1703.

Parks, J.M., Johs, A., Podar, M., et al., 2013. The genetic basis for bacterial mercury methylation. Science 339, 1332–1335.

Peñuelas, J., Asensio, D., Tholl, D., Wenke, K., Rosenkranz, M., Piechulla, B., Schnitzler, J.P., 2014. Biogenic volatile emissions from the soil. Plant Cell Environ. 37, 1866–1891. https://doi.org/10.1111/pce.12340.

Pierik, R., Ballaré, C.L., Dicke, M., 2014. Ecology of plant volatiles: taking a plant community perspective. Plant Cell Environ. 37, 1845–1853. https://doi.org/10.1111/pce.12330.

Prieto, I., Armas, C., Pugnaire, F.I., 2012. Water release through plant roots: new insights into its consequences at the plant and ecosystem level. New Phytol. 193, 830–841. https://doi.org/10.1111/j.1469-8137.2011.04039.x.

Prosser, J.I., 2015. Dispersing misconceptions and identifying opportunities for the use of 'omics' in soil microbial ecology. Nat. Rev. Microbiol. 13 (7), 439–446. https://doi.org/10.1038/nrmicro3468.

Read, D.B., Bengough, A.G., Gregory, P.J., Crawford, J.W., Robinson, D., Scrimgeour, C.M., et al., 2003. Plant roots release phospholipid surfactants that modify the physical and chemical properties of soil. New Phytol. 157, 315–326.

Redeker, K.R., Kalin, R., 2012. Methyl chloride isotopic signatures from Irish forest soils and a comparison between abiotic and biogenic methyl halide soil fluxes. Glob. Chang. Biol. 18, 1453–1467. https://doi.org/10.1111/j.1365-2486.2011.02600.x.

Redeker, K.R., Treseder, K., Allen, M.F., 2004. Ectomycorrhizal fungi: a new source of atmospheric methyl halides? Glob. Chang. Biol. 10 (6), 1009–1016.

Redeker, K.R., Davis, S., Kalin, R.M., 2007. Isotope values of atmospheric halocarbons and hydrocarbons from Irish urban, rural, and marine locations. J. Geophys. Res. 112, D16307. https://doi.org/10.1029/2006JD007784.

Redeker, K.R., Baird, A.J., The, Y.A., 2015. Quantifying wind and pressure effects on trace gas fluxes across the soil-atmosphere interface. Biogeosciences 12 (24), 7423–7434.

Ryu, C.-M., Farag, M.A., Hu, C.-H., Reddy, M.S., Wei, H.-X., Paré, P.W., Kloepper, J.W., 2003. Bacterial volatiles promote growth in Arabidopsis. Proc. Natl. Acad. Sci. U. S. A. 100 (8), 4927–4932. https://doi.org/10.1073/pnas.0730845100.

Saini, H.S., Attieh, J.M., Hanson, A.D., 1995. Biosynthesis of halomethanes and methanethiol by higher plants via a novel methyltransferase reaction. Plant Cell Environ. 18, 1027–1033.

Saito, T., Yokouchi, Y., 2008. Stable carbon isotope ratio of methyl chloride emitted from glasshouse-grown tropical plants and its implication for the global methyl chloride budget. Geophys. Res. Lett. 35, L08807. https://doi.org/10.1029/2007GL032736.

Schmidt, R., Etalo, D.W., de Jager, V., Gerards, S., Zweers, H., de Boer, W., Garbeva, P., 2015. Microbial small talk: volatiles in fungal–bacterial interactions. Front. Microbiol. 6, 1495. https://doi.org/10.3389/fmicb.2015.01495.

Schmidt, R., Etalo, D.W., de Jager, V., Gerards, S., Zweers, H., de Boer, W., Garbeva, P., 2016. Microbial small talk: volatiles in fungal–bacterial interactions. Front. Microbiol. 6, 1495. https://doi.org/10.3389/fmicb.2015.01495.

Schulz-Bohm, K., Zweers, H., de Boer, W., Garbeva, P., 2015. A fragrant neighborhood: volatile mediated bacterial interactions in soil. Front. Microbiol. 6, 1212. https://doi.org/10.3389/fmicb.2015.01212.

Serna-Chavez, H.M., Fierer, N., van Bodegom, P.M., 2013. Global drivers and patterns of microbial abundance in soil. Glob. Ecol. Biogeogr. 22, 1162–1172. https://doi.org/10.1111/geb.12070.

Stein, L.Y., Yung, Y.L., 2003. Production, isotopic composition, and atmospheric fate of biologically produced nitrous oxide. Annu. Rev. Earth Planet. Sci. 31, 329–356. https://doi.org/10.1146/annurev.earth.31.110502.080901.

Toljander, J.F., Lindahl, B.D., Paul, L.R., Elfstrand, M., Finlay, R.D., 2007. Influence of arbuscular mycorrhizal mycelial exudates on soil bacterial growth and community structure. FEMS Microbiol. Ecol. 61, 295–304. https://doi.org/10.1111/j.1574-6941.2007.00337.x.

Turlings, T.C.J., Hiltpold, I., Rasmann, S., 2012. The importance of root-produced volatiles as foraging cues for entomopathogenic nematodes. Plant Soil 358, 51–60. https://doi.org/10.1007/s11104-012-1295-3.
Ul Hassan, M.N., Zainal, Z., Ismail, I., 2015. Green leaf volatiles: biosynthesis, biological functions and their applications in biotechnology. Plant Biotechnol. J. 13, 727–739. https://doi.org/10.1111/pbi.12368.
Verstraeten, W.W., Veroustraete, F., Feyen, J., 2008. Assessment of evapotranspiration and soil moisture content across different scales of observation. Sensors 8, 70–117.
Virtanen, A.I., 1965. Studies on organic sulphur compounds and other labile substances in plants. Phytochemistry 4, 207–228.
von Fischer, J.C., Hedin, L.O., 2007. Controls on soil methane fluxes: tests of biophysical mechanisms using stable isotope tracers. Glob. Biogeochem. Cycles 21, GB2007. https://doi.org/10.1029/2006GB002687.
Wasmund, K., Mußmann, M., Loy, A., 2017. The life sulfuric: microbial ecology of sulfur cycling in marine sediments. Environ. Microbiol. Rep. 9, 323–344. https://doi.org/10.1111/1758-2229.12538.
Watling, R., Harper, D., 1998. Chloromethane production by wood-rotting fungi and an estimate of the global flux to the atmosphere. Mycol. Res. 102 (7), 769–787.
Yamada, T., Sekiguchi, Y., Hanada, S., Imachi, H., Ohashi, A., Harada, H., Kamagata, Y., 2006. *Anaerolinea thermolimosa* sp. nov., *Levilinea saccharolytica* gen. nov., sp. nov. and *Leptolinea tardivitalis* gen. nov., sp. nov., novel filamentous anaerobes, and description of the new classes Anaerolineae classis nov. and Caldilineae classis nov. in the bacterial phylum *Chloroflexi*. Int. J. Syst. Evol. Microbiol. 56 (6), 1331–1340. https://doi.org/10.1099/ijs.0.64169-0.
Yamada, T., Imachi, H., Ohashi, A., Harada, H., Hanada, S., Kamagata, Y., Sekiguchi, Y., 2007. *Bellilinea caldifistulae* gen. nov., sp. nov. and *Longilinea arvoryzae* gen. nov., sp. nov., strictly anaerobic, filamentous bacteria of the phylum *Chloroflexi* isolated from methanogenic propionate-degrading consortia. Int. J. Syst. Evol. Microbiol. 57, 2299–2306. https://doi.org/10.1099/ijs.0.65098-0.
Yuste, J.C., Baldocchi, D.D., Gershenson, A., Goldstein, A., Misson, L., Wong, S., 2007. Microbial soil respiration and its dependency on carbon inputs, soil temperature and moisture. Glob. Chang. Biol. 13, 1–18. https://doi.org/10.1111/j.1365-2486.2007.01415.x.
Zhang, H., Kim, M.S., Krishnamachari, V., Payton, P., Sun, Y., Grimson, M., et al., 2007. Rhizobacterial volatile emissions regulate auxin homeostasis and cell expansion in *Arabidopsis*. Planta 226, 839–851. https://doi.org/10.1007/s00425-007-0530-2.
Zhang, W., Li, Y., Xu, C., Li, Q., Lin, W., 2016. Isotope signatures of N_2O emitted from vegetable soil: ammonia oxidation drives N_2O production in NH_4^+-fertilized soil of North China. Sci. Rep. 6. Article number: 29257.
Zhu, B., van Dijk, G., Fritz, C., Smolders, A.J.P., Pol, A., Jetten, M.S.M., Ettwig, K.F., 2012. Anaerobic oxidization of methane in a minerotrophic peatland: enrichment of nitrite-dependent methane-oxidizing bacteria. Appl. Environ. Microbiol. 78 (24), 8657–8665. https://doi.org/10.1128/AEM.02102-12.

FURTHER READING

Sorek, R., Cossart, P., 2010. Prokaryotic transcriptomics: a new view on regulation, physiology and pathogenicity. Nat. Rev. Genet. 11, 9–16. https://doi.org/10.1038/nrg2695.
Wahman, D.G., Katz, L.E., Speitel Jr., G.E., 2005. Cometabolism of trihalomethanes by *Nitrosomonas europaea*. Appl. Environ. Microbiol. 71 (12), 7980–7986.

CHAPTER FIVE

Using Social Media for Biomonitoring: How Facebook, Twitter, Flickr and Other Social Networking Platforms Can Provide Large-Scale Biodiversity Data

Jon Chamberlain[1]
School of Computer Science and Electronic Engineering, University of Essex, Essex, United Kingdom
[1]Corresponding author: e-mail address: jchamb@essex.ac.uk

Contents

1. Introduction 134
2. Related Work 135
   2.1 Automatic Image Classification 136
   2.2 Collecting and Labelling Data With a Crowd 138
3. Examples of Biomonitoring Using Social Networking Platforms 141
   3.1 Identifying Species in an Image 142
   3.2 Identifying New Species 142
   3.3 Requesting Observations of a Species 143
   3.4 Coordinating Citizen Science Schemes 143
   3.5 Observations of Presence 144
   3.6 Proliferation of Changing Nomenclature 144
4. Analysis of Posts About Wildlife on Social Networking Platforms 144
   4.1 Data Collection and Preparation 145
   4.2 Data Analysis 148
   4.3 Accuracy of Image Labelling 153
   4.4 Comparison to Microworking 154
5. Discussion 155
   5.1 Data Acquisition and Annotation 155
   5.2 User Motivation 156
   5.3 Task Difficulty 157
   5.4 Social Learning and the Expert in the Crowd 158
   5.5 Harnessing Collective Intelligence on Social Networking Platforms 159
   5.6 Limitations of a Groupsourcing Approach 160

Advances in Ecological Research, Volume 59
ISSN 0065-2504
https://doi.org/10.1016/bs.aecr.2018.06.001

6. Applications 161
7. Future Directions 163
8. Conclusions 165
Acknowledgements 165
References 166

Abstract

In this chapter, social networking platforms are explored to see whether they can be a useful resource for biomonitoring; more specifically do they contain reliable biodiversity data and to what extent can we extract that information, both by analysing conversation threads and understanding how groups of people solve image classification problems.

A corpus of messages was analysed from Facebook containing 39,039 conversation threads. Social network groups that were set up specifically for users to exchange biodiversity information show a high workrate, fast response time, short message lifespan and more in-thread activity and discussion. Image classification tasks posted in these groups get a fast reply, elicit more data from users and are more likely to have the task completed. Users distribute work unevenly (the top 20% of users do 88.4% of the work), following a Zipf distribution.

This technology offers researchers a new opportunity to gather biodiversity data; however, it is not without its challenges. Tasks posted in such groups tend to be difficult to solve; however, the resulting labelling quality is very high when compared to experts and to other approaches. Automatic processing in some form for these types of data is essential given the rate of increase of data being added every day to social networking platforms; however, this is a complex problem due to informal language use and access to the data.

1. INTRODUCTION

By 2050 all coral reefs are estimated to be at risk from human activities including tourism, coral mining, pollution, overfishing, canal dredging and the warming and acidification of oceans (Hoegh-Guldberg et al., 2007; Wilkinson, 2008). Given the current global priorities of economic growth, energy security, threats of terrorism and pandemics[1] there is not likely to be a change in policy or increase in funding for conservation monitoring or research. It is apparent that we need more radical solutions for monitoring biodiversity that can collect vast amounts of data and process it for actionable knowledge without incurring the costs of traditional research.

[1] https://g20.org/wp-content/uploads/2014/12/brisbane_g20_leaders_summit_communique.pdf.

A solution is born from one of the threats itself: the increasing popularity of recreational SCUBA diving. Coupled with the affordability of underwater digital camera equipment, more data are being created and shared in informal ways, such as on social networking platforms,[2] with data being annotated by an enthusiastic community on a scale never been seen before. With more ecosystems being casually monitored by the public in this way a huge resource is being created and this research lays the foundations for developing a full-scale solution to the problem of global ocean biomonitoring with the collective data of social networking platforms.

In response to these challenges, this research investigates whether social networking platforms can be a useful resource to monitor wildlife; more specifically do they contain reliable biodiversity data and to what extent can we extract that information, both by analysing conversation threads and understanding how groups of people solve image classification problems. Section 2 discusses different approaches to using groups of people (referred to as a crowd) for data collection and labelling tasks and proposes that social networking platforms can be seen as such a system. Examples of how people are using these platforms for monitoring biodiversity are discussed in Section 3 before analysing data for a specific example of image labelling (Section 4). Using social networking platforms in this way is discussed in more depth in Section 5 and an example application is presented in Section 6.

2. RELATED WORK

Collecting the primary data to determine conservation management practices is a resource-intensive and time-consuming process in which traditionally the data are collected by a team of researchers in the field who subsequently spend numerous hours labelling the data, such as identifying and classifying habitats, species and behaviour. These efforts would perhaps be validated by other experts and inconsistencies (see Fig. 1) would be resolved. However, other methods can be used to create large-scale data resources, with high-quality labelling of information about the data, such as approaches that use groups of people to solve a particular problem or task. This can be achieved by developing structured systems for collecting data or by data mining and information extraction.

[2] Social networks in this context refer to software applications that allow Internet users to share information, photos and videos (referred to as social media).

Fig. 1 An image of a school of Red Sea Bannerfish, highlighting some of the difficulties when identifying and counting objects (in this case fish), such as partial objects (A), occlusion (B), rotation, contrast and depth of field (C).

Information (labels) can be added to data in a number of ways:

1. at the point of creation, most usually by the person who created the data, but also by the device that was used (for example, a camera will record metadata with every image taken which includes information about the manufacturer of the camera, the lens settings, GPS coordinates, etc.);
2. by automatic processing by applying algorithms that try to interpret the data (although this may be error-prone and require supervision from administrators);
3. manually after the data have been created via a labelling task.

Images can have different labels applied to them: the entire image can be labelled; regions can be labelled; or specific objects can be outlined and labelled (see Fig. 2).

2.1 Automatic Image Classification

Analysis of marine species in images has recently become important due to the increasing use of Autonomous Underwater Vehicles (AUV) that can collect data for many hours at a time. These images are either very numerous or very complex in their content, or perhaps both, making it impossible for human annotators to assess the contents of images on a large scale.

Categorising and classifying images, as well as the entities contained within them, have been the long-term goal for computer vision (Barnard et al., 2003); however, only in the last few decades have screen-based images

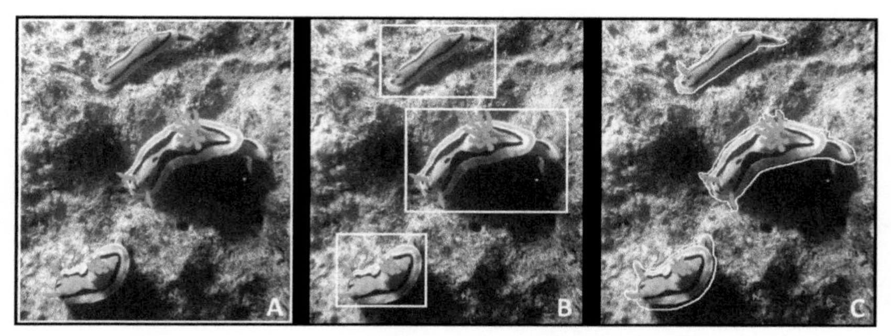

Fig. 2 Three images showing the different styles of image annotation: entire image labelled (A); regions labelled (B) and object outline labelled (C).

and digital photography made image classification so ubiquitous, and so important. It is therefore not surprising that automatic image annotation is an active area of research (Lu and Weng, 2007) with specific industry-supported tracks, such as Yahoo's Flickr-tag challenge at ACM Multimedia 2013.[3]

Gold standard image datasets, where the true label has been confirmed by expert annotators, exist for images of wildlife, such as Caltech-UCSD Birds 200, a repository of 200 species of birds displayed in 6033 images (Welinder et al., 2010a). Such datasets are used to train machine learning algorithms to identify visual features within images and map them to labels and concepts. More recently, there have been ImageCLEF challenges focused on automatically identifying species of fish from video still images.[4]

For easy-to-identify taxa, automatic systems achieve comparable results to that of experts; however, more difficult groups present problems for both approaches. Efforts to classify the habitat shown in an image by automatic recognition have shown some success at the broadest classification levels, such as *iSIS* which achieves 84% overall accuracy (Schoening et al., 2012) and *DiCANN* which achieves 90% accuracy for easy-to-classify images (Culverhouse et al., 2003). Other efforts have reported higher accuracy of 92%–95% with semisupervised classification (Beijbom et al., 2012).

It has been suggested that 'obtaining genus or species level data from even the highest quality digital images is very challenging and not without the possibility of human error' and that machine learning will be limited by the gold standards used for training (Henry and Roberts, 2014).

[3] http://acmmm13.org/submissions/call-for-multimedia-grand-challenge-solutions/yahoo-large-scale-flickr-tag-image-classification-challenge (accessed 27 August 2017).
[4] http://www.imageclef.org/2014/lifeclef/fish (accessed 27 August 2017).

2.2 Collecting and Labelling Data With a Crowd

Crowdsourcing is an approach to replace the work traditionally done by a single person by the collective action of a group of people via the Internet (Howe, 2008). Crowdsourcing has established itself in the mainstream of research methodology in recent years using a variety of approaches to engage humans to solve problems that otherwise could not be solved. One such task is the labelling of images for creating gold standard datasets for training machine learning algorithms.

2.2.1 Peer Production

Peer production is a way of completing tasks that relies on self-organising communities of individuals in which effort is coordinated towards a shared outcome (Benkler and Nissenbaum, 2006). The key aspects that make peer production so successful are the openness of the data resource being created and the transparency of the community that is creating it (Dabbish et al., 2014; Lakhani et al., 2007).

People who contribute information to Wikipedia are motivated by personal reasons such as the desire to make a particular page accurate, or the pride in one's knowledge in a certain subject matter (Yang and Lai, 2010). This motivation is also behind the success of *citizen science* projects, such as the *Zooniverse*[5] collection of projects where participants work on tasks including identifying astronomic objects, tagging wildlife in footage from motion sensitive cameras and transcribing ship's logs. The scientific research in these projects is conducted mainly by amateur scientists and members of the public (Clery, 2011). The costs of ambitious data labelling tasks are also kept to a minimum, with experts only required to validate a small portion of the data (which is also likely to be the data of most interest them).

Some citizen science projects get members of the public to classify objects in images taken from ROVs (Remotely Operated Vehicles),[6],[7],[8] while others require the users to supply the source data as well as the classification.[9],[10],[11],[12],[13] The quality of citizen scientist generated data has been

---

[5] https://www.zooniverse.org.
[6] http://www.planktonportal.org.
[7] http://www.seafloorexplorer.org.
[8] http://www.subseaobservers.com.
[9] http://www.projectnoah.org.
[10] http://www.arkive.org.
[11] http://www.brc.ac.uk/irecord.
[12] http://observation.org.
[13] https://www.inaturalist.org.

shown to be comparable to that generated by experts when producing taxonomic lists (Holt et al., 2013) even when the task is not trivial (He et al., 2013). Citizen science efforts at labelling marine images show high accuracy for complex tasks, such as measuring scallops in images from *Seafloorexplorer*, with contributions correlating to expert labels created under the same conditions.[14]

Question answering systems attempt to learn how to answer a question automatically from a human, either from structured data or from processing natural language of existing conversations and dialogue. Here we are more interested in *Community Question Answering (cQA)*, in which the crowd is the system that attempts to answer the question through natural language, such as StackOveflow[15] and Yahoo Answers.[16]

Image classification in a cQA format is common in SCUBA diving forums,[17] but can suffer from not having a broad enough community of users to answer the questions. Social networking platforms follow a similar cQA dialogue style in which threads may contain true classification tasks (a question is asked and is answered) or implied tasks (the post is augmented with additional data).

2.2.2 Microworking

Amazon Mechanical Turk[18] pioneered microwork crowdsourcing by using the Web as a way of reaching large numbers of workers (often referred to as turkers) who get paid to complete small items of work called human intelligence tasks (HITs). This is typically very little, in the order of 0.01–0.20 US$ per HIT.

Some studies have shown that the quality of resources created this way are comparable to that of resources created by experts, provided that multiple judgements are collected in sufficient number and that enough post-processing (removing poor quality judgements, spam, etc.) is done (Callison-Burch, 2009; Snow et al., 2008).

A reported advantage of microworking is that the work is completed very fast. It is not uncommon for a HIT to be completed in minutes, but this is usually for simple tasks. In the case of more complex tasks, or tasks

[14] http://archive.is/20150703012530/blog.seafloorexplorer.org (accessed 27 August 2017).
[15] http://stackoverflow.com.
[16] https://uk.answers.yahoo.com.
[17] https://www.scubaboard.com/community/forums/name-that-critter.360 (accessed 27 August 2017).
[18] http://www.mturk.com.

in which the worker needs to be more skilled, e.g. translating a sentence in an uncommon language, it can take much longer (Novotney and Callison-Burch, 2010). Microwork crowdsourcing is becoming more common for creating small-scale resources and addressing image labelling problems (Welinder et al., 2010b), but is prohibitively expensive on a large scale.

2.2.3 Gaming and Games-With-a-Purpose

Generally speaking, a game-based or game-with-a-purpose (GWAP) crowdsourcing approach uses entertainment rather than financial payment to motivate participation. The approach is motivated by the observation that every year people spend billions of hours playing games on the Web (von Ahn, 2006). The approach showed enormous initial potential, with the first, and perhaps most successful, image labelling game called the *ESP Game* attracting over 200,000 players who produced over 50 million labels (von Ahn, 2006). Players of this game were asked to provide a label for an image and if it matched the label provided by another player they both score points. Other image labelling games include the *Puzzle Racing* game (Jurgens and Navigli, 2014) and the two stage game (called *Infection* and *Knowledge Tower*) for validating image concepts (Vannella et al., 2014). A GWAP approach to classifying images of wildlife (moths) called *Happy Moths* also showed good accuracy (Prestopnik et al., 2014).

2.2.4 Social Computing and Social Networks

Social computing has been described as 'applications and services that facilitate collective action and social interaction online with rich exchange of multimedia information and evolution of aggregate knowledge' (Parameswaran and Whinston, 2007). It encompasses technologies that enable communities to gather online such as blogs, forums and social networking platforms, although the purpose is largely not to solve problems directly.

Increasingly, social networking platforms are being used to organise data, to pose problems and to connect with people who may have solutions that can be contributed in a simple and socially convenient fashion. Facebook[19] has been used as a way of connecting professional scientists and amateur enthusiasts with considerable success (Gonella et al., 2015; Sidlauskas et al., 2011). However, there are drawbacks with this method of knowledge sharing and problem solving: data may be lost to people interested in them in

[19] https://www.facebook.com.

the future and they are often not accessible in a simple way, for example, with a search engine.

Social network crowdsourcing is distinguished by several features:

1. data and tasks are created by the users;
2. input is unconstrained and developed in series while simultaneously validated by the users themselves;
3. users are inherently motivated, socially trained and work collaboratively;
4. the output is immediately accessible and beneficial to all, with users receiving recognition for their efforts.

This chapter focuses on using social networking platforms for collecting biodiversity information and performing related tasks, such as image labelling, on a large scale.

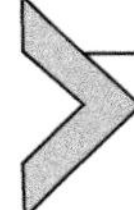

3. EXAMPLES OF BIOMONITORING USING SOCIAL NETWORKING PLATFORMS

Social networking platforms can be used in different ways to give insights into wildlife behaviour, distribution and morphology. Users of such networks leverage the functionality of media applications that are best suited for their purpose, if not originally intended by the developers of the systems.

Flickr[20] is a popular social networking platform for users to upload, share and tag images. The site is intended to be archival and media-rich, displaying users interesting images as if in an art gallery. The images, tags and location data can be accessed simply through the application programming interface (API), allowing analysis of species distribution to be performed (Barve, 2014; ElQadi et al., 2017). Additionally, sophisticated tagging tools allow users to develop libraries of images specifically for demonstrating morphological features.[21]

Conversely, Twitter[22] which is another well-used social network platform allows users to share "tweets" containing text and images. Only a small proportion of these data can be accessed through the API, with time restrictions, but can be filtered by location of the user and keywords making it a useful resource for monitoring up-to-date occurrences (Becken et al., 2017; Daume, 2016).

Although both Flickr and Twitter allow discussion between users in some form, the most popular platform is Facebook (over 79% of adults

[20] https://www.flickr.com.

[21] https://www.flickr.com/photos/56388191@N08/albums (accessed 24 August 2017).

[22] https://twitter.com.

use Facebook compared to 24% for Twitter).[23] Facebook has a vast resource of uploaded images from its community of users, with over 250 billion images, and a further 350 million posted every day. Images of things (rather than people or places) that have been given captions by users only represent 1% of these data, but it is still of the order of 2.6 billion images.[24] Facebook allows users to create complex threads of discussion which may be related to different posted media and the community uses this functionality in a number of ways.

3.1 Identifying Species in an Image

One of the first uses of social networking platforms in the context of monitoring wildlife was the identification of species by discussing morphological features and the tagging in experts to help (Sidlauskas et al., 2011). This is perhaps now the most common use, where a user posts an image and requests the community to provide an identification (and perhaps additional information), for example *Seasearch Identifications*.[25] Such requests may be made as part of a wider data collection scheme or out of personal curiosity. This behaviour is discussed and analysed in more detail in Section 4.

3.2 Identifying New Species

In the process of identifying species uploaded by users, occasionally a user may unknowingly upload an image of something that nobody can identify. These interesting cases spark much discussion regarding morphology and taxonomy, and may ultimately lead to further research to discover new species (Gonella et al., 2015). Unlike traditional monitoring schemes where a species classification must exist before an observation can be classified, the discussion can indicate to the community that the taxonomic classification of the species may not be certain and that a different way of classifying the image may be more appropriate (such as the use of common names, referring to a species complex or a reference to a previous unknown tag or morphospecies, e.g. Species A). This was the case for the common NE Atlantic opistobranch *Diaphorodoris luteocincta* (M. Sars, 1870) that was know to have a white variation and so users were instructed to record observations of the latter as *D. luteocincta var. alba*. Anatomical characteristics and

[23] http://www.pewinternet.org/2016/11/11/social-media-update-2016 (accessed 24 August 2017).
[24] http://www.insidefacebook.com/2013/11/28/infographic-what-types-of-images-are-posted-on-facebook (accessed 27 August 2017).
[25] https://www.facebook.com/groups/seasearch.identifications (accessed 27 August 2017).

molecular analyses eventually showed that the two colour morphs were indeed two different species (Furfaro et al., 2016) and previous classifications could be easily reconciled because the recording of uncertain observations had been coordinated, in part, via social networking platforms.

3.3 Requesting Observations of a Species

The work of users on social networking platforms is largely undirected and uncoordinated; however, there are some cases where the community can be tasked with a specific goal. Specific species that may be of interest due to rarity or conservation priority can be highlighted to the community while they are doing their normal activities, or to search through their records for observations that may fit the description. For example, the Group of Research on Opistobranchs in Catalonia (GROC)[26] requested observations for the pelagic opistobranch *Cymbulia peronii* Blainville 1818, using the simple post "Did you find this pelagic species?" and an accompanying image to assist with identification. In response their social network community had 3000 reads and provided four useful observations (pers. comm.).

3.4 Coordinating Citizen Science Schemes

Beyond requesting specific observations, social networking platforms can be used to coordinate the efforts of users who would traditionally be described as citizen scientists. A good example of this is the UK Hoverflies Facebook group[27] that is the Facebook presence of the UK Hoverfly Recording Scheme.[28] Users are requested to upload images with grid references using online resources[29] that allow each observation to be added to the national database. This labour-intensive manual process means the administrators are spending more time processing the data (estimated to be 25,000 observations in 2017 from all sources) than assisting with the community with identifications; however, it has led to some detailed insights into species distribution and phenology.[30]

[26] http://www.opistobranquis.org/en/home (accessed 24 August 2017).
[27] https://www.facebook.com/groups/609272232450940 (accessed 24 August 2017).
[28] http://www.hoverfly.org.uk (accessed 24 August 2017).
[29] http://www.gridreferencefinder.com (accessed 24 August 2017).
[30] http://stamfordsyrpher.blogspot.co.uk (accessed 24 August 2017).

3.5 Observations of Presence

Marine Ecological Solutions[31] recently developed an ID guide called Marine Signs of Life that showed images of indicators of species when they are not present such as eggs, tubes, siphons, casts, etc., for video analysts and in-field surveyors. The authors of the guide used archival social networking platforms such as Flickr to find images that were tagged with a species name but without the species actually being present in the image. Additionally, they used custom search functionality (see Section 6) for Facebook to search through posts of marine life (pers. comm.).

3.6 Proliferation of Changing Nomenclature

It is not uncommon for the underlying taxonomic hierarchy of species to change dramatically, for example, the update to the taxonomic group *Chromodorididae* (Johnson and Gosliner, 2012) that rendered many Web resources and books out of date. Social networking platforms can be useful for the proliferation and acceptance of a new taxonomy, although it is beyond the scope of this chapter to discuss this in any depth. It is notable that taxonomic discussion features frequently in online discussions with classifications corrected citing online references within the thread.

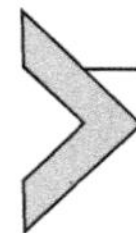

4. ANALYSIS OF POSTS ABOUT WILDLIFE ON SOCIAL NETWORKING PLATFORMS

The examples of how social networking platforms are used for monitoring biodiversity indicate that not only is there considerable interactivity among the community but also that the information that the users are acquiring is of a sufficiently high quality to keep them returning to contribute more. Here we investigate two key questions: do the posts on social networking platforms contain reliable biodiversity data and to what extent can we extract that information.

Similar to crowdsourcing, *groupsourcing* is defined as *completing a task using a group of intrinsically motivated people of varying expertise connected through a social network* (Chamberlain, 2014). A *group* in this context is a feature of a social network platform that allows a small subset of users to communicate through a shared message system. Groups on Facebook are initially set up in response to the needs of a few people and the community evolves as news from the group is proliferated around the network in feeds and user activity. The

[31] http://www.marine-ecosol.co.uk.

group title, description and 'pinned' posts usually give clear indications as to whom the group is aimed at and for what purpose. This research focused on three types of group motivation that were considered likely to contain images with classifications of wildlife:

1. *Task Request (TR)*—groups in which users were encouraged to post messages with a task, e.g. *ID Please (Marine Creature Identification)*
2. *Media Gallery (MG)*—groups in which users were encouraged to share media (image and video) for its artistic merit, e.g. *Underwater Macro Photographers*
3. *Knowledge Sharing (KS)*—groups used for coordination of activities or for distributing knowledge, research and news, e.g. *British Marine Life Study Society*

Groups can also be categorised into those that are *specific* to a taxonomic subject (e.g. *NE Atlantic Nudibranchs*, appended with -*S*) and those that are non-specific or *generalist* (e.g. *The Global Diving Community*, appended with -*G*).

A portion of messages is termed a *corpus*, and the complete dataset from a group (stored as multiple corpora) is called a *capture* (see Chamberlain (2014) for further technical details).

The *thread* of a typical post on a social network platform (such as Facebook, see Fig. 3) is structured:

1. A user posts a *message*.
2. Users (including the first user) can post a *reply*.
3. Users can *like* the message and/or replies including their own posts.

4.1 Data Collection and Preparation

In order to investigate image classification on social networking platforms, several social network (Facebook) groups were selected as they were thought likely to contain good examples. These groups were identified using the inbuilt search functionality on the platform, group recommendations and checking the group membership of prominent users in groups already found. Groups that were sufficiently mature (over 50 messages and over 50 members) were selected and were categorised according to purpose and generality.[32] The total cached message database includes 34 groups from Facebook containing 39,039 threads and a total of 213,838 messages and replies. The data were transformed into an anonymous database so users

[32] The group categorisation was done independently by Jon Chamberlain and two postgraduate researchers at the University of Essex. When there was not consensus on the categorisation (18%), a final decision was made by Jon Chamberlain after group discussion.

Fig. 3 Detail of a typical message containing an image classification task posted on a social networking platform (in this case Facebook).

cannot directly be associated with the data stored. This use of data is in line with Facebook's Data Use Policy.[33]

Messages posted to a group on Facebook can be one of six types: photo; link (URL); video; a question (in the form of an online poll); a scheduled event or just simply text (status)[34] although the majority of messages are either 'photo', 'link' or 'status' (see Fig. 4).

The Task Request (TR) and Media Gallery (MG) groups have more photo type messages posted in them compared to Knowledge Sharing (KS) groups both in the general and topic-specific categories (TR $n(6350)$ 62.5%, MG $n(17{,}831)$ 64.2%, KS $n(14{,}858)$ 38.0%, $P < 0.01$, z-test). This is not surprising as the primary motivation for posting a message

[33] https://www.facebook.com/full_data_use_policy (accessed 28 August 2017).

[34] http://fbrep.com//SMB/Page_Post_Best_Practices.pdf.

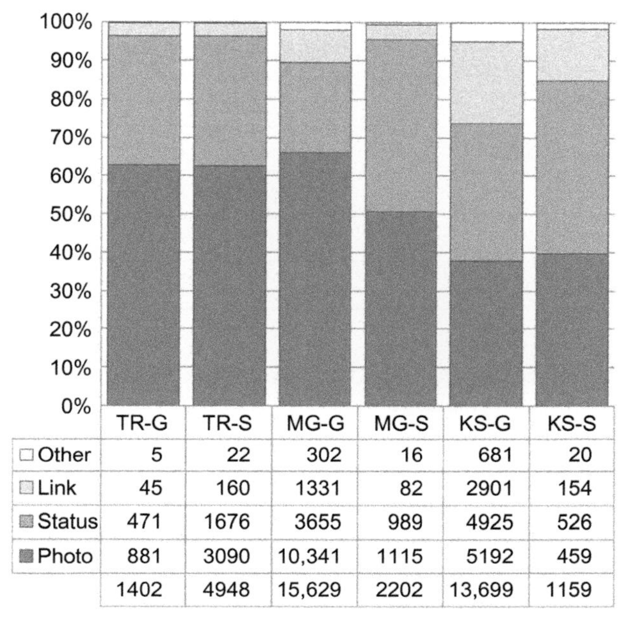

	TR-G	TR-S	MG-G	MG-S	KS-G	KS-S
Other	5	22	302	16	681	20
Link	45	160	1331	82	2901	154
Status	471	1676	3655	989	4925	526
Photo	881	3090	10,341	1115	5192	459
	1402	4948	15,629	2202	13,699	1159

Fig. 4 Distribution of thread types by group category. Groups were categorised by purpose: Task Request (TR); Media Gallery (MG) and Knowledge Sharing (KS), and by how specific they were to a taxonomic group (appended -G for general groups and -S for specific groups).

in TR and MG groups (seeking an identification or showing off a picture, respectively) requires an image to be attached. The KS groups show a more even distribution of message types as motivations for posting (arranging meetings, sharing research, posting information, etc.) do not require an image. This makes TR and MG groups better places to look for image classification tasks.

Image classification on social network platforms, much like Community Question Answering, occurs through the natural language of the message thread. For the purposes of this research, messages and replies were categorised by *inquisition* (question or statement) and *data load* (a solution to the task, see Table 1), although more detailed schemas (Bunt et al., 2012) and richer feature sets (Agichtein et al., 2008) have been used to describe cQA dialogue. The message and its replies form a thread that relates to what has been posted (photo, link, etc.). The thread may contain labels (or related data), irrespective of whether the poster requested them in the original message, as other users might augment or correct the posts (see Table 2).

Table 1 Categories of Posts With Examples of Content, Conditional on Inquisition (Question or Statement) and Data Load (in This Case the Scientific Name of a Species in the Image)

Category	Content
QUESTION	What is this?
CHECK	Is this *Chromodoris magnifica*?
NEUTRAL	Great photo from the trip!
ASSERTION	This is *Chromodoris magnifica*

Table 2 Categories of Threads When Viewed as a Task With Solutions

Category	Message	Reply
None	NEUTRAL	NEUTRAL
Unresolved	NEUTRAL	QUESTION
	QUESTION	QUESTION or NEUTRAL
Implied	NEUTRAL	CHECK or ASSERTION
	ASSERTION	Any
Suggestion	CHECK	Any
Resolved	QUESTION	CHECK or ASSERTION

4.2 Data Analysis

Analysis of a random sample of 1000 messages from the corpus showed a rapid drop in replies to messages after 4 weeks. Therefore, for the purposes of analysing thread activity, all messages less than 8 weeks old from the date of capture were ignored to reduce any bias in message activity of newly posted and currently active messages.

4.2.1 User Workload

The workload of each user was calculated as a total of all messages and replies they had posted. The users were then ranked by workload and fitted to a Zipf power law distribution ($R^2 = 0.957$, see Fig. 5).

In addition we find that the top 1% of users ($n = 79$) have contributed 41.6% of the work, the top 10% of users ($n = 792$) have contributed 79.2% of the work and the top 20% of users ($n = 1583$) have contributed 88.4% of the work. 53.5% of the 14,793 users who were members of the groups had contributed some form of work.

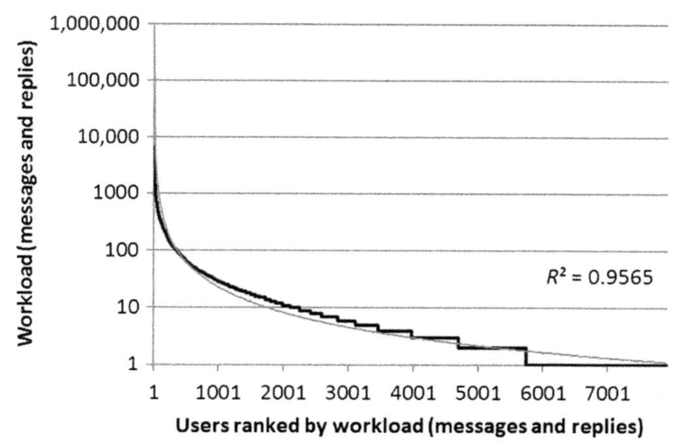

Fig. 5 Chart showing the workload (messages and replies) of users of the Facebook groups analysed, ranked by total workload and fitted to a Zipf power law distribution.

Table 3 A Table Summarising Groups' (Mean) Active Users (a User Who Has Posted a Message or Reply) and the Median and Mean Workrate (Messages/Replies per Active User)

	Active Users	Workrate (Median)	Workrate (Mean)
TR-G	28.0%	4	20.8
TR-S	36.5%	4	22.4
MG-G	20.3%	3	12.8
MG-S	32.4%	4	14.5
KS-G	18.4%	3	20.9
KS-S	38.3%	4	11.4

While the workload is unevenly distributed, social networking platforms have an active membership, perhaps because the barriers to contribution are lower than in other crowdsourcing systems.

4.2.2 User Activity

User activity was calculated as the proportion of group members that had posted a message or reply from the total membership at the time of the capture (see Table 3).

Topic-specific groups have more active users ($P < 0.05$, z-test, see Table 3), an indication that the community of users in these groups are more engaged with the subject matter and may even know each other personally (as specialist research areas tend to be quite small).

The TR groups have more active members who perform at a higher workrate ($P < 0.05$, z-test, see Table 3) than the MG groups, supporting the idea that users joining TR groups are more willing to participate actively in problem solving. Users of MG groups may be more passive by simply enjoying the images being shared.

It is clear that there is considerable information being added to social networking platforms such as Facebook with a rapid rise in new data being added each month (see Fig. 6).

4.2.3 Thread Response Time, Lifespan and Activity

The time to the first response (response time) and time to the last response (lifespan) were plotted on frequency graphs (see Table 4 and Figs. 7 and 8). 5%–10% of messages receive a reply in 8 min. The proportion of messages with replies beyond 1092:16:00 (6.5 weeks) from the time of the message being posted (outliers) is small so it makes an appropriate cut-off point for message analysis to ensure that messages have had a chance to receive all replies. The graphs show different profiles, indicating that response time is less predictable than lifespan.

General (-G) groups have a faster response rate and a shorter lifespan than topic-specific (-S) groups for MG and KS ($P < 0.05$, unpaired t-test, see

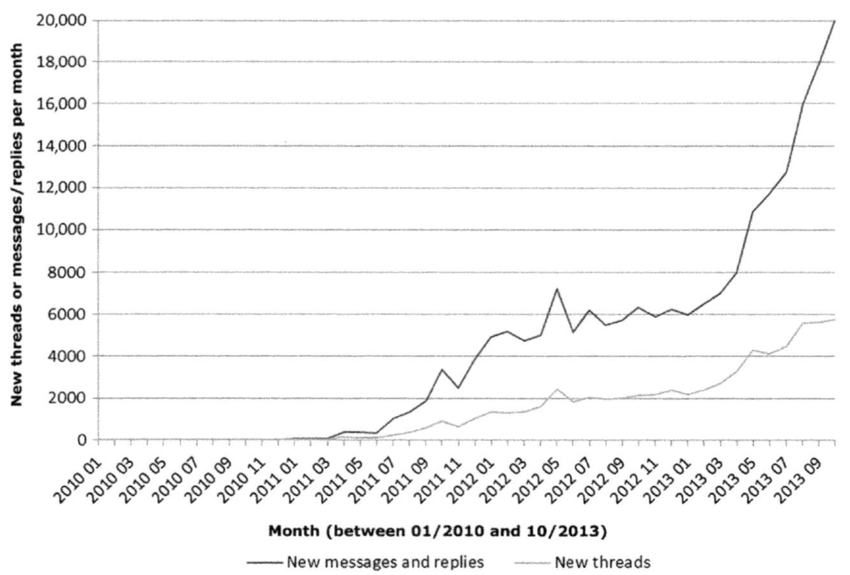

Fig. 6 Chart showing the amount of new threads and new messages/replies being added to the analysed Facebook groups each month.

Table 4 A Table Summarising Group Categories: the Proportion of Messages That Received a Reply; the Number of Replies (Median and Mean); the Response Time (Median) for the First Reply (hh:mm:ss); the Lifespan (Median) of the Thread (hh:mm:ss); and the Proportion of Outlier Replies Beyond 1092:16:00

	Received a Reply	Replies (Median)	Replies (Mean)	Response Time	Lifespan	Outliers
TR-G	81.5%	3	4.1	00:28:30	16:26:16	2.3%
TR-S	71.0%	2	3.2	00:48:57	11:55:09	1.5%
MG-G	42.7%	0	1.6	00:58:25	10:25:50	1.4%
MG-S	49.4%	0	1.8	01:59:46	16:39:43	4.0%
KS-G	50.5%	1	2.8	00:28:29	07:34:21	0.6%
KS-S	58.5%	1	2.2	01:24:45	18:12:20	3.1%

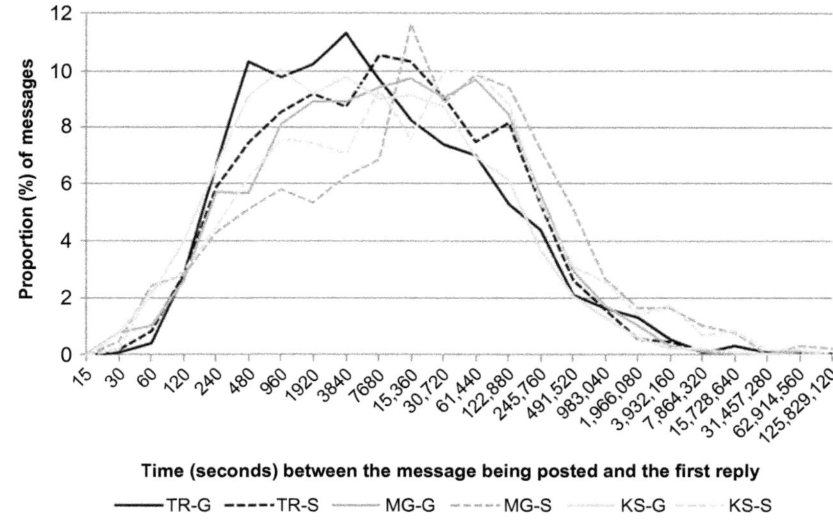

Fig. 7 Response time (seconds, log scaled) for a thread.

Table 4) perhaps indicating that users in general groups have a broad interest and make conversational replies that do not require a task to be solved.

Within topic-specific categories, the TR groups have a faster response time and shorter lifespan ($P < 0.05$, unpaired *t*-test, see Table 4) as users of these groups may anticipate task requests and are primed to submit a reply, especially if it is an opportunity to demonstrate their knowledge. This would be harder to achieve in general groups because the task may be too difficult for most users.

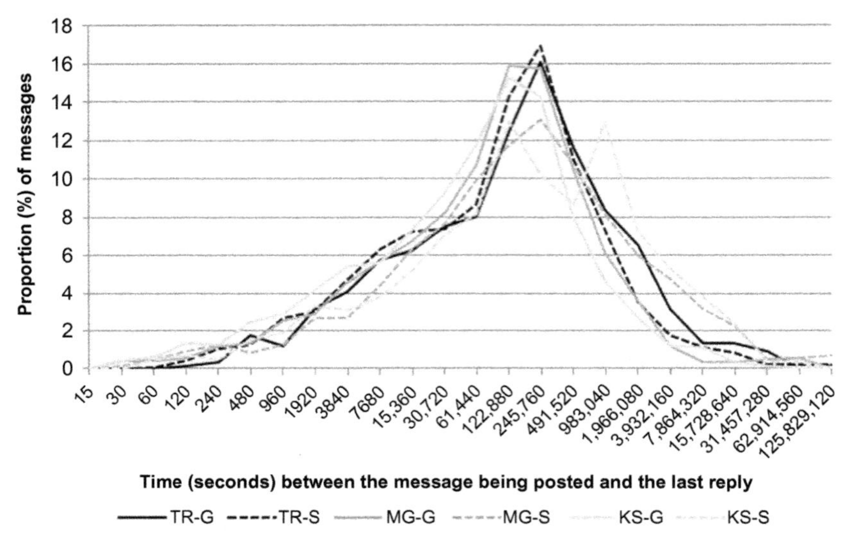

Fig. 8 Lifespan (seconds, log scaled) of a thread.

Response time and lifespan are influenced by the interface design of social networking platforms such as Facebook. When messages are first posted they appear on a user's news feed and/or notifications and the group wall. Over time they are replaced with other messages, move down the page until no longer visible and can only be accessed by clicking for older pages. If a message receives a reply it is moved back to the top of the page (termed 'bumping').

Messages posted in the TR groups have more replies than the other groups ($P < 0.05$, unpaired *t*-test, see Table 4). This is unsurprising as these groups are used for posting tasks that require a response, unlike the more passive nature of other groups. This makes the TR groups a good candidate for crowdsourcing because more users are potentially involved in the classification task.

4.2.4 Task Distribution

In order to assess the quality of data that could be extracted, and to investigate the distribution of the tasks within the group categories, 200 threads were selected at random from each category to form a subcorpus of 1200 threads.

The subcorpus was manually categorised in a random order for data load and inquisition by only viewing the thread text and author names, thus each thread could be classified as a task type (see Table 2).

Implied, Suggestion and Resolved tasks all contain data that could be extracted to solve the image classification tasks. TR groups have more data-loaded threads than MG or KS groups ($P < 0.05$, z-test) and it is not surprising due to the purpose of the groups (see Fig. 9). Additionally, tasks are more likely to be solved in the TR groups comparing resolved tasks to unresolved tasks ($P < 0.05$, z-test).

4.3 Accuracy of Image Labelling

Based on the previous findings it could be expected that the highest frequency of task requests and more accurate solutions would be found in the TR-S groups, although there are fewer explicit tasks compared to TR-G. A single topic-specific area of Opistobranchia (sea slugs or nudibranchs) was chosen in order to evaluate the accuracy of image classification. In this class of animals external morphology is often sufficient to confirm a classification from an image (unlike, for example, marine sponges) and this is also an active area on social media. In this research we analyse species level labels identified through morphology in photographs.

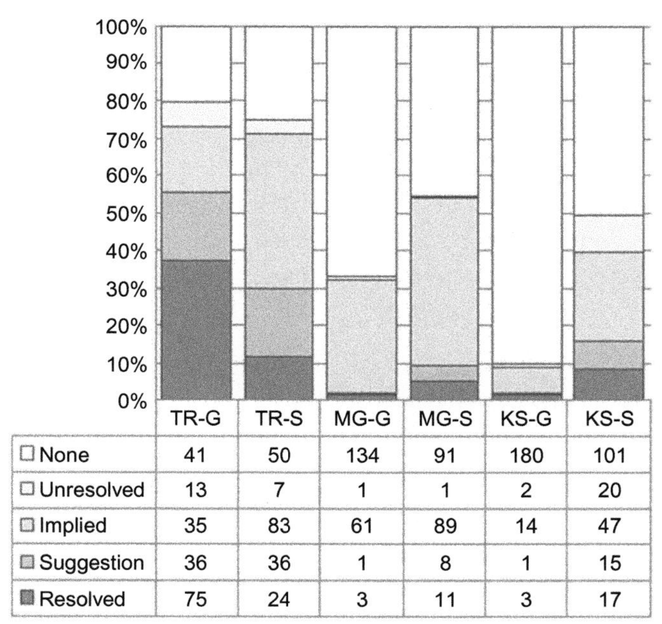

	TR-G	TR-S	MG-G	MG-S	KS-G	KS-S
None	41	50	134	91	180	101
Unresolved	13	7	1	1	2	20
Implied	35	83	61	89	14	47
Suggestion	36	36	1	8	1	15
Resolved	75	24	3	11	3	17

Fig. 9 Distribution of image classification tasks by group category. Groups were categorised by purpose: Task Request (TR); Media Gallery (MG) and Knowledge Sharing (KS), and by how specific they were to a taxonomic group (appended -G for general groups and -S for specific groups).

A random sample of threads from two groups (*Nudibase*[35] and *NE Atlantic Nudibranchs*[36]) from the TR-S subcorpus was taken. Only photo threads were selected and further threads removed if they were unsuitable for the image classification task (for example, not an Opistobranch, multiple species in an image, close-ups, words printed in the image, continuation and/or gallery threads). In total 61 threads were manually analysed using this method (called the *test set*).

The gold standard, created by examining eight resources (see (Chamberlain, 2014) for full details), was compared for inter-expert agreement using Fleiss' kappa, which allows for more than two annotators (unlike Cohen's kappa). This test showed an inter-annotator agreement of $\kappa = 0.61$, considered to be substantial agreement ($n(84)$, raters(8), $z = 34.1$, $\kappa = 0.61$, Fleiss' kappa). This is an underestimation of agreement between the resources as it accounts for all the images in the test set, including those when no classification was found.

By way of comparison the two resources that produced the most classifications (*Seaslug Forum* and *Nudipixel*) have very high agreement ($n(84)$, raters(2), $z = 9.2$, $\kappa = 0.84$, Cohen's kappa). Additionally, these two resources only disagreed with the classification on one occasion giving an inter-annotator agreement accuracy of 98.3% (counting only instances when both resources had a classification) which could be considered the top performance expected from any automatic aggregation of the groupsourcing data.

By using the gold standard to determine which answer from the subcorpus was correct, results show very high accuracy for the image classification task (0.93). This represents the upper limit of what could be expected from other categories of groups.

4.4 Comparison to Microworking

The images from the subcorpus (test set) were also classified using the microworking platform *Crowdflower*[37] to compare the accuracy. *Crowdflower* users were presented with an image and asked to provide a species name. Web resources were mentioned in the instructions, as well as the requirement for accurate spelling although minor capitalisation mistakes and synonyms were allowed. The microwork configuration selected the top 36% of users

[35] https://www.facebook.com/groups/206426176075326 (accessed 27 August 2017).
[36] https://www.facebook.com/groups/NE.Atlantic.nudibranchs (accessed 27 August 2017).
[37] http://www.crowdflower.com.

Table 5 Comparison of Image Classification Accuracy Between Different Crowdsourcing Methods

Crowdsourcing Method	Accuracy
Inter-expert (test set)	0.98
Groupsourcing (test set)	0.93
Crowdflower (training) @ \$0.05 $n = 10$	0.91
Crowdflower (test set) @ \$0.05 $n = 10$	0.49

on the system to work on the task who were offered \$0.05 per image annotated, with ten answers required for each image.

A *training set* of 20 images with known answers was created with the most common sea slugs found on the photo sharing website Flickr. This dataset was used both as a training gold standard (i.e. the users were told if their answers agreed with the known answer) and also as a benchmark dataset. Users were presented with images from both datasets, with high-performing (according to Crowdflower's assessment of performance against the gold standard) users' data being labelled as 'trusted'. In total 1525 labels were made, from 72 users, of which 701 labels were considered 'trusted'. The data collection cost \$104.

Results show that with microworking there was high accuracy in the training set, but the test set scored much lower accuracy (see Table 5). This is an indication of how hard the task was in the test set and if task difficulty is extrapolated to social network crowdsourcing it would achieve an accuracy of 0.99 on the training dataset.

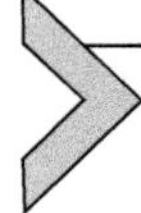

5. DISCUSSION

5.1 Data Acquisition and Annotation

Crowdsourcing approaches are typically used by a requester who has data they would like a task performed on; however, it may also be the case that the requester can acquire the data as part of the task, as seen in citizen science approaches, or even align with existing efforts, as has been seen with groupsourcing. It ultimately becomes a question of scalability: in order to scale up efforts every conventional bottleneck must be removed and this is why social networking platforms are so appealing. Users are motivated to answer the same type of questions as the requester and moderate themselves to ensure the resource is of high quality. Directing such a community of users is not straightforward and attempts at central control may give rise to resentment from some quarters. This makes the groupsourcing approach difficult when there is a shortage of skills or little general interest in the wider community.

Coupled with the bias of what data users want to work on Troudet et al. (2017) is the issue of data sparsity in general. For example in the citizen science project *GalaxyZoo*, only a few rare instances of unusual features are of real interest, although the general classification work assists with creating a large resource. In the context of images of marine species, some animals are more charismatic and easy to find than others, or are physically more common and well distributed, therefore there will be more images of these posted on social networking platforms.

Allowing participants in scientific activities a wider range of input may be the key to knowledge discovery. An unconstrained approach allows data to evolve in a way that interests the community, for example in the case of marine life, labelling interactions with other species, population dynamics, geographic distribution and other niche dimensions that could be indicators of ecosystem changes caused by pollution, overfishing or climate change.

A groupsourcing approach challenges what is known about a topic to cast a more realistic (although likely to be biased) view. This relates to the idea of a functional niche (i.e. the maximum parameters under which the concept as a whole could exist) compared to the realised niche (i.e. under what parameters individuals of the concept have been observed) (Hutchinson, 1957).

5.2 User Motivation

Crowds can be motivated in different ways, dependent on the system, task and goals, and the effectiveness of incentives can be measured by how many people participate and how much they contribute. Users that contribute nothing are consuming site resources (from bandwidth to interaction with administrators), as well as potentially producing spam or malicious content. Noncontributing members of a collective effort are commonplace on social networking platforms, with most users simply viewing the content rather than contributing new content or commenting on existing content. Social networking platforms have a very low barrier for participation, in that a user can simply 'like' content rather write a comment.

Reported numbers of recruitment and participation mask a large disparity between how much contribution individual participants make. It is common for individual contributions to follow a Zipf power law distribution (Zipf, 1949), in which only a few users make the majority of the contributions. A similar proposal[38] is suggested in the 90-9-1 rule (or the 1% rule in Internet culture) and has been shown to hold across a number of other domains including social networking platforms (van Mierlo, 2014).

On social networking platforms, groups that are set up specifically for users to share biodiversity information have users working at a higher rate with more in-thread activity and discussion than more general groups, an indication that this is inherently motivating for the community, although a wider study would be required to generalise this finding.

An additional motivation for users to collect and share data is the idea of pan-species listing,[39] a competitive ranking of naturalists and citizen scientists who spot the most species in a specific area. At the risk of trivialising natural history, the lists encourage people to observe nature by employing game-like elements in the data collection process.

5.3 Task Difficulty

Social networking platforms can produce high-quality data if communities of users can be found doing the task. Image classification is not simple and, although the majority of tasks were not hard, it is the uncommon, difficult tasks that require the skill of human labellers. A social network environment allows these difficult tasks to be solved in more organic ways.

The community of users in the groups examined on social networking platforms performed image classification tasks at near-expert levels on difficult tasks, considerably outperforming the same set of tasks on the Crowdflower microworking platform. In comparison to other approaches to wildlife image classification it also outperforms a gaming approach (Prestopnik et al., 2014). One of the greatest strengths of social network communities is the potential to target tasks to specific people who may be able to help, in a similar way to how the platforms themselves do targeted advertising.

A task may be difficult because the correct answer is difficult, but not impossible, to determine or that the answer is genuinely ambiguous and there is more than one plausible solution. The language used on social networking platforms creates even more ambiguity, with ill-formed grammar, misspelling, concatenation, contextual referencing and sentiment.

It has been shown that moderately diverse groups are better at solving tasks and have higher collective intelligence (termed *c*) than more homogeneous or very diverse groups. A balanced gender ratio within a group also

[38] http://www.nngroup.com/articles/participation-inequality.

[39] http://www.brc.ac.uk/psl/about (accessed 27 August 2017).

produces a higher *c* as females demonstrate higher social sensitivity towards group diversity and divergent discussion (Woolley et al., 2010).

Facebook is reported to have more female users[40] although, in the case of the social network groups investigated, there was a clear bias towards male users (Chamberlain, 2014). It may be that males prefer image-based tasks to word-based problems to solve (Mason and Watts, 2009), or that the topic is a male-dominated interest (66% of PADI diving certifications in 2010 were for men).[41]

5.4 Social Learning and the Expert in the Crowd

One of the distinct advantages of social networking over other approaches is that the participants learn from each other, not only how to contribute to the system, but also the knowledge to participate. This interaction is led by more experienced and knowledgeable members of the community in an open and transparent way, meaning that when a user receives an answer from an expert, many more may be passively learning from it. Outreach and communicating knowledge to the general public is a core objective of many citizen science projects and social networking platforms can be used to facilitate these aims. *Social learning* (Bandura and Walters, 1963), in which users on the platform teach and support each other in an ad-hoc manner, encourages users to engage in the learning process to an extent that suits their interests and time constraints. There are dangers of convergence towards the opinions of charismatic members or the majority; however, for difficult tasks a degree of discussion and consensus is preferable.

Users within groups typically learn how to interact with each other and how to post questions and replies by observation of the group's message feed. Administrators of the group set the rules of engagement in a short description of the group or with a pinned post, as well as advising members directly. These rules tend to proliferate across the group so over time the administrative load is reduced and the members become self-regulating. As an example, a common explicit guideline within marine species identification groups is to specify the location where the image was taken as this may have an important bearing on what the species might be (some species have limited geographical distribution).[42]

[40] http://royal.pingdom.com/2009/11/27/study-males-vs-females-in-social-networks (accessed 27 August 2017).

[41] http://www.padi.com/scuba-diving/about-padi/statistics/pdf.

[42] With other social media sharing sites such as Instagram and Flickr the image may be automatically geotagged in the metadata; however, with underwater images this is often not the case.

The advantage of having an expert in the crowd is that their knowledge is spread throughout the community and ultimately reduces their workload to only the most unique and difficult cases, which is a motivation for the expert to contribute in the first place. Some novice users will learn enough to be able to answer other users' questions reducing the traditional bottleneck of a few experts having to do the majority of the work. Small groups of annotators will not have the breadth of knowledge required to answer difficult, niche questions (Henry and Roberts, 2014), but a social network community allows experts from other groups to be drafted in by tagging (Sidlauskas et al., 2011).

The issue of expert bias has also been raised, when systems can be manipulated (intentionally or otherwise) by the perceived ability of an expert to answer a question due to their reputation (Alon et al., 2015). This is a long-standing issue in research areas of reputation management, expert finding and recommender systems; however, it is intuitive to believe that the expert in social networks is beneficial to the community.

In addition to experts in the crowd, the idea of crowd-powered experts has also been proposed. By using an approach in which the crowd deal with the majority of the easy work and experts focus on the difficult images, considerable improvements in an image classification system were made (Eickhoff, 2014). These results are comparable to what could be achieved by social networking platforms and can be considered a similar scenario in which the majority of group users take on the bulk of the work solving easy tasks, leaving the experts to focus on what is of most interest to them.

An issue with all crowdsourcing systems is how to gauge the user's ability to complete tasks, as well as have the internal knowledge required to solve problems. The distinction between a nonexpert and expert is often not clear cut (Brabham, 2012) and, over time, user abilities and biases will change the way they perform tasks which does not make them a consistent, long-term tool (Culverhouse et al., 2003).

5.5 Harnessing Collective Intelligence on Social Networking Platforms

Harnessing the collective intelligence of communities on social networking platforms is not straightforward, but the rewards are high. If a suitable community can be found to align with the task of the requester and the data can be extracted from the network, it has been shown here to be a useful approach. Aggregating the social network data in a similar way to crowdsourcing will allow for the automatic extraction of knowledge and sophisticated crowd aggregation techniques (Raykar et al., 2010) can be used to gauge the confidence of data extracted from threads on a large scale.

A validation model is intuitive to users and features in some form on most social network platforms. Typically a 'like' or 'upvote' button can be found on messages and replies, allowing the community to show favour for particular solutions. Other forms of voting exist, such as full validation (like and dislike) or graded voting (using a five star vote system) allowing for more fine-grained analysis of the community's preference; however, further research is needed to assess whether this is actually a waste of human effort and a simple like button proves to be the most effective (Chamberlain et al., 2018).

5.6 Limitations of a Groupsourcing Approach

Despite the many benefits of social networking platforms, there are also some significant limitations. The constantly changing underlying technology, as well as the popularity with users, means that long-term projects need to spend more time adjusting their software to maintain compliance. Although fairly mature with a high take-up rate, social networking platforms are still an emerging technology, and changes are made to the terms of service, access and software language that could swiftly render a dependent platform redundant.

Another drawback to using social networking platforms is that people use them in different ways and there is no formally correct way. Additionally, our understanding of such systems is only through observation as the algorithms for presenting information are never published and are constantly changing. There are also a proportion of user accounts used for spreading advertising or for spamming, although this is common in all crowdsourcing approaches. Users have different expectations that may lead to segregation in groups and data not being entered in a fashion that is expected. Users can also change a post after it has received replies, meaning a user can make a task request and then change the message once a solution has been offered, even deleting replies from the thread dialogue. This is not malicious or ungrateful behaviour, but simply a different way of using groups to organise data.

In a wider sense this raises the issue of whether an observation shared on a public forum or social network should have some level of permanence, to allow the community to refer back to the information in the future and build on an evolving resource. Clearly the image copyright belongs to the photographer but the knowledge contributed and shared by the community should be public domain. Social networking platforms have such a vast amount of data that this is not a priority; however, if such data are to be used for biodiversity research the fundamental questions of data privacy and ownership must be resolved.

It is unclear in the long term how social networking will continue as a popular pastime, and maintaining a community's interest in a project over time will need to be carefully managed. There may also be a saturation point of how many projects can be implemented to existing communities and this is also a problem for other peer production approaches.

A significant challenge for groupsourcing as a methodology is the automatic processing of the threads. There are a large quantity of data associated with threads and removing this overhead is essential when processing on a large scale (Maynard et al., 2012). The natural language processing needs to cope with ill-formed grammar and spelling, and sentences for which only context could make sense of the meaning, for example (taken from the subcorpus):

'And my current puzzle ...'
'Need assistance with this tunicate please.'
'couldn't find an ID based on these colours'
'Sven please talk Latin to me ;-)'

The image classification task that was investigated here uses natural language to solve the task; however, machine learning could use the image itself to classify the content. Much like the language of social networking platforms, images also vary in quality and there is little control over what is posted. Poor quality images or images with low illumination, unusual poses, clutter, occlusion, different viewpoints and low resolution will all make the image processing much more difficult.

6. APPLICATIONS

The data derived from the experiments in groupsourcing have been made available through a prototype website called *Purple Octopus*.[43]

In order to explore the ecological data, all text elements of the threads (messages and replies) were parsed for text strings representing marine species entities using the World Register of Marine Species (WoRMS) taxonomy[44] (see Fig. 10). In the same way a database of location names[45] was used

[43] http://www.purpleoctopus.org.
[44] http://www.marinespecies.org (accessed September 2012).
[45] http://www.dbis.informatik.uni-goettingen.de/Mondial/\#SQL (accessed September 2012).

Fig. 10 Detail of a typical message containing an image classification task having been analysed for named entities.

to find locations mentioned within the text. There were problems caused by the structure of the ontology, the informal reporting of locations in the thread text and disambiguation with other entities (it is also the case that marine species are not usually found in terrestrial locations and more usual for a location to be referenced by a locally known dive site name). However, using this simple text pattern matching, the prototype website can visualise the images and conversation threads of social networking platforms with marine species and locations represented in several ways:

- On the species page, all messages related to a species are listed along with a gallery of photographic examples of the species (see Fig. 11).[46]
- The species page also shows associated species, i.e. other species named in the same threads, which indicate interaction (for example, predation or symbiosis) or morphological similarity;
- On the species page, a map of comentioned locations for a species, representing its geographical distribution;
- On the explore page, a map showing species richness (total number of individual species comentioned with a country name) with a link to view all of the species comentioned with a particular country (see Fig. 12);
- Groups in which the data were extracted and top contributors from each group, ordered by the number of posts made.

The prototype interface allowed a degree of informal testing to investigate the information extraction and to see what kind of problems that were likely to be encountered if the groupsourcing approach were to be utilised for

[46] Only links to the images were stored, the images themselves are hosted on the social network platform. Each image was credited with the author's name.

Fig. 11 Screenshot of the *Purple Octopus* aggregated image gallery.

marine conservation in future work. The ultimate goal is to create an accurate database of information derived from social networking platforms that can be explored to provide actionable knowledge.

7. FUTURE DIRECTIONS

This chapter has detailed a number of approaches to using a crowd to monitor wildlife and all have their advantages and disadvantages. If initiatives to monitor biodiversity using public contributions are to increase on a scale that can accommodate huge data needs, they will need to engage a much larger audience. Several projects already implement lessons from social networking sites and weave this functionality into citizen science platforms to try and engage their audience more; however, this approach highlights a fundamental problem: most potential participants are not inherently motivated to contribute and need to be coerced with incentives. Users on social networking platforms such as Facebook, Flickr and Twitter are doing similar tasks, although frequently undirected, on a much larger scale because they are inherently motivated.

Fig. 12 Screenshot of species richness across the groupsourced dataset. *Darker blue* areas indicate areas of relatively high biodiversity (over 100 mentions).

To capitalise on this phenomenon future projects need to combine data analysis techniques to address the problems caused by the unconstrained interface with additional functionality to make software more useful to users, for example plugin apps that make suggestions for wildlife automatically identified in images, to more easily manage metadata (in particular location data) and to provide supplemental knowledge to help the users learn more about what they can see in the images. Additionally, with the development of virtual and augmented reality systems, the immediacy of such data could allow a much richer experience for users while they are in the environment where they are collecting data or to immerse the user in an environment that is normally inaccessible.

Not only would such an approach have the potential to deliver biodiversity data on a scale never been seen before, but it would also engage and educate the public in the marine environment and its conservation priorities. The ease and ubiquity of viral user-made content in a rapidly changing socio-technological landscape makes a social network citizen science approach an exciting and innovative direction for future research, education and exploration.

8. CONCLUSIONS

The goal of this research was to discover if social networking platforms could be used to create large-scale data resources, with high-quality labelling of information about the data. Social networking platforms can be viewed as a system for collecting and labelling biodiversity information; however, the benefits of using a crowd in this way are tempered with the many challenges.

In comparison to other methods of crowdsourcing, social networking platforms offer a high-accuracy, data-driven and low-cost approach. Users are self-organised and intrinsically motivated to participate, with open access to the data. By archiving social network data they can be categorised and explored in meaningful ways. There are significant challenges to automatically process and aggregate data generated from social networking platforms; however, this research shows the huge potential for this type of approach.

ACKNOWLEDGEMENTS

This research was partially funded by an EPSRC Doctoral Training Allowance granted by the University of Essex. The author would like to thank Professors Udo Kruschwitz, David J. Smith and Massimo Poesio for their advice in the production of this work.

REFERENCES

Agichtein, E., Castillo, C., Donato, D., Gionis, A., Mishne, G., 2008. Finding high-quality content in social media. In: Proceedings of the 1st ACM International Conference on Web Search and Data Mining (WSDM'08) , pp. 183–194.

Alon, N., Feldman, M., Lev, O., Tennenholtz, M., 2015. How robust is the wisdom of the crowds? In: Proceedings of the 24th International Joint Conference on Artificial Intelligence, IJCAI'15, pp. 2055–2061.

Bandura, A., Walters, R.H., 1963. Social Learning and Personality Development. Holt, Rinehart and Winston, New York.

Barnard, K., Duygulu, P., Forsyth, D., de Freitas, N., Blei, D.M., Jordan, M.I., 2003. Matching words and pictures. J. Mach. Learn. Res. 3, 1107–1135. ISSN 1532-4435.

Barve, V., 2014. Discovering and developing primary biodiversity data from social networking sites: a novel approach. Ecol. Inform. 24, 194–199. ISSN 1574-9541.

Becken, S., Stantic, B., Chen, J., Alaei, A.R., Connolly, R.M., 2017. Monitoring the environment and human sentiment on the great barrier reef: assessing the potential of collective sensing. J. Environ. Manage. 203, 87–97. ISSN 0301-4797.

Beijbom, O., Edmunds, P.J., Kline, D.I., Mitchell, B.G., Kriegman, D., 2012. Automated annotation of coral reef survey images. In: Proceedings of the 25th IEEE Conference on Computer Vision and Pattern Recognition (CVPR'12), June, Providence, Rhode Island.

Benkler, Y., Nissenbaum, H., 2006. Commons-based peer production and virtue. J. Polit. Philos. 14 (4), 394–419. ISSN 1467-9760.

Brabham, D.C., 2012. The myth of amateur crowds. Inform. Commun. Soc. 15 (3), 394–410.

Bunt, H., Alexandersson, J., Choe, J.W., Fang, A.C., Hasida, K., Petukhova, V., Popescu-Belis, A., Traum, D., 2012. ISO 24617-2: a semantically-based standard for dialogue annotation. In: Proceedings of the 8th International Conference on Language Resources and Evaluation (LREC'12), May.

Callison-Burch, C., 2009. Fast, cheap, and creative: evaluating translation quality using Amazon's Mechanical Turk. In: Proceedings of the 2009 Conference on Empirical Methods in Natural Language Processing (EMNLP'09).

Chamberlain, J., 2014. Groupsourcing: distributed problem solving using social networks. In: Proceedings of 2nd AAAI Conference on Human Computation and Crowdsourcing (HCOMP'14).

Chamberlain, J., Kruschwitz, U., Poesio, M., 2018. Optimising crowdsourcing efficiency: amplifying human computation with validation. Inform. Technol. 60 (1), 41–49.

Clery, D., 2011. Galaxy evolution. Galaxy zoo volunteers share pain and glory of research. Science 333 (6039), 173–175.

Culverhouse, P.F., Williams, R., Reguera, B., Herry, V., González-Gil, S., 2003. Do experts make mistakes? A comparison of human and machine identification of dinoflagellates. Mar. Ecol. Prog. Ser. 247, 17–25.

Dabbish, L., Stuart, H.C., Tsay, J., Herbsleb, J.D., 2014. Transparency and coordination in peer production. Commun. Res. Rep. abs/1407.0377.

Daume, S., 2016. Mining twitter to monitor invasive alien species—an analytical framework and sample information topologies. Ecol. Inform. 31, 70–82. ISSN 1574-9541.

Eickhoff, C., 2014. Crowd-powered experts: helping surgeons interpret breast cancer images. In: Proceedings of the 1st International Workshop on Gamification for Information Retrieval (GamifIR'14).

ElQadi, M.M., Dorin, A., Dyer, A., Burd, M., Bukovac, Z., Shrestha, M., 2017. Mapping species distributions with social media geo-tagged images: case studies of bees and flowering plants in Australia. Ecol. Inform. 39, 23–31.

Furfaro, G., Picton, B., Martynov, A., Mariottini, P., 2016. Diaphorodoris alba Portmann & Sandmeier, 1960 is a valid species: molecular and morphological comparison with D. luteocincta (M. Sars, 1870) (Gastropoda: Nudibranchia). Zootaxa 4193 (2), 304–316.
Gonella, P., Rivadavia, F., Fleischmann, A., 2015. Drosera magnifica (Droseraceae): the largest New World sundew, discovered on Facebook. Phytotaxa 220 (3), 257–267. ISSN 1179-3163.
He, J., van Ossenbruggen, J., de Vries, A.P., 2013. Do you need experts in the crowd?: a case study in image annotation for marine biology. In: Proceedings of the 10th Open Research Areas in Information Retrieval (OAIR'13), pp. 57–60.
Henry, L., Roberts, J.M., 2014. Recommendations for best practice in deep-sea habitat classification: Bullimore et al. as a case study. ICES J. Mar. Sci. 71 (4), 895–898.
Hoegh-Guldberg, O., Mumby, P.J., Hooten, A.J., Steneck, R.S., Greenfield, P., Gomez, E., Harvell, C.D., Sale, P.F., Edwards, A.J., Caldeira, K., 2007. Coral reefs under rapid climate change and ocean acidification. Science 318, 1737.
Holt, B.G., Rioja-Nieto, R., MacNeil, A.M., Lupton, J., Rahbek, C., 2013. Comparing diversity data collected using a protocol designed for volunteers with results from a professional alternative. Methods Ecol. Evol. 4 (4), 383–392.
Howe, J., 2008. Crowdsourcing: Why the Power of the Crowd Is Driving the Future of Business. Crown Publishing Group.
Hutchinson, G.E., 1957. Concluding remarks. Cold Spring Harb. Symp. Quant. Biol. 22, 415–427.
Johnson, R.F., Gosliner, T.M., 2012. Traditional taxonomic groupings mask evolutionary history: a molecular phylogeny and new classification of the chromodorid nudibranchs. PLoS ONE 7 (4).
Jurgens, D., Navigli, R., 2014. It's all fun and games until someone annotates: video games with a purpose for linguistic annotation. Trans. Assoc. Comput. Linguist. 2, 449–464.
Lakhani, K.R., Jeppesen, L.B., Lohse, P.A., Panetta, J.A., 2007. The value of openness in scientific problem solving. Harvard Business School. Working Paper 07-050.
Lu, D., Weng, Q., 2007. A survey of image classification methods and techniques for improving classification performance. Int. J. Remote Sens. 28 (5), 823–870.
Mason, W., Watts, D.J., 2009. Financial incentives and the "performance of crowds" In: Proceedings of the ACM SIGKDD Workshop on Human Computation (HCOMP'09).
Maynard, D., Bontcheva, K., Rout, D., 2012. Challenges in developing opinion mining tools for social media. In: Proceedings of the 8th International Conference on Language Resources and Evaluation (LREC'12) Workshop.
Novotney, S., Callison-Burch, C., 2010. Cheap, fast and good enough: automatic speech recognition with non-expert transcription. In: Proceedings of the 11th Annual Conference of the North American Chapter of the Association for Computational Linguistics (NAACL HLT 2010).
Parameswaran, M., Whinston, A.B., 2007. Social computing: an overview. Commun. Assoc. Inform. Syst. 19, 37.
Prestopnik, N., Crowston, K., Wang, J., 2014. Exploring data quality in games with a purpose. In: Proceedings of the 2014 iConference, Berlin, Germany.
Raykar, V.C., Yu, S., Zhao, L.H., Valadez, G.H., Florin, C., Bogoni, L., Moy, L., 2010. Learning from crowds. J. Mach. Learn. Res. 11, 1297–1322. ISSN 1532-4435.
Schoening, T., Bergmann, M., Ontrup, J., Taylor, J., Dannheim, J., Gutt, J., et al., 2012. Semi-automated image analysis for the assessment of megafaunal densities at the arctic deep-sea observatory HAUSGARTEN. PLoS ONE 7 (6), e38179.
Sidlauskas, B., Bernard, C., Bloom, D., Bronaugh, W., Clementson, M., Vari, R.P., 2011. Ichthyologists hooked on Facebook. Science 332 (6029), 537.
Snow, R., O'Connor, B., Jurafsky, D., Ng, A.Y., 2008. Cheap and fast - but is it good?: Evaluating non-expert annotations for natural language tasks. In: Proceedings of the 2008 Conference on Empirical Methods in Natural Language Processing (EMNLP'08).

Troudet, J., Grandcolas, P., Blin, A., Vignes-Lebbe, R., Legendre, F., 2017. Taxonomic bias in biodiversity data and societal preferences. Sci. Rep. 7 (1), 9132.
van Mierlo, T., 2014. The 1% rule in four digital health social networks: an observational study. J. Med. Internet Res. 16 (2), 33.
Vannella, D., Jurgens, D., Scarfini, D., Toscani, D., Navigli, R., 2014. Validating and extending semantic knowledge bases using video games with a purpose. In: Proceedings of the 52nd Annual Meeting of the Association for Computational Linguistics (ACL'14), pp. 1294–1304.
von Ahn, L., 2006. Games with a purpose. Computer 39 (6), 92–94. ISSN 0018-9162.
Welinder, P., Branson, S., Mita, T., Wah, C., Schroff, F., Belongie, S., Perona, P., 2010. Caltech-UCSD Birds 200. California Institute of Technology. CNS-TR-2010-001.
Welinder, P., Branson, S., Perona, P., Belongie, S.J., 2010. The multidimensional wisdom of crowds. In: Lafferty, J.D., Williams, C.K.I., Shawe-Taylor, J., Zemel, R.S., Culotta, A. (Eds.), Advances in Neural Information Processing Systems 23. Curran Associates, Inc., pp. 2424–2432
Wilkinson, C., 2008. Status of Coral Reefs of the World 2008. Global Coral Reef Monitoring Network & Reef and Rainforest Research Centre.
Woolley, A.W., Chabris, C.F., Pentland, A., Hashmi, N., Malone, T.W., 2010. Evidence for a collective intelligence factor in the performance of human groups. Science 330, 686–688. ISSN 1095-9203.
Yang, H., Lai, C., 2010. Motivations of Wikipedia content contributors. Comput. Hum. Behav. 26, 1377–1383.
Zipf, G.K., 1949. Human Behavior and the Principle of Least Effort. Addison-Wesley.

CHAPTER SIX

A Vision for Global Biodiversity Monitoring With Citizen Science

Michael J.O. Pocock*[,1], Mark Chandler[†], Rick Bonney[‡], Ian Thornhill[§,¶], Anna Albin[||], Tom August*, Steven Bachman[#], Peter M.J. Brown, Davi Gasparini Fernandes Cunha[††], Audrey Grez[‡‡], Colin Jackson[§§], Monica Peters[¶¶], Narindra Romer Rabarijaon[||||], Helen E. Roy*, Tania Zaviezo[##], Finn Danielsen[||]**

*Centre for Ecology & Hydrology, Wallingford, United Kingdom
[†]Earthwatch Institute, Boston, MA, United States
[‡]Cornell Lab of Ornithology, Ithaca, NY, United States
[§]Earthwatch Institute, Oxford, United Kingdom
[¶]College of Liberal Arts (CoLA), Bath Spa University, Bath, United Kingdom
[||]NORDECO, Copenhagen, Denmark
[#]Royal Botanic Gardens, Kew, Richmond, United Kingdom
**Applied Ecology Research Group, Department of Biology, Anglia Ruskin University, Cambridge, United Kingdom
[††]Departamento de Hidráulica e Saneamento, Escola de Engenharia de São Carlos, Universidade de São Paulo, Avenida Trabalhador São-Carlense, São Carlos, Brazil
[‡‡]Facultad de Ciencias Veterinarias y Pecuarias, Universidad de Chile and Director of Kauyeken, Santiago, Chile
[§§]A Rocha Kenya, Watamu, Kenya
[¶¶]People + science, Hamilton, New Zealand
[||||]Kew Madagascar Conservation Center, Antananarivo, Madagascar
[##]Facultad de Agronomía e Ingeniería Forestal, Pontificia Universidad Católica de Chile, Santiago, Chile
[1]Corresponding author: e-mail address: michael.pocock@ceh.ac.uk

Contents

1. Introduction 170
2. Citizen Science for Biodiversity Monitoring 171
   2.1 The Definition of Citizen Science 171
   2.2 Participatory Monitoring as a Citizen Science Approach 172
   2.3 Locally Based, Yet Global, Citizen Science 175
3. The Global Need for Biodiversity Monitoring 177
4. The Global Potential for Biodiversity Monitoring With Citizen Science 179
5. Approaches for Biodiversity Monitoring With Citizen Science: Who, What and How? 183
   5.1 Who Are the Potential Volunteers? 183
   5.2 How Should Biodiversity Be Recorded? 184
   5.3 What Should Be Recorded? 186
   5.4 How Can Technology Support Recording? 189
   5.5 How Should the Data Be Used to Produce Relevant Outputs? 193

Advances in Ecological Research, Volume 59
ISSN 0065-2504
https://doi.org/10.1016/bs.aecr.2018.06.003

6. Case Studies of Steps Towards Global Biodiversity Monitoring With Citizen Science 194
6.1 Assessing Opportunities for Biodiversity Monitoring in Chile With Citizen Science 195
6.2 eBird: Being Relevant to Local Participants While Global in Ambition 198
6.3 Citizen Science With Semistructured Recording: Kenya Bird Map 200
6.4 Codesign of Monitoring Protocols: New Zealand Environmental Community Groups 202
6.5 Zavamaniry Gasy: Making Recording Accessible Through Investment in Training and Internet Platforms 203
6.6 The Importance of Local Advocates to Support Participants: FreshWater Watch 205
6.7 A Proposal for Global Monitoring of Pollinators With Citizen Science 208
7. Conclusion 210
Acknowledgements 211
References 211

Abstract

Global biodiversity monitoring is urgently needed across the world to assess the impacts of environmental change on biodiversity. One way to increase monitoring is through citizen science. 'Citizen science' is a term that we use in this chapter to describe the diverse approaches that involve people in monitoring in a voluntary capacity, thus including participatory monitoring in which people work collaboratively with scientists in developing monitoring. There is great unrealised potential for citizen science, especially in Asia and Africa. However, to fulfil this potential citizen science will need to meet local needs (for participants, communities and decision makers, including people's own use of the data and their motivations to participate) and support global needs for biodiversity monitoring (including the United Nations' Sustainable Development Goals and the Aichi Biodiversity Targets). Activities should be feasible (for participants to provide scientifically rigorous data) and useful (for data users, from local to global scales). We use examples from across the world to demonstrate how monitoring can engage different types of participants, through different technologies, to record different variables according to different sampling approaches. Overall, these examples show how citizen science has the potential to provide a step change in our ability to monitor biodiversity—and hence respond to threats at all scales from local to global.

1. INTRODUCTION

Global biodiversity losses are escalating (Butchart et al., 2010; Chapin et al., 2000; Dobson et al., 2006) and monitoring is urgently needed across the world to assess the impacts of environmental change on biodiversity, to evaluate the impact of policy and management interventions

(Balmford et al., 2005; Johnson et al., 2017; Proença et al., 2017) and to assess the benefits that people gain from biodiversity (Millenium Ecosystem Assessment, 2005). The relatively scarce information on biodiversity change (at a global scale) results largely from a lack of taxonomic and spatial coverage (Boakes et al., 2010; Proença et al., 2017). We therefore need a step change in the way we undertake biodiversity monitoring (Schmeller et al., 2017). One way to accelerate monitoring is through the application of new technologies, for example, remote sensing (Pettorelli et al., 2014) or eDNA (Bohan et al., 2017). A complementary approach is to use citizen science, which engages nonprofessionals in a voluntary capacity to undertake monitoring (Dickinson et al., 2012). Environmental monitoring conducted by volunteers has a long history in many developed countries, especially in northern Europe and North America, with some initiatives having taken place for more than a century (Cooper, 2016; Dickinson et al., 2012; Pocock et al., 2015b; Schmeller et al., 2009). In addition to benefiting from the knowledge and capabilities of local participants, citizen science (or volunteer monitoring) can be less expensive than monitoring conducted by contracted professional staff (although the two approaches are not mutually exclusive) because it has the benefit of up-scaling more cost efficiently (Roy et al., 2012; Theobald et al., 2015; Tulloch et al., 2013a).

Overall, citizen science is making a substantial contribution to global biodiversity data (Amano et al., 2016; Chandler et al., 2017; Theobald et al., 2015), but there is considerable potential for increasing its contribution. Citizen science is valuable at the global scale because it not only collects large amounts of data, it also contributes to public engagement with the environment, potentially leading to behavioural change (McKinley et al., 2017) and production of partnerships between scientists and local people (Funder et al., 2013; Toomey and Domroese, 2013). In this chapter, we provide perspectives on how citizen science can meet the demands of global biodiversity monitoring. In particular we consider that for citizen science to be sustainable, it must meet local needs (for participants, communities and decision makers) as well as supporting global needs for biodiversity monitoring.

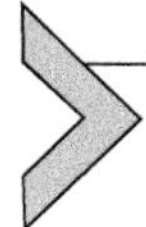

2. CITIZEN SCIENCE FOR BIODIVERSITY MONITORING

2.1 The Definition of Citizen Science

Firstly, it is important to define what we mean by 'citizen science' in the context of this chapter. 'Citizen science' is a useful encompassing term to describe the diverse range of approaches that involve people in science and monitoring in a noncontracted or voluntary capacity (Bonney et al., 2009b). However, we note that not all activities falling under this broad description would define themselves as 'citizen science' and it is important to be sensitive to the concerns of those practitioners. There has been rapidly growing interest in the use of 'citizen science', especially in North America, Europe, Australia and New Zealand (Bonney et al., 2014; Pocock et al., 2017; Theobald et al., 2015), and interest is growing elsewhere: in Africa (Citizen Science Association, 2017), Central/South America (Cunha et al., 2017b; Fundación Ciencia Ciudadana, 2017) and Asia (http://www.citizenscience.asia).

It is helpful to consider the different approaches in citizen science. One broad distinction is between contributory approaches (where participants are primarily involved as data collectors) and collaborative/cocreated approaches (where participants are involved in additional stages of the scientific process, including identifying the question of interest, designing methodologies, analysing data and using the results) (Bonney et al., 2009a; Shirk et al., 2012). Contributory approaches comprised the vast majority of 500 projects surveyed in a recent review (Pocock et al., 2017). However, the distinctions are not clearcut: even if some people are involved in the scope and design of activities (collaborative), many others can subsequently be involved in contributing data (contributory). Citizen science also includes both mass participation activities in which anyone can get involved, and those engaging interest groups and volunteer experts (Pocock et al., 2017; Tulloch et al., 2013b), including 'biological recording', for which there is a long history of recording by volunteer expert naturalists in some north European countries (e.g. Pocock et al., 2015b).

2.2 Participatory Monitoring as a Citizen Science Approach

While there continues to be debate about the use of the term 'citizen science' (Eitzel et al., 2017), we find it helpful here as a 'term of convenience', thus including activities that may not define themselves as 'citizen science'. One such activity is participatory monitoring (also called community-based environmental monitoring). Participatory monitoring is focused on participation by local people with a strong stake in their local environment, with the aim that the monitoring is defined by local people rather than being 'top down', scientist-led activities (Conrad and Hilchey, 2011; Danielsen et al., 2005a). It is mostly found in developing countries and the Arctic where community members are dependent on living resources for their livelihood and cultural identity (Danielsen et al., 2000; Johnson et al., 2015).

A focus is on direct benefits to the local participants because the information informs their role as resource managers (Danielsen et al., 2010; Evans and Guariguata, 2008). Following Chandler et al. (2017), we include participatory monitoring within our definition of citizen science, and henceforth use the term 'citizen science' with this broad sense. We note that other interpretations are possible: Kennett et al. (2015) contrasted participatory monitoring with 'citizen science': '[participatory] monitoring could benefit from the large-scale databases and knowledge integration pioneered by citizen science… [while] citizen science could benefit from the community-based monitoring practices used to build data-collection methods, analytical tools, communication networks and skilled workforces in culturally appropriate, place-based governance structures'. Specifically, they seemed to reference 'contributory' citizen science in developed countries when making valid points about how different approaches can gain benefit from each other.

Citizen science and participatory monitoring are each growing as global communities with international communities of practice for both participatory monitoring (http://www.pmmpartnership.com/) and citizen science (Bonney et al., 2014; Citizen Science Association, 2017). Historically, 'citizen science' (typically contributory) and participatory monitoring have tended to focus on different types of participants (the general public in contributory citizen science vs people whose livelihoods depend on natural resource management in participatory monitoring), in different parts of the world (developed vs developing countries), and for different purposes (e.g. regional and national monitoring vs data for local people to benefit from). However, some activities, e.g., community-based monitoring of illegal resource use, such as poaching and logging, have been described as both collaborative citizen science and participatory monitoring (Danielsen et al., 2010; Stevens et al., 2013). The nascent Citizen Science Global Partnership (https://www.wilsoncenter.org/article/concept-note-citizen-science-global-partnership) seeks, in part, to bring these communities together.

Danielsen et al. (2009) provided a conceptual description of five different types of monitoring in discussion about participatory monitoring that was subsequently verified by statistical analysis of published monitoring programmes (Danielsen et al., 2014b). It is helpful to compare these with citizen science approaches (Table 1). Specifically, the approaches where local people are involved in data collection (categories 2, 3 and 4 in Table 1) have potential to scale-up to contribute to global biodiversity monitoring.

Table 1 Different Types of Participatory Monitoring and How These Relate to Types of Citizen Science

Category of Monitoring	Primary Data Gatherers	Primary Users of Data	Comments
1. Externally driven, professionally executed	Professional researchers	Professional researchers	Costly, does not scale cost efficiently
2. Externally driven with local data collectors	Professional researchers, local people	Professional researchers	For example, 'contributory citizen science'. Has potential to be cost efficient compared to monitoring by professionals. Has potential to contribute to global biodiversity monitoring, but relies upon the enthusiasm of volunteers
3. Collaborative monitoring with external data interpretation	Local people with professional researcher advice	Local people and professional researchers	For example, 'collaborative citizen science' and participatory monitoring. Has potential to contribute to global biodiversity monitoring but also produces locally relevant outputs
4. Collaborative monitoring with local data interpretation	Local people with professional researcher advice	Local people	For example, 'collaborative citizen science' and participatory monitoring. Produces locally relevant outputs, and when the data is shared this could contribute to global biodiversity monitoring
5. Autonomous local monitoring	Local people	Local people	Does not serve additional benefits for larger-scale monitoring (if data and information are shared it becomes defined as category 3 or 4)

The first three columns are adapted from Danielsen, F., Burgess, N.D., Balmford, A., Donald, P.F., Funder, M., Jones, J.P.G., Alviola, P., Balete, D.S., Blomley, T., Brashares, J., Child, B., Eenghoff, M., Fjeldså, J., Holt, S., Hübertz, H., Jensen, A.E., Jensen, P.M., Massao, J., Mendoza, M.M., Ngaga, Y., Poulsen, M.K., Rueda, R., Sam, M., Skielboe, T., Stuart-Hill, G., Topp-Jørgensen, E., Yonten, D., 2009. Local participation in natural resource monitoring: a characterization of approaches. Conserv. Biol., 23, 31–42.

2.3 Locally Based, Yet Global, Citizen Science

When fulfilling the vision of global biodiversity monitoring with citizen science, we suggest that it needs to be 'locally based, yet global' (Chandler et al., 2012; He and Tyson, 2017) (Fig. 1). The challenging question is what this means in practice. Throughout this chapter, we contend that

1. The local perspective is *essential* to the *success* of any biodiversity monitoring activity. This is the scale at which people choose to participate and benefits would be directly experienced by people. Participation will therefore be influenced by people's interests, and the application of data.
2. The larger scale (national to global) perspective is *important* to increase the *impact* of biodiversity monitoring. This larger scale is often where policy and management decisions are made and where funding may be obtained. The international scale can give impetus and focus to monitoring, via international targets such as the Aichi Biodiversity Targets (SCBD, 2010)

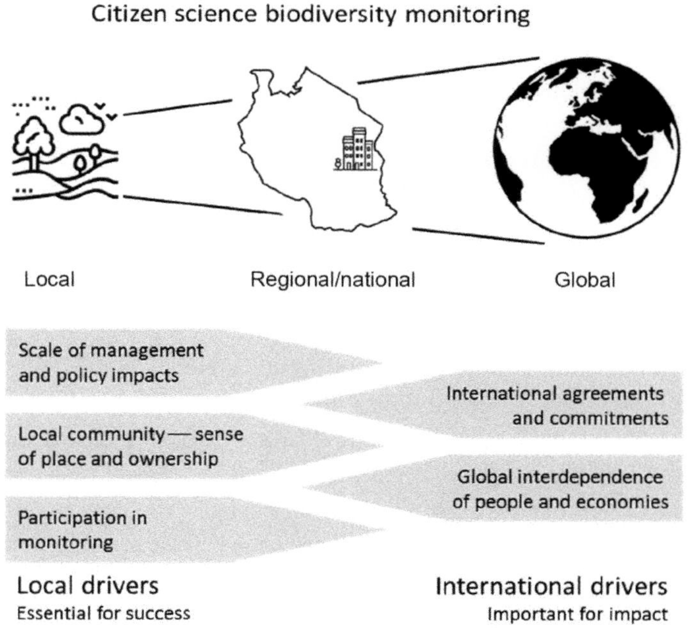

Fig. 1 There are different needs and drivers for biodiversity monitoring at different scales, as exemplified in this diagram. We contend that when considering citizen science biodiversity monitoring it is essential to meet both the needs of participants (at the local level) and it is important to meet the needs of funders and data users (at the regional to global level). *Icons CC-BY from thenounproject.com: 'landscape' by Becris, 'Tanzania' by Fatemah Manji, 'city' by Made by Made, 'earth' by David.*

and the United Nation's Sustainable Development Goals (UNGA, 2015) and via aspirations such as the essential biodiversity variables (EBVs) (Pereira et al., 2013) (see Section 3).

A different way of considering this is the need to focus on both the *product* (i.e. observations added to databases and analysed) and the *process* of volunteer biodiversity monitoring (the way in which people are recruited, retained and motivated) (Lawrence, 2006). Focussing solely on the national to international scale can lead to a 'demand-driven' process that prioritises the global needs for biodiversity monitoring information, and ignores local relevance, which can exacerbate asymmetric power dynamics (Ayensu et al., 1999; Lawrence, 2006). The actors who make 'demands' for data (e.g. government environmental bodies, international nonprofit organisations and professional scientists) often hold power in terms of funding, database ownership and access to resources. In contrast, local people (typically participating in a voluntary capacity) may hold less power for setting up the monitoring programs, but their involvement is essential for data collection, for the activities to generate local benefits and for them to be sustainable. In developing countries, participatory monitoring often serves the purpose of advocacy for socio-environmental justice (Stevens et al., 2013), empowerment (Lawrence, 2006), governance (Liu et al., 2014) or sustainable management of daily-use resources (Danielsen et al., 2005a), rather than regional or national benefits (Staddon et al., 2015). The Manaus Letter (Participatory Monitoring and Management Partnership (PMMP), 2015), which sought to raise the profile of participatory monitoring, explained this further. Key messages from the Letter include:

- Initiatives should be constructed from the bottom up, incorporating local as well as academic visions and knowledge;
- It is important to ensure high data quality and to standardise data collection at the necessary scales (among monitors, among communities and among initiatives, if the scale of monitoring is regional or global);
- Monitoring initiatives must reconcile and balance the interests and motivations of local, regional and global actors involved in the initiative and
- When monitoring initiatives are designed for use at the regional or global scale, they should ensure the return of information and results to participating local communities.

These recommendations align well with the subsequent Ten ECSA Principles for Citizen Science (ECSA, 2017) including the importance of feedback, data quality and involvement of participants.

3. THE GLOBAL NEED FOR BIODIVERSITY MONITORING

The aspiration for global biodiversity monitoring stems in large part from the demands of policy instruments that have regional, national and global scope. These 'green' instruments incorporate biodiversity data as part of existing frameworks (e.g. European Common Agricultural Policy), to ensure the implementation of conventions dedicated to managing species (Convention on Biological Diversity, Convention on Migratory Species of Wild Animals), and as part of global collaborative platforms that stipulate broad national level assessments (e.g. Intergovernmental Panel on Biodiversity and Ecosystem Services: IPBES). Beyond these dedicated policies, biodiversity data can be, and is, used by governments at all levels to assess the effects of other policy decisions and investments (e.g. infrastructure development) on biodiversity (Millenium Ecosystem Assessment, 2005).

Currently there is a gap between the biodiversity information available (e.g. in the Global Biodiversity Information Facility (GBIF)) and the information required to adequately assess the impact of conservation-oriented policies (Collen et al., 2008; Joppa et al., 2016; Tittensor et al., 2014), but could citizen science monitoring help to meet these needs? Recently, Danielsen et al. (2014b) looked at indicators underpinning 12 major international agreements and mapped onto these different types of monitoring approaches, ranging from scientist-driven to those undertaken by local people (Table 1). Overall, 63% of the existing 186 indicators in the 12 agreements were found suitable for some form of 'citizen science' (including 'participatory monitoring') (Danielsen et al., 2014b). Nine agreements (in the quadrants on the right side of the graph in Fig. 2) are well suited to involving local stakeholders in collecting monitoring data (with professional input for analysis), or both collecting and analysing monitoring data. The study did not look at local or subnational environmental policies, which might have even more potential for public involvement (Haklay, 2015), but there is clearly considerable opportunity for involving local stakeholders in collecting relevant data.

The findings of Danielsen et al. (2014b) showed that citizen science and community-based monitoring could enhance monitoring progress within global environmental conventions. However, they also have the potential

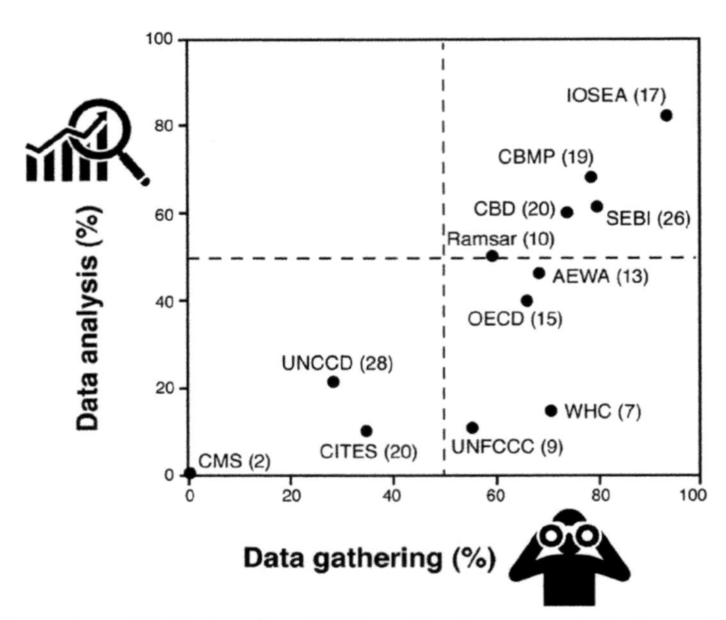

Fig. 2 Percentage of indicators from 12 international agreements that are suitable for local stakeholder involvement in data gathering and data analysis. For each agreement, we show the proportion of indicators that can be measured with the involvement of local stakeholders as data collectors (*x*-axis) and the proportion that can be measured by local stakeholders, who not only collect data but also process and interpret the data and present the findings to decision makers (*y*-axis). *Dashed lines* indicate 50% values for each axis. The number of indicators of each agreement is shown in *brackets*. *AEWA*, Agreement on the Conservation of African-Eurasian Migratory Waterbirds; *CBD*, Convention on Biological Diversity; *CBMP*, Circumpolar Biodiversity Monitoring Program; *CITES*, Convention on International Trade in Endangered Species; *CMS*, Convention on Migratory Species; *IOSEA*, Indian Ocean—South-East Asian Marine Turtle Memorandum of Understanding; *OECD*, Organisation for Economic Cooperation and Development; *Ramsar*, Convention on Wetlands of International Importance; *SEBI*, Streamlining European 2010 Biodiversity Indicators; *UNCCD*, United Nations Convention to Combat Desertification; *UNFCCC*, United Nations Framework Convention on Climate Change; *WHC*, World Heritage Convention. *From Danielsen, F., Pirhofer-Walzl, K., Adrian, T.P., Kapijimpanga, D.R., Burgess, N.D., Jensen, P.M., Bonney, R., Funder, M., Landa, A., Levermann, N., Madsen, J., 2014b. Linking public participation in scientific research to the indicators and needs of international environmental agreements. Conserv. Lett. 7, 12–24, reproduced with permission. Icons CC-BY from thenounproject.com: 'binoculars' by Luis Prado, 'analysis' by Arafat Uddin.*

to raise awareness, scientific literacy and enhance decision making for resource management (Pretty and Smith, 2004). Consequently, the social ambitions, as well as the monitoring needs, of the United Nations Sustainable Development Goals (UNGA, 2015) and the Convention on Biological Diversity's

Aichi Targets (SCBD, 2010) could also be supported by local stakeholder involvement in monitoring (West and Pateman, 2017).

Making progress internationally will benefit when biodiversity data can be harmonised globally (e.g. the EBVs; Pereira et al., 2013). For this, data from disparate sources would be harmonised for a minimum set of critical variables required to monitor biodiversity change and the impacts of interventions (Kissling et al., 2018). Citizen science can successfully contribute data into approaches such as the EBVs (Chandler et al., 2017), e.g., its role for monitoring alien species occurrence, status and impact (Latombe et al., 2017; McGeoch and Squires, 2015).

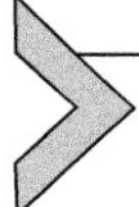

4. THE GLOBAL POTENTIAL FOR BIODIVERSITY MONITORING WITH CITIZEN SCIENCE

At present, there is a heterogeneous distribution of biodiversity monitoring and recording across the world (Beck et al., 2014; Chandler et al., 2017; Proença et al., 2017; Schmeller et al., 2009). Furthermore, there is pressing need to assess data biases and gaps in existing databases, and improve coverage (taxonomic, spatial and temporal) ensuring openness and accessibility to facilitate analysis of trends (Edwards, 2000; Hortal et al., 2008; Meyer et al., 2016; Troudet et al., 2017).

Here, we undertook a simple analysis to describe the global potential for citizen science and community-based monitoring to contribute to biodiversity monitoring. We are rapidly approaching a global human population of 8 billion which is distributed unevenly across the world (Fig. 3A). We considered the capacity for species to be recorded within a location (i.e. a 5 arc-minute grid cell in our analysis) to be a function of the number people and the number of species at that location. We used the species richness of mammals, birds and amphibians as a proxy for total species richness (Jenkins et al., 2013), which is heterogeneously distributed (Fig. 3B). The product of human population and species richness gives the potential for people to record species: the observation potential (Fig. 3C). Specifically, we use log-transformed human population and species richness [$y=\log_{10}(x+1)$ to account for zeros in the data], and scaled these to lie within the range (0,1) [scaled $y=y/y_{\max}$]:

$$\text{Observer potential} = \text{human population (scaled and log transformed)} \times \text{species richness (scaled and log transformed)}$$

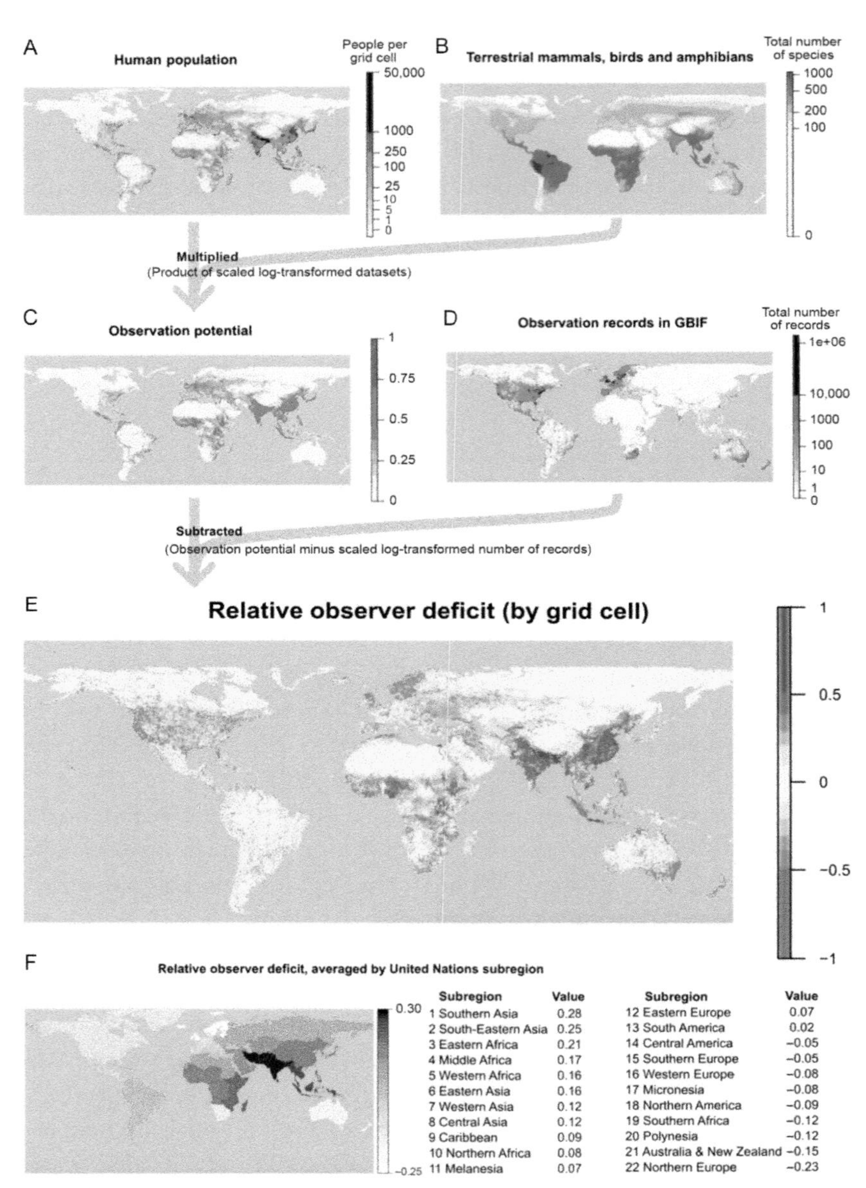

Subregion	Value	Subregion	Value
1 Southern Asia	0.28	12 Eastern Europe	0.07
2 South-Eastern Asia	0.25	13 South America	0.02
3 Eastern Africa	0.21	14 Central America	−0.05
4 Middle Africa	0.17	15 Southern Europe	−0.05
5 Western Africa	0.16	16 Western Europe	−0.08
6 Eastern Asia	0.16	17 Micronesia	−0.08
7 Western Asia	0.12	18 Northern America	−0.09
8 Central Asia	0.12	19 Southern Africa	−0.12
9 Caribbean	0.09	20 Polynesia	−0.12
10 Northern Africa	0.08	21 Australia & New Zealand	−0.15
11 Melanesia	0.07	22 Northern Europe	−0.23

Fig. 3 The global distribution of (A) human population and (B) total species richness of mammals, birds and amphibians. The product of these, once log transformed and scaled to lie in the range (0,1) was (C) the observation potential. The observation potential minus (D) the number of records in GBIF (log transformed and scaled to lie in the range (0,1)) gives the index of (E) the relative observer deficit, which is (F) aggregated (using arithmetic mean) to the United Nations subregion (*darker colours* indicate higher relative observer deficit). All data are shown for 5 arc-minute grid cells.

(Continued)

There are nearly 1 billion records in GBIF (https://www.gbif.org) from many biodiversity monitoring activities. Each record represents a single species observation at a specified time and place. Over half of records are from citizen science, although with a strong bias towards birds (especially via eBird, a dominant provider) and towards Europe and North America (Amano et al., 2016; Chandler et al., 2017). Comparing the observer potential with the number of records in GBIF (Fig. 3D) gives an indication of how well the observer potential is being met. To quantify this, we calculated the observer deficit as the difference, for each location, between the observer potential and the count of GBIF records (Fig. 3E). As with human population and species richness, the count of GBIF records was log transformed ($y = \log_{10}(x+1)$) and scaled to lie in the range (0,1).

$$\text{Relative observer deficit} = \text{observer potential} - \text{number of GBIF records (scaled and log transformed)}$$

Observer deficit therefore ranged from +1 (high observer deficit: few GBIF records relative to the number of people and the richness of biodiversity, and greatest potential to increase records, e.g. through citizen science) to −1 (low observer deficit: many GBIF records relative to the number of people and the richness of biodiversity). Of course, whatever the value of observer deficit, there is still potential to increase the amount of observation effort in that location. Relative observer deficit provided a standardised measure to compare locations and showed that there is substantial unevenness in its distribution across the world (Fig. 3E and F; Table 2). The highest relative observer deficits are southern and south-eastern Asia and sub-Saharan Africa (except Southern Africa). South America has moderate observer deficit because relatively few people live in the most species-rich

Fig. 3—Cont'd *Data on vertebrate species was downloaded from BiodiversityMapping.org and is based on Jenkins, C.N., Pimm, S.L., Joppa, L.N., 2013. Global patterns of terrestrial vertebrate diversity and conservation. Proc. Natl. Acad. Sci. U. S. A. 110, E2602–E2610. Data were provided by IUCN (mammals and amphibians) and BirdLife International and NatureServe (birds). Human population data were obtained from GPWv4 (Center for International Earth Science Information Network—CIESIN—Columbia University, 2016). Note that the human population data are aggregated at different spatial resolutions depending on availability of data in each country. We thank Tim Robertson at GBIF for support in downloading the data.*

Table 2 The Countries in the World With the Highest and Lowest Average Value of Observer Deficit

Rank of Country	Highest Observer Potential (Highest First)	Lowest Observer Potential (Lowest First)
1	Bangladesh	Sweden
2	Burundi	UK + Isle of Man
3	Nigeria	Denmark
4	Rwanda	Norway
5	India	Luxembourg
6	North Korea	Estonia
7	Moldova	New Zealand
8	Vietnam	Portugal
9	Uganda	Costa Rica
10	Sierra Leone	Belgium
11	Cote d'Ivoire	South Africa
12	Togo	Spain

Averages were calculated as the mean across 5 arc-minute grid cells covering each country. Very small countries (less than 10 grid cells) were excluded.

areas and has relatively good data in GBIF (although see Section 6.1 for discussion about opportunities and barriers to citizen science biodiversity monitoring in Chile).

While this is a simplistic measure of the potential of citizen science, it demonstrates the unrealised potential for citizen science (i.e. highest observer deficit). This could help in identifying where to focus efforts for greatest gains in global knowledge of biodiversity, especially when considering other cultural factors, including access to technology (Section 5.4). For instance, the burgeoning middle class in emerging economies in Asia (Kharas, 2017) are likely to have good access to technology and are in regions with high observer deficit, meaning that it could be fruitful to engage with them as potential participants. Also, taxa of functional importance, for example, many plants and pollinating insects, are underrepresented in GBIF (Chandler et al., 2017) and could be fruitful focus of future effort (Troudet et al., 2017).

5. APPROACHES FOR BIODIVERSITY MONITORING WITH CITIZEN SCIENCE: WHO, WHAT AND HOW?

We have discussed the global-to-local approach required for sustainable and impactful biodiversity monitoring with citizen science (Section 2), the international need for biodiversity information (Section 3) and the potential for citizen science to meet this need (Section 4). The question is how this can be achieved. Here, we discuss different issues relating to the implementation of biodiversity monitoring with citizen science. Ultimately any recording, whether initiated by local participants or by programme organisers, should meet two important requirements:

1. Recording should be *feasible* for participants: it should engage participants, fit with their motivation and interests (which could include their own use of the data) and it should enable them to contribute with confidence.
2. Recording should be *useful* by being of sufficient quality for its intended scientific purpose. Quality can be considered in two ways. Firstly, individual records should be accurate (within prescribed limits of acceptability). Efforts should be put in place during the project design to support accuracy, and appropriate verification should be used to prove that accuracy that is fit-for-purpose (Kosmala et al., 2016). Secondly, the dataset as a whole should (after analysis) be able to provide valuable, unbiased and reliable information, e.g. on the status of an individual species, or as a measure/proxy of habitat quality (Buckland and Johnston, 2017).

In this section we discuss: who is recording, what they are recording and how the information should be used.

5.1 Who Are the Potential Volunteers?

Broadly we suggest that biodiversity monitoring could involve three types of participants: people who are already interested and have expertise in recording wildlife (Pocock et al., 2015b); local stakeholders who are involved in participatory monitoring to protect land and resources (Danielsen et al., 2005a,b); and the general public who become engaged with existing activities. Opportunities and challenges are associated with each potential audience, and all are key to scaling up to increase coverage of global biodiversity monitoring.

Understanding what motivates people to engage in citizen science is important (Conrad and Hilchey, 2011; Geoghegan et al., 2016;

Lawrence, 2006). It may be helpful to distinguish between intrinsic and extrinsic motivations (Blackmore et al., 2013). Extrinsic frames are those that relate to self-interest; so for citizen science this would include an appeal to monitoring the benefits we get from nature. However, long-term participation is better supported through reference to intrinsic frames, which are about connections with nature and people, positive action, the appreciation of beauty and self-discovery (August et al., in review). Participatory citizen science approaches can result in personal transformation and sustain long-term motivation among participants (Lawrence, 2006), but more participatory approaches may have a stronger impact in terms of the volunteers' engagement, interest and empowerment (Rotman et al., 2012; West and Pateman, 2016; Wilderman et al., 2004). Motivations can also be influenced by level of education and the expectation of a reward, as found in a recent study on the potential of citizen science for agricultural research (Beza et al., 2017). The long-term motivation of participants in the United States, India and Costa Rica was found to be affected by many different aspects, but poor communication and inadequate technical infrastructure were found to be strong demotivators (Rotman et al., 2014). Overall, there is much to be learned about how people in cultures and demographics across the world are motivated to engage with citizen science. However, for individual activities, the best approach is to codesign monitoring activities along with potential participants.

5.2 How Should Biodiversity Be Recorded?

Related to the motivations of the participants is the way in which biodiversity is recorded. This can be done in many different ways (Pocock et al., 2017) (Table 3). Whichever approach is used, it is important to be clear about the aims of the project (Buckland and Johnston, 2017; Lindenmayer and Likens, 2010; Pocock et al., 2015a).

Fully structured recording has advantages because the data will be consistent and can easily be analysed and aggregated. As the sites are selected by the organisers of the activity, they can be chosen strategically or randomly, and the resulting data is representative, rather than being biased by the participant's choice (Fragoso et al., 2016; Newson et al., 2005). However, participants are required to travel and visit specific locations so this requires dedication and is most suited to participation by volunteer experts. It can require a lot of investment by volunteer coordinators to support

Table 3 Summary of Different Approaches for Recording Suitable for Biodiversity Monitoring, as Discussed in the Text

Monitoring Approach	Choice of Location and Time	Use of Sampling Protocol
Fully structured recording	Organiser's choice	Following a protocol at all times and places selected by project organisers (the scientific ideal being randomised locations)
Semistructured recording	Participant's choice	Following a monitoring protocol but at times and/or places of the participant's choice
Unstructured recording, with assessment of effort	Participant's choice	None, although 'effort' should be recorded in a consistent way
Unstructured recording	Participant's choice	None
Participatory assessment	In the participant's locality	Focus group discussions, interviews, documentation of oral history

the retention and recruitment of volunteers to ensure sufficient spatial and temporal coverage.

Semistructured recording also uses specific monitoring protocols, but the time and/or location of the sampling is at the participant's choosing. This ranges from high-skilled ecological surveys through to mass participation projects where a structured protocol is followed, e.g., making observations in one location for a fixed period of time. The benefit of having a structured protocol is that it helps to standardise sampling effort, thus making the results more comparable across surveys. It can also be suitable for participation by less-skilled volunteers because the protocol is very clear, although for ensured success, protocols should be designed in collaboration with professionals and potential participants. Selection bias can be expected where volunteers choose sites and/or times for recording because, for example, recorders tend to favour locations closer to home or in protected areas (Boakes et al., 2010; McGoff et al., 2017; Tulloch et al., 2013a), or where there are currently presences (Buckland and Johnston, 2017). This can make it more difficult to generalise about the state of the overall environment. One step to help account for selection bias is the incorporation of species distribution modelling, by combining remotely sensed habitat information with species occurrence data (Coxen et al., 2017) or statistical correction (Robinson et al., 2018).

Unstructured recording is typical of biological recording and 'mass participation' citizen science (Pocock et al., 2017). For expert naturalists, this is beneficial because they are not constrained by specific protocols, but there is the same risk of selection bias as semistructured recording. New statistical approaches can be used to extract information from these unstructured data (Hill, 2012; Isaac et al., 2014; Maes et al., 2015; van Strien et al., 2013) including on the impacts of environmental change, e.g., the impacts of pesticide use (Woodcock et al., 2016). However, the lack of clarity and structure could be demotivating for public contributors (August et al., in review). The quality of the information can be enhanced by including an assessment of effort, e.g., distance travelled, time spent observing or using list-based recording (Sullivan et al., 2014; van Dyck et al., 2009).

All the methods above are primarily to collect direct observations of the current state of biodiversity. There is also value in considering approaches from social sciences for gathering information, e.g., focus group discussions (Danielsen et al., 2014a,c), interviews (Jones et al., 2008; Topp-Jørgensen et al., 2005), oral history documentation (Fernández-Llamazares and Cabeza, 2017; Mustonen, 2015) and participatory mapping (Rich et al., 2015). These are probably particularly useful for gaining accurate information on change in state over time (i.e. trends in abundance).

5.3 What Should Be Recorded?

5.3.1 Different Variables That Can Be Recorded

Biodiversity recording traditionally has involved recording the presence of an organism at a certain place and time, and undertaking analysis on the collation of such occurrence records (Hochachka et al., 2012; Powney and Isaac, 2015). In parts of the world and for many species groups this is the only quantitative biodiversity information available and so many decision makers rely on this information, e.g., in creating Red Lists of threatened species (Gardiner and Bachman, 2016). The simplicity of these records means that they can be easily harmonised and collated in large-scale databases such as GBIF (Chandler et al., 2017) and used for EBVs (Kissling et al., 2018; Pereira et al., 2013). However, other variables can also be recorded and may be as, or more, useful for biodiversity monitoring (Table 4).

Table 4 Variables That can be Recorded to Contribute to Monitoring Biodiversity

Biodiversity Variables	Description
Uniquely identifiable individuals	Recording individually identifiable organisms across space and time
Occurrence	Reporting when a species has been recorded in a given area. This is presence-only data, so there is no information on nondetections
Presence/absence	Reporting whether a species has been detected or not in a given area (including recording against a checklist)
Abundance	Total number of individuals of a species recorded over a discrete period of time in a given area
Physiological attributes	Assessment of health or quality of an organism, e.g., measurement of size
Phenology	Reporting periodic biological events for selected taxa/phenomena at a given location, e.g., timing of breeding, leaf coloration, flowering, migration
Interactions	Interactions between species, which could be long-lasting or short-term, e.g., insects visiting flowers
Habitat quality	Presence/absence of indicator species, or assessment of habitat attributes, e.g., vegetation height
Ecosystem function	Measurement of a specific aspect of ecosystem functioning, e.g., fruit and seed set or decomposition

See text for further description.

Interpreting occurrence records is challenging because observer coverage varies across space and time (Isaac and Pocock, 2015): i.e., does the absence of a record mean the absence of the species or the absence of an observer? More informative analysis can be undertaken when we have knowledge of, or can infer, nondetections so creating 'presence–absence' datasets (Table 4). These can be available in structured and semistructured recording (Table 3), although for single species the motivation to report a presence is different to the motivation to reporting an absence. One simple approach to achieve this is encouraging observers to report a full list of sightings (Szabo et al., 2010), as can be facilitated through reporting systems such as eBird (Sullivan et al., 2014).

Another commonly collected biodiversity variable is abundance. This is most valuable when it is standardised and the effort is recorded (see Sections 6.2 and 6.3), e.g., the number of individuals in a location over a set time period (trends in time) or revisiting set transects (Dennis et al., 2017; Vianna et al., 2014). For mobile or cryptic species the number seen is often taken as an index of relative abundance so changes in counts are often taken as an index of change in true abundance, although this can be confounded with seasonal or long-term changes in detectability.

For a few animal species and some plants, each individual can be uniquely identified or marked, allowing true population size and other demographic parameters to be estimated using mark-recapture analysis, e.g., using photographic records from volunteers for large whales and sharks (Davies et al., 2013) or zebras (Parham et al., 2017), or mark-recapture or mark-resighting which is especially popular for birds (Greenwood, 2007). This is also relevant for recording plant phenology, especially for trees, for which individuals can be tracked through seasons and across years.

The presence and/or abundance of specific indicator species or communities of species are often used as a measure of habitat quality, including to assess the impact of interventions. Measures of other attributes of individuals, e.g., fish size, tree size or disease status (Danielsen et al., 2014c) are valuable because they provide more information about the quality of the individuals, and so are especially relevant where species are being harvested (although care needs to be taken to account for sampling biases when interpreting the data). Species interactions are also important because these generate many ecosystem services and disservices. The individual interactions can be long-lasting, e.g., epiphytic lichens or fungi on host plants, or short term, e.g., predation and flower–pollinator interactions. Finally, as well as recording functionally important groups, it can also be useful to monitor ecosystem function directly, e.g., decomposition (Keuskamp et al., 2013), pollination (Birkin and Goulson, 2015) or pest control.

5.3.2 Species-Level Recording: A Benefit or a Constraint?

For many biodiversity scientists, species-level data are of key interest, but for citizen science, species-level recording could be constraining because we are limited (i) by the availability of people who can, or want to be able to, accurately identify taxa to the species level and (ii) to the taxa where this could be feasibly achieved (thus excluding the majority of invertebrate groups). Birds, in particular, are extremely well recorded by volunteers, but even for them there is still patchy coverage of recording by volunteers (see Section 6.2). We should be careful not to 'cling on to the sanctity of the species' (quoting Raffaelli, 2007 who describes this in a different context) when developing a vision for global biodiversity monitoring with citizen science.

Rather than recording to the species level, it may be suitable to record aggregated groups (e.g. 'morphotypes' comprising individuals that look similar). This is used by 'parataxonomists' in rapid biodiversity assessments in highly biodiverse locations (Krell, 2004; Schmiedel et al., 2016). Morphotypes are also used as indicators of habitat quality where there is knowledge of their sensitivity to particular drivers of environmental change, e.g., air quality (Seed et al., 2013) and water quality (Graham et al., 2004; Wright et al., 1998). Morphotype classifications could be codesigned with professional scientists (who can ensure that the groups are functionally informative for the intended purpose; Lawler et al., 2003) and potential participants (who can test whether distinguishing these groups is practicable; Roy et al., 2016), ensuring the groups are morphologically and functionally distinctive (e.g. Ullmann et al., 2010). For instance, trends in the numbers of bright blue butterflies in European grasslands could be just as informative as the trends in the counts of the individual species (comprising bright blue males and brown females, but with subtle differences between species). However, effective sharing of morphotype data is a challenge, especially if it is not known which species are in each morphotype.

5.4 How Can Technology Support Recording?

The technological revolution of the late 20th and early 21st century has brought dramatic changes to citizen science for biodiversity monitoring: shaping what is possible, and how monitoring can be undertaken (August et al., 2015). Data of many types can now be captured, including images, videos and sounds. These new data can support better verification and allow new questions to be addressed, e.g., Pipek et al. (2018). Networks of technologies allow these data to be shared, via the internet, in globally accessible datasets. Finally, the data and their analysis can be made available locally, to further motivate participants and inform them, e.g., about the impacts of local resource management.

5.4.1 Data Collection

When used in citizen science, technology must be *accessible*, *useable* and *useful*, all of which have increased in recent time. The gradual reduction in cost of hardware has increased accessibility while advances in design and sensors have increased usability and usefulness. Modern sensors allow a range of data to be collected in the field by volunteers. These technologies include, digital cameras, acoustic recording devices, GPS, drones and eDNA among many others. At a local scale, technological developments can lead to greater potential for local action. For example, small and low-cost pollution sensors can be used in participatory monitoring to detect local pollution events and prompt local government action (Borghi et al., 2017; Glasgow et al., 2004; Toivanen et al., 2013).

Mobile phones have created an amazing resource for citizen science (Fig. 4). Smart phones combine connectivity technologies (i.e. mobile

Fig. 4 One of the authors (H.E.R.) as a participant in citizen science demonstrating the way technology facilitates biodiversity recording. She is using a smartphone to upload a ladybird beetle sighting in the high Andes in Chile (see Section 6.1) to the Chinita arlequín project (http://www.chinita-arlequin.uchile.cl/).

network and internet) with sensor technologies such as cameras, GPS and microphones. The ubiquity of mobile technology in much of the world (Pew Research Center, 2016), and the ease with which data can be captured and transmitted to central data stores has made them a popular tool for both locally focused participatory monitoring, e.g., EpiCollect (Aanensen et al., 2009), and global citizen science, e.g., eBird (Sullivan et al., 2014). However, mobile phone use varies: in a study of farmers in India, Honduras and Ethiopia more than 90% of people owned mobile phones, but less than 60% were always connected to mobile phone networks, and very few (<10%) used the internet via their phones (Beza et al., 2017). Smartphone ownership varies; it is currently very high in developed countries, high in China and Turkey (58%–59%), low in India, Indonesia, Kenya and Nigeria (17%–28%) and very low in other countries, including Ethiopia (4%) (Pew Research Center, 2016). There will also be substantial variation within countries and across demographics due to access to mobile networks and the interrelated variables of income, education and location (e.g. rural vs urban). There are few studies available on the role of technology in locally based biodiversity monitoring in real-life cases (Brammer et al., 2016), but a recent study in Cambodia's Prey Lang forest (Brofeldt et al., in press) was able to successfully collect large amounts of high-quality data using smartphone apps. Multiple entry points along a decreasing technological hierarchy such as smartphone, web application and face-to-face communication is one way to address generational and cultural differences in use of new technology.

Social media, such as Facebook and Twitter, provides a valuable tool for the organisation and support of monitoring by acting as a forum. Through social media, disparate people can come together and share knowledge and discuss topics of interest. It can also be used for groups to self-organise, such as the Garden Bioblitz in the United Kingdom (http://www.gardenbioblitz.org/) which was initiated and established via social media. Despite its prevalence, social media does not facilitate the efficient databasing of biodiversity observations: platforms designed for biodiversity recording are better because they provide consistency for taxonomic names, locations and dates.

5.4.2 Making Global Databases From Local Datasets

Greater global connectivity, facilitated by technology, has given rise to greater sharing of data. Centralisation of data, e.g., in GBIF, allows large-scale analyses to be undertaken. There are practical barriers to the flow of data from local to global scales. These include the time and resources required to submit the data in a standardised format (Wieczorek et al., 2012). The submission of local data to global datasets often does not have a tangible benefit for individual organisers, so if data are collated locally then sharing the data can be low on the list of priorities of busy local organisers and could even threaten the sustainability of participation (Pearce-Higgins et al., 2018). Additionally, sensitivities may mean that data are shared with reduced locational precision or with limits on who can view the data (Groom et al., 2016). These barriers to data sharing are not technological, but economical, institutional and motivational (Thessen and Patterson, 2011), but they must be addressed to realise the potential of the local to global scaling in citizen science.

Some projects allow individuals to submit directly to a central global platform, bypassing the need for local organisers to summarise, format and submit data themselves. This also saves the cost of maintaining individual databases for each project. This process directly links the local to the global but may bypass established data quality processes and personal interactions that may exist with local coordinators. Also where databases are maintained centrally (e.g. in North America or Europe) there can be issues with perceived ownership of the activity and its data (Pulsifer et al., 2011), and there is the risk of developing 'one size fits all' solutions. One solution is the development of international databases with locally relevant data portals, developed and run by local organisations. This approach has been adopted by eBird (see Section 6.2), iNaturalist (e.g. its Mexican version Naturalista http://www.naturalista.mx/), iSpot (e.g. its South African version https://www.ispotnature.org/communities/southern-africa) and for the Kenyan Bird Map (see Section 6.3). This can also be done locally, with projects such as iRecord in the United Kingdom (http://www.brc.ac.uk/irecord/) being a single data infrastructure but with local instances for specific user groups (e.g. those interested in recording bees or fungi). The City Nature Challenge is an example of a project benefitting on the local to global potential of citizen science. It was a distributed global bioblitz (>60 cities on four continents in 2018) enabled through the use of a common app using the iNaturalist platform.

5.4.3 Data Analysis and Feedback to Participants

The centralisation of data has allowed for analyses and data visualisation at large spatial scales. These can be shared with widely with participants via web and mobile interfaces. While traditionally analyses would be carried out by a project organiser or third party analyst, interactive websites now allow users to control the visualisation of data and in some instances the nature of the analysis (Belbin and Williams, 2016). These advances help to close the data flow cycle, enriching the experience of participants and empowering them in local resource management. This has been shown to increase the retention of volunteers as well as the quality of their data (Blake et al., 2012).

Data collated by projects can be used to create systems that aid participants in future data collection, including systems that validate new records based on expectations from previous records. For example, these systems are typically able to identify records that come from outside the known geographic distribution of the species (e.g. the United Kingdom's 'record cleaner' developed by the National Biodiversity Network) and more verified images of species aids iNaturalist's species recognition software to more accurately propose identifications to users. More advanced systems are able to use collected data to predict which species a participant is likely to observe, given their location and time of year, and present them with a guide to those species (Goldsmith et al., 2016). These technologies can create a virtuous cycle whereby data collected are used to improve the quality of future data collection and the motivation of those collecting the data (van der Wal et al., 2016).

5.5 How Should the Data Be Used to Produce Relevant Outputs?

Ultimately one of the aims of monitoring is its impact on natural resource management and conservation (Danielsen et al., 2007). There can be impact at a local scale, with the participants being users of the data, as for some participatory monitoring and collaborative citizen science projects (Earthwatch Institute, 2017). However, often it is valuable for the information to be collated at a larger scale to influence national policy making through provision of trends in 'indicator' groups or calculated metrics (e.g. of ecosystem health). This requires two things: that the data are analysed and interpreted; and that the information is collated, and this can be done in either order (Fig. 5).

The most intuitive way of collating information is to collect data into a single database or database framework at a regional, national or international level. This is the approach of projects such as iNaturalist, eBird (see Section 6.2) and the Freshwater Information Platform (FIP; see Section 6.6),

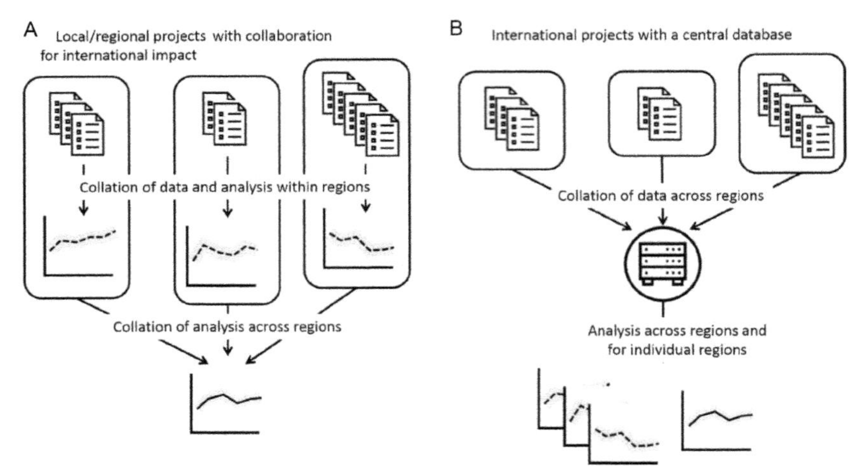

Fig. 5 Examples of two different approaches for adding value to individual monitoring projects by collating information at a larger scale (regional, national or global). *Icons CC-BY from thenounproject.com: 'list' by unlimicon, 'database' by Chameleon Design. Graphs by the author.*

which also then share their data with GBIF as a global repository of biodiversity data. Alternatively, meta-analytic approaches can be used to collate analysis, rather than collate the data themselves, e.g., the collation of country-level trends to create supranational indices for butterflies (van Swaay et al., 2008). These outputs can be used to report on biodiversity trends and monitor the impact of management interventions, including through policy.

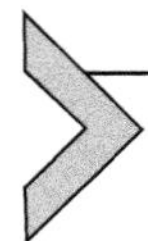

6. CASE STUDIES OF STEPS TOWARDS GLOBAL BIODIVERSITY MONITORING WITH CITIZEN SCIENCE

Thus far we have discussed some principles related to the potential for global biodiversity monitoring with citizen science. In the following section we illustrate these principles with reference to current and proposed activities. We explore the benefits and barriers of citizen science in a country (Chile) with no long history of citizen science (Section 6.1). We compare two projects based on birds showing how: activities can have local and global perspectives (eBird: Section 6.2); and how structured monitoring adds value to the data (Kenya Bird Map: Section 6.3).

We discuss how working with local participants supports sustainable recording with examples from community monitoring in New Zealand (Section 6.4), developing technology for plant recording in Madagascar (Section 6.5) and working locally but in a global network through Freshwater Watch (FWW) (Section 6.6). Finally, we bring these together in a proposal for global monitoring of pollinators with citizen science (Section 6.7).

6.1 Assessing Opportunities for Biodiversity Monitoring in Chile With Citizen Science

Chile is considered to be one of the 35 biodiversity hotspots in the world (Mittermeier et al., 2011), and about 25% of the described species are endemic (Gligo, 2016; Ministerio del Medio Ambiente, 2014). Biodiversity monitoring is important here, especially because of the potential impact of major environmental change, including volcano eruptions, forest fires and land use change (Martinez-Harms et al., 2017; Sala, 2000). There is a recognised need for biodiversity information across Chile (Gligo, 2016; Ministerio del Medio Ambiente, 2014), but, despite government efforts, information is still incomplete. Participation in recording from Chile's 17.5 million inhabitants would greatly increase our knowledge.

A 1-day meeting on citizen science on biodiversity was held at Pontificia Universidad Católica de Chile in January 2017 to assess opportunities, barriers and topics for using citizen science to gather biodiversity data in Chile (using the same approach as implemented previously in East Africa; Pocock et al., 2018). This meeting involved 31 professional participants (http://www.kauyeken.cl/con-exito-se-realizaron-el-taller-y-los-seminarios-sobre-ciencia-ciudadana/) from a range of organisations with an interest in citizen science including universities, colleges, NGOs, the Ministry of the Environment, the National Forestry Corporation (CONAF) and the Chilean Navy. In summary: (1) participants completed a short questionnaire in advance of the workshop to list the barriers, opportunities and topics they considered to be a priority for advancing citizen science in Chile; (2) responses were compiled and refined through discussion during the workshop; and (3) participants anonymously voted for priorities in each category (scoring 3, 2 and 1 for their top, second and third priority) and these scores were summed to rank the barriers and opportunities (Table 5) and the priority topics (Table 6) for biodiversity citizen science in Chile.

Table 5 The Top Opportunities (a) and Barriers (b) to Implementation of Citizen Science for Biodiversity Conservation in Chile

a.

Overall Rank	Opportunities	Total Number of Votes	Total Score
1	Increase in sampling (improved spatial and temporal coverage of data)	19	45
2	Improved knowledge of the natural environment for citizens and scientist	17	36
3	Sense of belonging with our natural and social environment	14	33
4	Improved networks between citizens, scientists and government agencies for managing environmental problems	14	22
5	Produce useful information for decision making	12	21
6	Empowerment of local communities	11	18
7	Strengthening the relation between citizens and science	6	11
8	Free access to engage and participle in science projects	3	6

b.

Overall Rank	Barrier	Total Number of Votes	Total Score
1	Disconnection between scientists and citizens	26	63
2	Lack of adequate resources (e.g. funding, time)	15	26
3	Citizens disconnection with the natural environment	8	17
4	People do not find the research problem a priority	7	15
5	Lack of validation of citizen science from the scientific community	6	12
6	Data quality and its management	5	12
7	Lack of commitment from volunteers and scientists	4	10
8	Limited outreach and publicity for citizen science projects	5	9
9	Technology limitations for some groups of citizens (e.g. lack of access to internet; lack of technology literacy by older people)	5	6
10	Complexity of the research methods	3	6

Table 6 Topics Suggested as Priorities for Citizen Science for Biodiversity Conservation in Chile

Overall Rank	Topic	Total Number of Votes	Total Score
1	Species distribution and diversity (both native and alien)	25	66
2	Monitoring the effects of humans on ecosystems	16	28
3	Assessing abiotic variables important for biodiversity conservation	14	23
4	Illegal or harmful activities (poaching, close season times, etc.)	9	17
5	Assessing ecosystem services (e.g. pollination)	8	16
6	Assessing habitat quality through indicator species	9	14
7	Human dimension of conservation biology (e.g. native biodiversity use and management)	6	14
8	Emerging wildlife diseases	5	8
9	Species phenology (e.g. blooming, migration)	4	6

A number of overarching themes emerged from the collaborative prioritisation (Tables 5 and 6). For example, scientists were seen as disconnected from citizens, but citizen science was seen as an opportunity to improve communication between citizens, scientists and government agencies. It was also suggested that there is a need to improve the scientific literacy of people and consequently increase their confidence to participate in science and engage in scientific debates. In order to identify organisms to the species there is a need for additional resources and training for experts and nonexperts alike, which would support the gathering of information on the distribution and diversity of species (Table 6). However, monitoring functional morphotypes (rather than species) may be sufficient for addressing several of the topics, including human impact on ecosystems (see Section 5.3.2).

There has been a recent growth in citizen science activities in Chile developed by different organisations, for example, to monitor rainfall (http://milluvia.dga.cl/index.php), beach litter (http://www.cientificosdelabasura.cl/) and an invasive species of ladybird (http://www.chinita-arlequin.uchile.cl/).

Social media has a major role in promoting citizen science activities (see Section 5.4), e.g., Facebook groups for Murciélagos de Chile (monitoring bats), Moscas Florícolas de Chile (monitoring flower-visiting flies) and Salvemos Nuestro Abejorro (monitoring an endangered bumblebee). The growth of new initiatives has led to the creation of Fundación Ciencia Ciudadana (http://cienciaciudadana.cl/), a Chilean organisation that promotes citizen science and has recently published a Citizen Science Guide (Fundación Ciencia Ciudadana, 2017). Participatory monitoring was piloted in protected areas of Patagonia (Aysén; Region XI) by CONAF and the National History Museum of Santiago in 2003 (Danielsen et al., 2005b).

Citizen science is a relatively new concept in Chile, meaning it serves as a useful case study in exploring the potential for global biodiversity monitoring with citizen science. Assessments of the potential for citizen science as in Chile (as discussed here) and in East Africa (Pocock et al., 2018) provide evidence to prioritise issues and raise the profile of citizen science.

6.2 eBird: Being Relevant to Local Participants While Global in Ambition

As we have discussed in this chapter, one way for citizen science to contribute to global biodiversity monitoring is to collate observations globally via a single project. Birdwatching is a popular pastime across the world, and the observations that people make have great value for scientists if they can be collated and analysed. eBird is one project collating information on sightings of birds and is the largest source of global information about the distribution and abundance of bird species currently available, with most of the data from volunteers (Amano et al., 2016; Sullivan et al., 2009, 2014). As of May 2018, more than 400,000 unique observers had submitted over 29,000,000 complete checklists to eBird, including 98% of the world's bird species and representing over 30 million hours of time in the field.

eBird is based in the United States, and this is where it has its highest levels of participation. However, eBird enables project coordinators to tailor the infrastructure to meet their specific needs, including language. In Mexico, for example, the local version of eBird Averaves (http://ebird.org/content/averaves/acerca/) allows for more relevant local programming (Ortega-Álvarez et al., 2012). Currently the eBird website is available in 11 languages, and its app in 26 languages. There continues to be discussion how a platform like eBird meshes with existing well-supported platforms and projects elsewhere in the world (e.g. BirdTrack http://www.birdtrack.net or Trektellen https://www.trektellen.nl/).

One of the reasons that eBird is successful in collecting data is that it serves the needs of a community by providing a tool for birders to store their records, both locally and when travelling (which shows the value of combining ecotourism with citizen science, although spatial coverage by travelling birders will be biased towards ecotourism hotspots). However, by providing some structure to the data that people submit, the records have added value for analysis and use. Participants in eBird generate bird abundance and distribution data at high spatial and temporal resolution by submitting checklists of birds they have seen. They also can choose to add additional information such as breeding status and to attach photos. Although participants can enter records made anytime and at any place, there are simple ways in which eBird requires people to record their effort (see Section 5.2). Firstly, people record when, where and for how long (both time and distance) they spent birding, and eBird provides various options of the protocol used for data gathering, such as point counts, transects and area searches. Secondly, observers are encouraged to fill out checklists of all the birds they have seen (not just the 'special' species), thereby providing information on nondetections as well as presence. The checklists are automatically created in real time based on the likelihood data for the time and region where the observations are being made. This is an excellent example of a way in which a citizen science project can encourage people to provide more informative data, without making it a burden. When checklists of sightings are submitted, automated filters provide an instantaneous first layer of screening. Flagged records are sent for verification by one of the approximately 1500 regional editors spread around the globe. Once verified, flags are removed and the data enter the permanent database.

eBird data summaries are available on the project website along with data exploration tools that allow anybody to create maps and frequency distributions of bird species around the world. The data are used to make distribution maps, model migration of individual species, indicate species in decline, demonstrate relationships between bird populations and habitat quality, and inform management and policy decisions affecting birds (Callaghan and Gawlik, 2015; Hochachka et al., 2012). It is also used for bird atlases in three US states: Wisconsin, Virginia and Maine. Data users include national and state agencies, nongovernmental organisations, academic researchers, birders and students of all ages. Data products are available in a variety of forms and data visualisations. As of early 2017, 150 scientific publications used eBird data, appearing in a variety of fields including ecology, statistics, computer science and public policy.

Other projects use a similar approach to eBird, with all the data being submitted to a single data infrastructure, but also having local versions of data portals to make the project locally relevant to participants. Examples include iNaturalist (e.g. its Mexican version Naturalista http://conabio.inaturalist.org) and iSpot (e.g. its South African version https://www.ispotnature.org/communities/southern-africa).

6.3 Citizen Science With Semistructured Recording: Kenya Bird Map

Kenya has long been known to have a rich and diverse avifauna (Bennun and Njoroge, 1999). However, vast areas of many of its habitats have been dramatically altered by human impact over the past 40–50 years. This has, without a doubt, affected the distribution and occurrence of Kenya's rich biodiversity—but to what extent? It is important to know this for effective conservation management to take place (Loftie-Eaton, 2015), but currently there is not effective monitoring of the trends in biodiversity. The Kenya Bird Map project seeks to address these questions (Wachira et al., 2015). A key factor in the success of the Kenya Bird Map (http://kenyabirdmap.adu.org.za/) is the carefully designed protocol that was adopted from the Southern Africa Bird Atlas Project 2 (SABAP2). This was designed to reduce observer bias due to, e.g., observer effort and skill, time of day, observation conditions and conspicuousness of a species (Underhill, 2016), while maximising ease and the observer's enjoyment of the data collection process (Fig. 6). Both of these factors are critical to a successful citizen science project—and the results so far for SABAP2 (10,688,000 records) and the bird atlases in both Kenya (167,000 records over 4 years from 230 observers) and Nigeria (48,000 records over 2 years from 76 observers) suggest that balance of structured vs unstructured and ease vs rigour of the protocol has been successful.

Feedback on the success of the protocol indicates the importance of the 'customer satisfaction' that the participant receives from their involvement and contribution to the project. This may be self-pursued, in that one can check for oneself the status of her/his contribution, or through active feedback from the organisers of the project to provide updates on project progress. A primary feature of this project is that the atlas is based on a grid system for the country coverage that is combined with real-time updates

Fig. 6 (A) Young Kenyan birders taking part in citizen science bird surveys, Malindi, Kenya. Although here they are using a notebook to record sightings, the use of a smartphone app has supported increased participation in the Kenya Bird Map project. (B) A birdwatcher following a structured protocol to contribute to the Kenya Bird Map around Lake Mikimba in pentad 0305_4000 just inland from Malindi, Kenya.

of that coverage on the atlas website, so that observers can choose to focus their effort on underrecorded squares and have the satisfaction of seeing their contribution to the overall project. This is supported by project Facebook pages where discussion takes place among recorders.

One of the challenges at the start of the project in 2014 was whether it would work in Kenya, where most birders are young Kenyans who do not have their own vehicles and often not even their own computers, compared to South Africa, where most participants came from the more affluent white communities. Initially in Kenya there was participation mainly from a small group of keen birders. However, the development of a Kenyan version of the BirdLasser app in 2013 (developed by a participant in the Kenya Bird Map project), coincided with a rapid increase in the availability of smartphones in Kenya and this led to a substantial leap in participation. The design of BirdLasser focussed on allowing birders to do the thing that they enjoy: birding. One benefit of the smartphone app was that it automated the detection of an observer's location, allowing birders to focus on birdwatching, rather than map reading. The app was designed to function without mobile network coverage, and subsequent submission of data is a simple process when mobile coverage is gained. The focus on the participant's needs is believed to have been vital in its contribution to the growth of bird atlassing not just in southern Africa but now across the whole continent.

6.4 Codesign of Monitoring Protocols: New Zealand Environmental Community Groups

Within the past 700 years, human colonisation and the arrival of exotic biota (both intentionally and unintentionally introduced) have resulted in substantial loss of biodiversity across New Zealand. Now, over 500 community environmental groups protect and restore flora, fauna and habitat in diverse ecosystems: forests, rivers and streams, freshwater and saline wetlands and coastlines (Peters et al., 2015). These groups are generally small (<20 active volunteer participants), but many have been active for over a decade and some larger projects employ coordinators. Partnerships, mostly with land management agencies, play a crucial role visiting sites and providing labour (e.g. pest and weed control), technical advice (e.g. monitoring design, species identification) and funding (Hardie-Boys, 2010; Peters et al., 2015). Many groups have achieved important biodiversity conservation gains through sustained control of animal pests and weeds, revegetation with native species, constructing predator-proof fenced sanctuaries, and translocating native flora and fauna species to managed sites (Campbell-Hunt and Campbell-Hunt, 2013; Cromarty and Alderson, 2012; Hardie-Boys, 2010; Sullivan and Molles, 2016).

Many groups carry out environmental monitoring to quantify their restoration management activities and to a lesser degree, the outcomes of their management (Peters et al., 2016). Ecosystem monitoring toolkits have been designed to provide protocols for monitoring the health of forests, wetlands, streams, coastal areas, rivers and estuaries, e.g., Biggs et al. (2002), Handford (2004), Robertson and Peters (2006), Tipa and Teirney (2003). These toolkits help community groups because they facilitate the collection of standardised data. In some cases, new protocols have been developed to meet community needs or existing protocols simplified to enhance their usability. Uptake of these toolkits, however, has generally been low, mostly due to a lack of ongoing technical and logistical support (e.g. from land management agencies and nongovernment agencies) as well as online platforms to facilitate analysing, storing and reporting on findings (Peters et al., 2016).

However, the bewildering array of biomonitoring protocols available is confusing for groups. This has prompted one local government (Auckland Council) to consolidate the most suitable methods into a technical guide specifically for nonspecialist community users in preparation. Wide consultation with agency staff as well as a cross-section of community groups resulted in 11 criteria for assessing monitoring methods (including level of skill, resourcing, scientific robustness) of about 30 established protocols, including simplified methods drawn from the aforementioned ecosystem monitoring toolkits. The wider rationale for the guide is simple: to build consistency among groups' monitoring efforts and enable agency staff to deliver uniform advice to community groups. This directly addresses a lack of technical expertise, which is a known barrier for groups to undertake monitoring. Currently, monitoring data are mostly used to shape the management of groups' own project sites, support funding applications and report back to funders (Peters et al., 2016). A future prospect is to aggregate groups' data for State of the Environment reporting with the guide playing a pivotal role in directing groups to use a limited but cohesive suite of methods. Therefore, working with community groups to equip them to undertake scientifically rigorous monitoring of the impact of their ecosystem management supports their efforts and gives wider benefit through sharing data.

6.5 Zavamaniry Gasy: Making Recording Accessible Through Investment in Training and Internet Platforms

Madagascar has a rich diversity of more than 11,000 native plants species, 80% of which are endemic (Goyder et al., 2017; Madagascar Catalogue, 2017) and despite hundreds of years of botanical collecting and study, new species are still being added to its flora (Andriamihajarivo et al., 2016; Darbyshire et al., 2017; Vorontsova et al., 2013). Documenting the distribution and ecology of species has become increasingly urgent as threats to biodiversity in Madagascar increase (Hannah et al., 2008; Harper et al., 2007), especially because only about 9% of plants native to Madagascar have been assessed for their level of extinction risk according to the IUCN Red List Categories and Criteria (IUCN, 2016).

Responding to this lack of information requires innovative solutions and citizen science offers great potential for species discovery and monitoring in Madagascar. The contribution of public participation, via technology, is exemplified by the story of a palm enthusiast, resident in Madagascar, who posted a photo of an unidentified species on a web-based forum. The photo was eventually examined by experts from the Royal Botanic Gardens, Kew, who were able to relocate the population in wild. When the plants were analysed they turned out to be not only a new species (*Tahina spectabilis*), but an entire new genus of palm (Dransfield et al., 2008). Its restricted range, threats from fire and grazing, small known population size and lack of protection immediately warranted a classification of Critically Endangered on the IUCN Red List (Rakotoarinivo and Dransfield, 2012).

Acknowledging the potential role that georeferenced photos of plants could play in contributing towards the Madagascar plant inventory and the assessment of extinction risk, a project was initiated in 2014 by the Kew Madagascar Conservation Centre (KMCC) called Zavamaniry Gasy (Plants of Madagascar) with funding from the JRS Biodiversity Foundation. Utilising the existing infrastructure of iNaturalist (http://www.inaturalist.org) the project was quickly and easily established and started to accumulate plant observations. The project was coordinated by KMCC who undertake regular expeditions to the field and their aim was to supplement collections of voucher specimens with observations which are quicker and easier to generate. Plant species could then be mapped across their ranges more accurately, and could also be used to report on localised threats, both of which support the production Red List assessments of extinction risk.

The iNaturalist observation workflow was conceived in developed countries where it relies on technology such as GPS-enabled smartphones and good internet connectivity; however, this presented challenges in developing countries like Madagascar. To bridge this gap, development by iNaturalist improved the functionality of the app to make records where internet connectivity is low and smartphones were purchased for local participants. KMCC curated observations and used their taxonomic expertise to verify observations as 'Research Grade'. Research grade observations are subsequently harvested from iNaturalist by the GBIF. It was important that the observations contributed to globally important datasets like GBIF, but at the same time were freely accessible for more localised research and conservation activities. To date, nearly 14,000 observations have been made representing 2692 species (nearly a quarter of native plant species), 5052 research grade observations have been submitted to GBIF and records are beginning to be used for Red List assessments.

Efforts were made to engage other members of the plant science community in Madagascar by conducting BioBlitz events, which are sustained 2–3-day biological surveys of a particular area with multiple participants. These events lead to a noticeable increase in observations (Fig. 7). More recently, growth in observations has been supported by tourists, who are already engaged with iNaturalist. They are adding observations that can be subsequently verified by the KMCC team or other Madagascar plant experts. While records from tourists are valuable, future work must focus on engaging with the custodians of Madagascar's parks and reserves such as guides and park rangers who can undertake repeat visits so that observations can be used to document population trends, not just presence or absence, and support people's engagement with their local environment.

6.6 The Importance of Local Advocates to Support Participants: FreshWater Watch

Freshwater ecosystems are among the most degraded of habitats on the planet owing to land use intensification, point and nonpoint pollution sources, river channel modification, and over exploitation (Vörösmarty et al., 2010). They occupy only 0.8% of the world's surface but support disproportionally high biodiversity (Dudgeon et al., 2006; WWF, 2016). They also provide a wide range of ecosystem services, such as flood regulation, food provision and cultural importance, which are intrinsically linked to the diversity of functions provided by the organisms present (Diaz and Cabido, 2001; Millenium Ecosystem Assessment, 2005) and so can unite human societies with their environment (Vörösmarty et al., 2015).

FWW comprises of a core standardised global method for the assessment of nutrient pollution (nitrate and phosphate concentrations) and turbidity by citizen scientists and a suite of questions related to ecosystem condition (e.g. presence of aquatic vegetation, water level and land use). Protocols for water sampling and testing and ecosystem evaluation are delivered through a consistent training approach developed by Earthwatch Institute. However, regional partners (research institutions, NGOs or governmental departments) specify additional parameters that support a local defined research priority. Such local priorities vary from anthropogenic litter at Great Lakes sites in the United States (Vincent et al., 2017) to algal blooms in urbanised Brazilian streams (Castilla et al., 2015; Cunha et al., 2017a) or nutrient

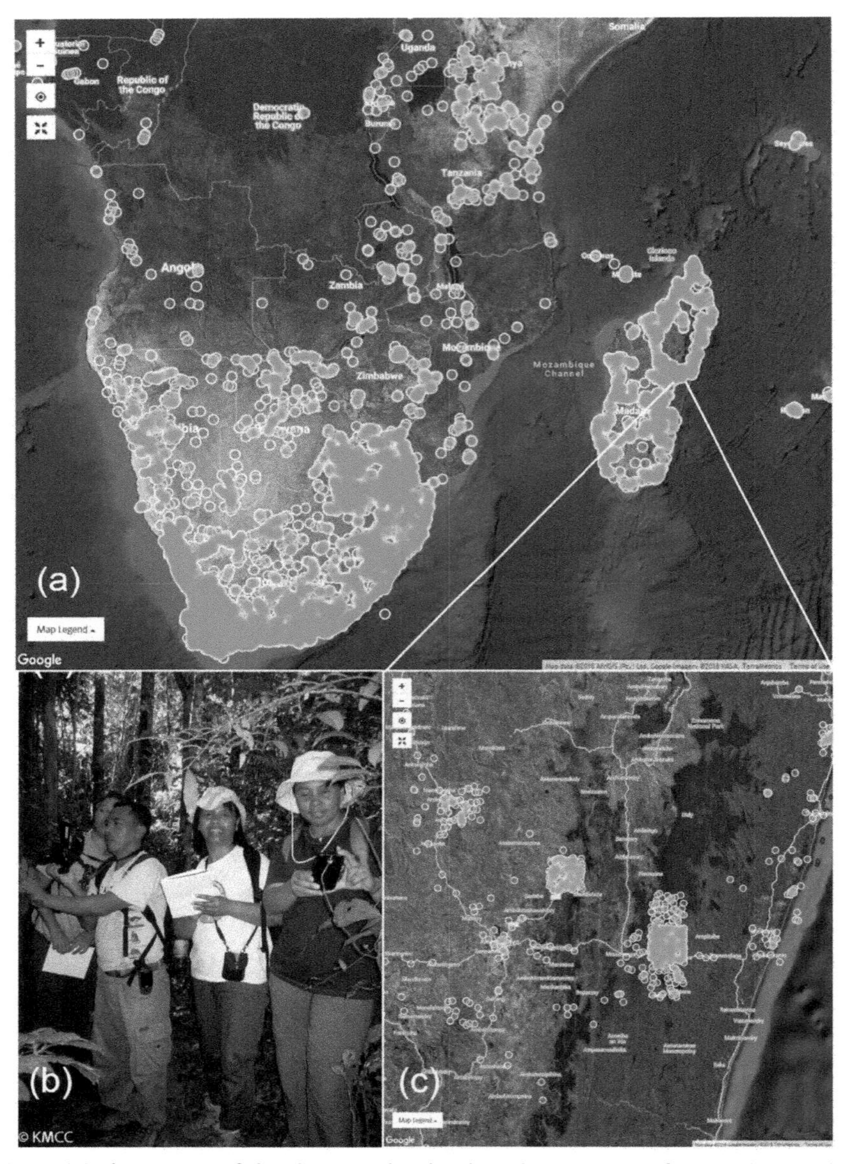

Fig. 7 (A) The impact of developing a localised implementation of iNaturalist can be clearly seen because records for Madagascar are more abundant than other countries in mainland Africa (excluding South Africa). (B) Local 'bioblitz' participants contribute many records, as demonstrated by (C) the density of records from Mantadia/Andasibe National Park and Analamazaotra Reserve (with records shown randomly within 20 km grid squares). *Map data: Search of 'research grade' and 'verifiable' records of plants from www.iNaturalist.org, accessed 20 June 2018 (A) ©2018 AfriGIS (Pty) Ltd., Google Imagery ©2018 NASA Terrametrics; (B) ©2017 Google Imagery ©2017 Terrametrics. (C) © Kew Madagascar Conservation Centre, used with permission.*

pollution in small waterbodies in the United Kingdom (McGoff et al., 2017). This dual track approach allows both data to be globally harmonisable as well as locally relevant and all projects feed into a single online platform. In this way, all core data are both high resolution and comparable across local projects, e.g., in the Americas (Loiselle et al., 2016) or across regions in China (Thornhill et al., 2017).

Over the first 5 years (2012–2016), FWW focused primarily upon corporate volunteers, training over 8000 citizen scientists and generated more than 15,000 datasets. The presence of such local 'champions' or environmental ambassadors helped spread freshwater conservation lessons, and recruit new volunteers (see https://freshwaterwatch.thewaterhub.org/volunteer_day) and the most active and dedicated volunteers are recognised through online tools and awards. Participants and principal investigators valued being able to contribute to global research via consistent methods (Earthwatch Institute, unpublished data).

A challenge to using a global approach to nutrient testing can be compatibility with approaches that are already established in the country or region. However, the automated feedback function within the FWW website and app, and the capacity to contrast local measures of nitrate with global values empowered participants making records from a Brazilian spring to alert the principal investigator to high nitrate values, who then alerted the local authority, and informed the community of the implications of drinking this water (Earthwatch Institute, 2017). Being part of the decision-making process can embed stewardship and overcome the feeling of 'monitoring for the sake of monitoring' that has been reported as a barrier in some citizen science initiatives (Ballard et al., 2017; Sharpe and Conrad, 2006; Sinclair and Diduck, 2001).

In summary, FWW is both a global and local project, and so provides an example for biodiversity monitoring. The FWW core dataset allows comparisons across the globe, while additional parameters support local researchers answer specific questions. The project demonstrates the potential for citizen science to complement professional monitoring through data with high spatiotemporal resolution (Krasny et al., 2014), to generate social capital (Overdevest and Stepenuck, 2004) and support behaviour change (Toomey and Domroese, 2013). This global–local approach has resulted in a wide range of scientific publications, e.g., Castilla et al. (2015), Loiselle et al. (2017), Thornhill et al. (2016), local actions and increased environmental awareness (Earthwatch Institute, 2017).

6.7 A Proposal for Global Monitoring of Pollinators With Citizen Science

Drawing on the discussion in the chapter so far, we now apply this to create a conceptual approach for how citizen science could be utilised to address the global challenge to monitor pollinators. The recently published record on pollinators and the ecosystem service of pollination by the Intergovernmental Science-policy Platform on Biodiversity and Ecosystem Services (IPBES 2016) demonstrated the importance of pollinators and pollination and showed that there is limited information on trends in insect pollinators at large spatial scales. Specifically, the only information is on declines in the occurrence and range size of individual species that have been detected in parts of northern Europe and North America (Biesmeijer et al., 2006; Cameron et al., 2011), although information was also gathered from indigenous honey hunters (Tengö et al., 2017). However, "although there is some evidence for changes [in abundance], this is a topic for which much additional work is needed before we have a clear picture for trends on a global scale" and this is "because of a lack of baseline datasets and monitoring schemes" (IPBES 2016). Another important report stresses that "there is the need for a global monitoring program to track trends in pollinator diversity and abundance" (LeBuhn et al., 2016).

Standardised methods have been proposed to develop global pollinator monitoring, e.g., using insect traps with experts paid to undertake the identification of insects. It will cost an estimated $2 million per region to generate trends in abundance (LeBuhn et al., 2013, 2016). While this may be modest compared to the value of pollination to agriculture (globally >$200 billion per year): (1) it is expensive in absolute terms, (2) it does not scale efficiently (doubling the number of samples roughly doubles the cost), (3) it requires large amounts of long-term funding and (4) it does not engage nonspecialist communities in valuing pollinators. However, if we develop methods (i.e. citizen science or participatory monitoring) that are suitable to be used by people who are not skilled in insect identification or ecological sampling techniques, then many of these issues will be ameliorated.

We suggest that a standardised approach of identifying and counting taxa at specific 'lure' plants for a set period of time could be a valuable approach to be scaled-up to global pollinator monitoring (Fig. 8). This has been used at small scale in citizen science projects (Roy et al., 2016) and forms a citizen science component of the UK Pollinator Monitoring Scheme (Carvell et al., 2016; https://www.ceh.ac.uk/our-science/projects/pollinator-monitoring). The focal monitoring plants could be selected from a small set of widespread

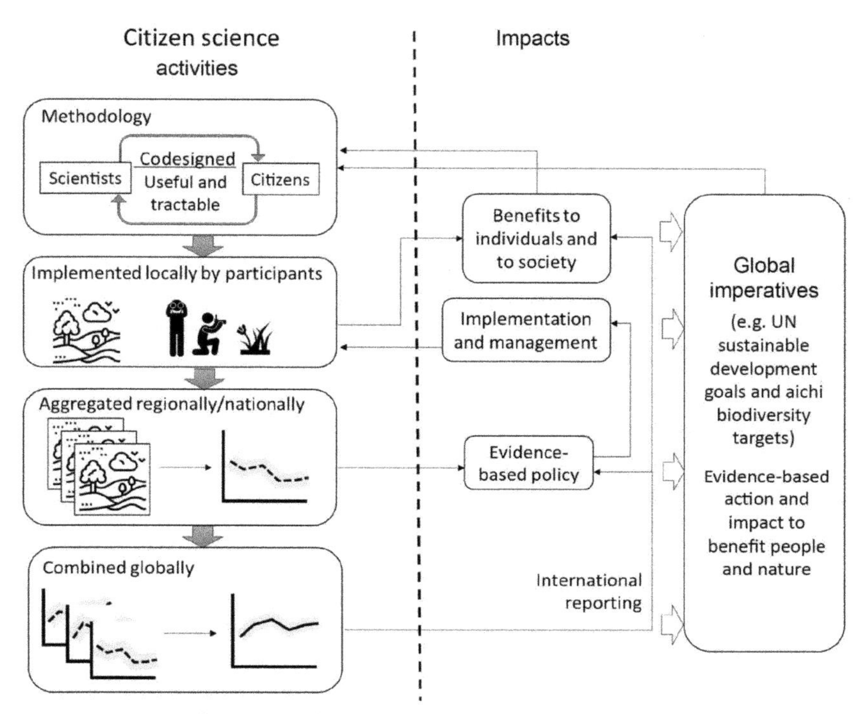

Fig. 8 A proposed framework for global pollinator monitoring, which is based on the principles discussed in this paper to create local–global monitoring. *Icons CC-BY from www.thenounproject.com: 'landscape' by Becris, 'binoculars' and 'man taking picture' by Gan Khoon Lay, 'plants' by Hamish. Graphs by the author.*

(wild, cultivated or ornamental) lure plants, with the lists of species created regionally. One crucial aspect is how the insects should be classified. We suggest that functional-indicator morphotypes should be used, because they are informative and tractable (see Section 5.3). This will require interaction between pollinator experts (who can define the useful functional categories) and potential participants (who can assess whether they can be accurately distinguished without much expertise). A different, or complementary, approach would be to monitor the benefits we gain from pollinators, i.e., pollination. This has been done by recording seed set in a plant requiring cross-fertilisation through the Great Sunflower Project (Domroese and Johnson, 2017) in the United States. Being an indirect measure of pollinators, if would only be able to detect change below a certain threshold, but if the same assay species were used across regions then the results could be used for monitoring in space and time, and it does provide information of direct relevance to people growing their own food (i.e. whether pollination is limiting food production).

Any global pollinator monitoring would have to be 'locally based, yet global' (He and Tyson, 2017), so while a global core methodology would be valuable, activities must be designed collaboratively so that the information is useful to and useable by local participants. Our proposal has similarities with The Global Mosquito Alert as a 'locally based, yet global platform' through linking successful projects in individual countries including Spain, the Netherlands, the United States, Indonesia, Hong Kong and Colombia (He and Tyson, 2017), where knowledge on mosquitoes as vectors of disease has direct bearing on people's health. Facilities should also be created so that data are easily shared and made globally accessible, so contributing to our current lack of information on pollinator trends; this could include adopting 'older' technologies such as SMS (text messages) to facilitate participation in regions where smartphones and the internet are less accessible. Paradoxically, if those participating are empowered and informed to go and improve their local environment for pollinators, based on their monitoring evidence, then this does create a problem for the global monitoring. This is because the sites monitoring would be an improving subset of the wider environment, rather than being representative (see Buckland and Johnston, 2017): this is an important issue that would need to be solved.

7. CONCLUSION

Citizen science is being increasingly promoted as a tool for global solution to many different problems. For biodiversity monitoring, there are many people living in regions where there is relatively little data on biodiversity and its trends, and there is a long history of biological recording by volunteers in some parts of the world; both of these show the potential for citizen science to contribute to global biodiversity monitoring. However, we will never achieve this potential by only adopting 'top down' control of citizen science activities, because local motivations and participation are essential for the success of any activity: we need a 'local to global' perspective (Chandler et al., 2017), based on learning from the communities of practice in citizen science and participatory monitoring. Ultimately, the need is great, the potential is great and together citizen science, and related activities, could provide a step change in our ability to monitor biodiversity—and hence respond to its threats in the lights of the benefits we gain locally, regionally and globally.

ACKNOWLEDGEMENTS

M.J.O.P., T.A. and H.E.R. receive support through the Natural Environment Research Council via national capability funding to the Centre for Ecology & Hydrology. The workshop in Chile was funded through FONDECYT (project 1140662) with additional contributions from Kauyeken Association, Doctorate in Agricultural and Veterinary Sciences (University of Chile), College of Agriculture and Forest Sciences (Pontificia Universidad Católica de Chile) and the British Ecological Society. We are grateful to all the participants at the workshop: http://www.kauyeken.cl/con-exito-se-realizaron-el-taller-y-los-seminarios-sobre-ciencia-ciudadana/. eBird: Thanks to Gretchen LeBuhn and Andrea Wiggins for contributions to the case study. Zavamaniry Gasy was supported by JRS Biodiversity Foundation. Kenya Bird Map: Thanks to Les Underhill and the Animal Demography Unit for supporting us in Kenya to use the SABAP2 system, and Michael Brooks for database design, implementation and maintenance. Thanks also to the National Museums of Kenya and Tropical Biology Association for their partnership with A Rocha Kenya on the bird atlas and for support from NatureKenya. Kenya Bird Map has been made possible through funding received from the People Programme (Marie Curie Actions) of the European Union's Seventh Framework Programme FP7/2007–2013/under REA Grant agreement no 317184 and the Natural History Museum of Denmark. FreshWater Watch: DGFC thanks CNPq (Conselho Nacional de Desenvolvimento Científico e Tecnológico) for the research productivity Grant (Process Number 300899/2016-5) and Earthwatch Institute/Neville Shulman for the Earthwatch Shulman Award. Pollinator monitoring: Thanks to Wanja Kinuthia (National Museums of Kenya) and Rosie Trevelyan (Tropical Biological Association) for discussions about pollinators.

REFERENCES

Aanensen, D.M., Huntley, D.M., Feil, E.J., Al-Own, F., Spratt, B.G., 2009. EpiCollect: linking smartphones to web applications for epidemiology, ecology and community data collection. PLoS One 4, e6968.

Amano, T., Lamming, J.D.L., Sutherland, W.J., 2016. Spatial gaps in global biodiversity information and the role of citizen science. Bioscience 66, 393–400.

Andriamihajarivo, T.H., Porter, P.L., Schatz, G.E., 2016. Endemic families of Madagascar. XIV. A new restricted range species of Pentachlaena H. Perrier (Sarcolaenaceae) from Central Madagascar. Candollea 71, 167–172.

August, T., Harvey, M., Lightfoot, P., Kilbey, D., Papadopoulos, T., Jepson, P., 2015. Emerging technology for biological recording. Biol. J. Linn. Soc. 115, 731–749.

August, T., West, S.E., Robson, H., Lyon, J., Huddart, J., Velasquez, L.F., Thornhill, I., Freshwater and Citizen Science Research Hackathon Group, 2018. Citizen meets social science: predicting volunteer involvement in a global freshwater monitoring experiment. Freshw. Sci. (in review).

Ayensu, E., van Claasen, D.R., Collins, M., Dearing, A., Fresco, L., Gadgil, M., Gitay, H., Glaser, G., Juma, C., Krebs, J., Lenton, R., Lubchenco, J., McNeely, J.A., Mooney, H.A., Pinstrup-Andersen, P., Ramos, M., Raven, P., Reid, W.V., Samper, C., Sarukhán, J., Schei, P., Tundisi, J.G., Watson, R.T., Guanhua, X., Zakri, A.H., 1999. International ecosystem assessment. Science 286, 685–686.

Ballard, H.L., Dixon, C.G.H., Harris, E.M., 2017. Youth-focused citizen science: examining the role of environmental science learning and agency for conservation. Biol. Conserv. 208, 65–75.

Balmford, A., Crane, P., Dobson, A., Green, R.E., Mace, G.M., 2005. The 2010 challenge: data availability, information needs and extraterrestrial insights. Philos. Trans. R Soc. Lond. B Biol. Sci. 360, 221–228.

Beck, J., Böller, M., Erhardt, A., Schwanghart, W., 2014. Spatial bias in the GBIF database and its effect on modeling species' geographic distributions. Eco. Inform. 19, 10–15.

Belbin, L., Williams, K.J., 2016. Towards a national bio-environmental data facility: experiences from the Atlas of living Australia. Int. J. Geogr. Inf. Sci. 30, 108–125.

Bennun, L.A., Njoroge, P., 1999. Important Bird Areas in Kenya, first ed. Nature Kenya, Nairobi, Kenya.

Beza, E., Steinke, J., van Etten, J., Reidsma, P., Fadda, C., Mittra, S., Mathur, P., Kooistra, L., 2017. What are the prospects for citizen science in agriculture? Evidence from three continents on motivation and mobile telephone use of resource-poor farmers. PLoS One 12, e0175700.

Biesmeijer, J.C., Roberts, S.P.M., Reemer, M., Ohlemüller, R., Edwards, M., Peeters, T., Schaffers, A.P., Potts, S.G., Kleukers, R., Thomas, C.D., Settele, J., Kunin, W.E., 2006. Parallel declines in pollinators and insect-pollinated plants in Britain and the Netherlands. Science 313, 351–354.

Biggs, B.J.F., Kilroy, C., Mulcock, C.M., Scarsbrook, M.R., 2002. New Zealand Stream Health Monitoring and Assessment Kit. Stream Monitoring Manual. Version 2. NIWA, Christchurch, New Zealand.

Birkin, L., Goulson, D., 2015. Using citizen science to monitor pollination services. Ecol. Entomol. 40, 3–11.

Blackmore, E., Underhill, R., McQuilkin, J., Leach, R., 2013. Common Cause for Nature: Finding Values and Frames in the Conservation Sector. Public Interest Research Centre, Machynlleth, Wales.

Blake, S., Siddharthan, A., Nguyen, H., Sharma, N., Robinson, A.-M., O'Mahony, E., Darvill, B., Mellish, C.S., Van Der Wal, R., 2012. In: Kay, M., Boitet, C. (Eds.), Natural language generation for nature conservation: automating feedback to help volunteers identify bumblebee species. International Conference on Computational Linguistics: Proceedings of COLING 2012: Technical Papers, pp. 311–324. Mumbai, India.

Boakes, E.H., McGowan, P.J.K., Fuller, R.A., Chang-qing, D., Clark, N.E., O'Connor, K., Mace, G.M., 2010. Distorted views of biodiversity: spatial and temporal bias in species occurrence data. PLoS Biol. 8, e1000385.

Bohan, D.A., Vacher, C., Tamaddoni-Nezhad, A., Raybould, A., Dumbrell, A.J., Woodward, G., 2017. Next-generation global biomonitoring: large-scale, automated reconstruction of ecological networks. Trends Ecol. Evol. 32, 477–487.

Bonney, R., Ballard, H., Jordan, R., Mccallie, E., Phillips, T., Shirk, J., Wilderman, C., 2009a. Public Participation in Scientific Research: Defining the Field and Assessing Its Potential for Informal Science Education. A CAISE Inquiry Group Report. Center for Advancement of Informal Science Education (CAISE), Washington, D.C.

Bonney, R., Cooper, C.B., Dickinson, J., Kelling, S., Phillips, T., Rosenberg, K.V., Shirk, J., 2009b. Citizen science: a developing tool for expanding science knowledge and scientific literacy. Bioscience 59, 977–984.

Bonney, R., Shirk, J.L., Phillips, T.B., Wiggins, A., Ballard, H.L., Miller-Rushing, A.J., Parrish, J.K., 2014. Citizen science. Next steps for citizen science. Science 343, 1436–1437.

Borghi, F., Spinazzè, A., Rovelli, S., Campagnolo, D., Del Buono, L., Cattaneo, A., Cavallo, D.M., 2017. Miniaturized monitors for assessment of exposure to air pollutants: a review. Int. J. Environ. Res. Public Health 14, 909.

Brammer, J.R., Brunet, N.D., Burton, A.C., Cuerrier, A., Danielsen, F., Dewan, K., Herrmann, T.M., Jackson, M.V., Kennett, R., Larocque, G., Mulrennan, M., Pratihast, A.K., Saint-Arnaud, M., Scott, C., Humphries, M.M., 2016. The role of digital data entry in participatory environmental monitoring. Conserv. Biol. 30, 1277–1287.

Brofeldt, S., Argyriou, D., Turreira-Garcia, N., Meilby, H., Danielsen, F., Theilade, I., 2018. Community-based monitoring of tropical forest crimes and forest resources using information and communication technology: experiences from Prey Lang. Cambodia. Citiz. Sci. Theory Pract. (in press).

Butchart, S.H.M., Walpole, M., Collen, B., van Strien, A., Scharlemann, J.P.W., Almond, R.E.A., Baillie, J.E.M., Bomhard, B., Brown, C., Bruno, J., Carpenter, K.E., Carr, G.M., Chanson, J., Chenery, A.M., Csirke, J., Davidson, N.C., Dentener, F., Foster, M., Galli, A., Galloway, J.N., Genovesi, P., Gregory, R.D., Hockings, M., Kapos, V., Lamarque, J.-F., Leverington, F., Loh, J., McGeoch, M.A., McRae, L., Minasyan, A., Morcillo, M.H., Oldfield, T.E.E., Pauly, D., Quader, S., Revenga, C., Sauer, J.R., Skolnik, B., Spear, D., Stanwell-Smith, D., Stuart, S.N., Symes, A., Tierney, M., Tyrrell, T.D., Vié, J.-C., Watson, R., 2010. Global biodiversity: indicators of recent declines. Science 328, 1164–1168.

Buckland, S.T., Johnston, A., 2017. Monitoring the biodiversity of regions: key principles and possible pitfalls. Biol. Conserv. 214, 23–34.

Callaghan, C.T., Gawlik, D.E., 2015. Efficacy of eBird data as an aid in conservation planning and monitoring. J. Field Ornithol. 86, 298–304.

Cameron, S.A., Lozier, J.D., Strange, J.P., Koch, J.B., Cordes, N., Solter, L.F., Griswold, T.L., 2011. Patterns of widespread decline in North American bumble bees. Proc. Natl Acad. Sci. U.S.A. 108, 662–667.

Campbell-Hunt, D.M., Campbell-Hunt, C., 2013. Ecosanctuaries: Communities Building a Future for New Zealand's Threatened Ecologies. Otago University Press, Dunedin, New Zealand.

Carvell, C., Isaac, N.J.B., Jitlal, M., Peyton, J., Powney, G.D., Roy, D.B., Vanbergen, A.J., O'Connor, R.S., Jones, C.M., Kunin, W.E., Breeze, T.D., Garratt, M.P.D., Potts, S.G., Harvey, M., Ansine, J., Comont, R.F., Lee, P., Edwards, M., Roberts, S.P.M., Morris, R.K.A., Musgrove, A.J., Brereton, T., Hawes, C., Roy, H.E., 2016. Design and Testing of a National Pollinator and Pollination Monitoring Framework. Final Summary Report to the Department for Environment, Food and Rural Affairs (Defra), Scottish Government and Welsh Government: Project WC1101. Defra, London, UK.

Castilla, E.P., Cunha, D.G.F., Lee, F.W.F., Loiselle, S., Ho, K.C., Hall, C., 2015. Quantification of phytoplankton bloom dynamics by citizen scientists in urban and peri-urban environments. Environ. Monit. Assess. 187, 690.

Center for International Earth Science Information Network—CIESIN—Columbia University, 2016. Gridded Population of the World Version 4 (GPWv4): Population Density, Beta Release. CIESIN, Palisades, NY.

Chandler, M., Bebber, D.P., Castro, S., Lowman, M.D., Muoria, P., Oguge, N., Rubenstein, D.I., 2012. International citizen science: making the local global. Front. Ecol. Environ. 10, 328–331.

Chandler, M., See, L., Copas, K., Bonde, A.M.Z., López, B.C., Danielsen, F., Legind, J.K., Masinde, S., Miller-Rushing, A.J., Newman, G., Rosemartin, A., Turak, E., 2017. Contribution of citizen science towards international biodiversity monitoring. Biol. Conserv. 213, 280–294.

Chapin, F.S., Zavaleta, E.S., Eviner, V.T., Naylor, R.L., Vitousek, P.M., Reynolds, H.L., Hooper, D.U., Lavorel, S., Sala, O.E., Hobbie, S.E., Mack, M.C., Díaz, S., 2000. Consequences of changing biodiversity. Nature 405, 234–242.

Citizen Science Association, 2017. Local Actions, Global Connections: Advancing Citizen Science with UN-Environment (UNEP). Citizen Science Association, Ithaca, USA. http://citizenscience.org/2017/12/27/local-actions-global-connections-advancing-citizen-science-with-un-environment-unep/.

Collen, B., Ram, M., Zamin, T., McRae, L., 2008. The tropical biodiversity data gap: addressing disparity in global monitoring. Trop. Conserv. Sci. 1, 75–88.

Conrad, C.C., Hilchey, K.G., 2011. A review of citizen science and community-based environmental monitoring: issues and opportunities. Environ. Monit. Assess. 176, 273–291.

Cooper, C.B., 2016. Citizen Science. The Overlook Press, New York.

Coxen, C.L., Frey, J.K., Carleton, S.A., Collins, D.P., 2017. Species distribution models for a migratory bird based on citizen science and satellite tracking data. Glob. Ecol. Conserv. 11, 298–311.

Cromarty, P., Alderson, S., 2012. Translocation statistics (2002 – 2010), and the revised Department of Conservation translocation process. Notornis 60, 55–62.

Cunha, D.G.F., Casali, S.P., de Falco, P.B., Thornhill, I., Loiselle, S.A., 2017a. The contribution of volunteer-based monitoring data to the assessment of harmful phytoplankton blooms in Brazilian urban streams. Sci. Total Environ. 584–585, 586–594.

Cunha, D.G.F., Marques, J.F., De Resende, J.C., De Falco, P.B., De Souza, C.M., Loiselle, S.A., 2017b. Citizen science participation in research in the environmental sciences: key factors related to projects' success and longevity. An. Acad. Bras. Cienc. 89, 2229–2245.

Danielsen, F., Balete, D.S., Poulsen, M.K., Enghoff, M., Nozawa, C.M., Jensen, A.E., 2000. A simple system for monitoring biodiversity in protected areas of a developing country. Biodivers. Conserv. 9, 1671–1705.

Danielsen, F., Burgess, N.D., Balmford, A., 2005a. Monitoring matters: examining the potential of locally-based approaches. Biodivers. Conserv. 14, 2507–2542.

Danielsen, F., Jensen, A.E., Alviola, P.A., Balete, D.S., Mendoza, M., Tagtag, A., Custodio, C., Enghoff, M., 2005b. Does monitoring matter? A quantitative assessment of management decisions from locally-based monitoring of protected areas. Biodivers. Conserv. 14, 2633–2652.

Danielsen, F., Mendoza, M.M., Tagtag, A., Alviola, P.A., Balete, D.S., Jensen, A.E., Enghoff, M., Poulsen, M.K., 2007. Increasing conservation management action by involving local people in natural resource monitoring. Ambio 36, 566–570.

Danielsen, F., Burgess, N.D., Balmford, A., Donald, P.F., Funder, M., Jones, J.P.G., Alviola, P., Balete, D.S., Blomley, T., Brashares, J., Child, B., Eenghoff, M., Fjeldså, J., Holt, S., Hübertz, H., Jensen, A.E., Jensen, P.M., Massao, J., Mendoza, M.M., Ngaga, Y., Poulsen, M.K., Rueda, R., Sam, M., Skielboe, T., Stuart-Hill, G., Topp-Jørgensen, E., Yonten, D., 2009. Local participation in natural resource monitoring: a characterization of approaches. Conserv. Biol. 23, 31–42.

Danielsen, F., Burgess, N., Funder, M., Blomley, T., Brashares, J., Akida, A., Jensen, A., Mendoza, M., Stuart-Hill, G., Poulsen, M.K., Ramadhani, H., Sam, M.K., Topp-Jørgensen, E., 2010. Taking stock of nature in species-rich but economically poor areas: an emerging discipline of locally based monitoring. In: Lawrence, A. (Ed.), Taking Stock of Nature: Participatory Biodiversity Assessment for Policy, Planning and Practice. Cambridge University Press, Cambridge, pp. 88–112.

Danielsen, F., Jensen, P.M., Burgess, N.D., Coronado, I., Holt, S., Poulsen, M.K., Rueda, R.M., Skielboe, T., Enghoff, M., Hemmingsen, L.H., Sørensen, M., Pirhofer-Walzl, K., 2014a. Testing focus groups as a tool for connecting indigenous and local knowledge on abundance of natural resources with science-based land management systems. Conserv. Lett. 7, 380–389.

Danielsen, F., Pirhofer-Walzl, K., Adrian, T.P., Kapijimpanga, D.R., Burgess, N.D., Jensen, P.M., Bonney, R., Funder, M., Landa, A., Levermann, N., Madsen, J., 2014b. Linking public participation in scientific research to the indicators and needs of international environmental agreements. Conserv. Lett. 7, 12–24.

Danielsen, F., Topp-Jørgensen, E., Levermann, N., Løvstrøm, P., Schiøtz, M., Enghoff, M., Jakobsen, P., 2014c. Counting what counts: using local knowledge to improve Arctic resource management. Polar Geogr. 37, 69–91.

Darbyshire, I., Tripp, E.A., Onjalalaina, G.E., 2017. Ruellia domatiata (Acanthaceae), a striking new species from Madagascar. Kew Bull. 72, 13.
Davies, T.K., Stevens, G., Meekan, M.G., Struve, J., Rowcliffe, J.M., 2013. Can citizen science monitor whale-shark aggregations? Investigating bias in mark? Recapture modelling using identification photographs sourced from the public. Wildl. Res. 39, 696.
Dennis, E.B., Morgan, B.J.T., Brereton, T.M., Roy, D.B., Fox, R., 2017. Using citizen science butterfly counts to predict species population trends. Conserv. Biol. 31, 1350–1361.
Diaz, S., Cabido, M., 2001. Vive la difference: plant functional diversity matters to ecosystem processes. Trends Ecol. Evol. 16, 646–655.
Dickinson, J.L., Shirk, J., Bonter, D., Bonney, R., Crain, R.L., Martin, J., Phillips, T., Purcell, K., 2012. The current state of citizen science as a tool for ecological research and public engagement. Front. Ecol. Environ. 10, 291–297.
Dobson, A., Lodge, D., Alder, J., Cumming, G.S., Keymer, J., McGlade, J., Mooney, H., Rusak, J.A., Sala, O., Wolters, V., Wall, D., Winfree, R., Xenopoulos, M.A., 2006. Habitat loss, trophic collapse and the decline of ecosystem services. Ecology 87, 1915–1924.
Domroese, M.C., Johnson, E.A., 2017. Why watch bees? Motivations of citizen science volunteers in the Great Pollinator Project. Biol. Conserv. 208, 40–47.
Dransfield, J., Rakotoarinivo, M., Baker, W.J., Bayton, R.P., Fisher, J.B., Horn, J.W., Leroy, B., Metz, X., 2008. A new Coryphoid palm genus from Madagascar. Bot. J. Linn. Soc. 156, 79–91.
Dudgeon, D., Arthington, A.H., Gessner, M.O., Kawabata, Z.-I., Knowler, D.J., Lévêque, C., Naiman, R.J., Prieur-Richard, A.-H., Soto, D., Stiassny, M.L.J., Sullivan, C.A., 2006. Freshwater biodiversity: importance, threats, status and conservation challenges. Biol. Rev. Camb. Philos. Soc. 81, 163–182.
Earthwatch Institute, 2017. Water: Our Precious Resource. A Report on the Impact of Five Years of FreshWater Watch. Oxford.
ECSA, 2017. Ten Principles of Citizen Science. ECSA. Vienna.
Edwards, J.L., 2000. Interoperability of biodiversity databases: biodiversity information on every desktop. Science 289, 2312–2314.
Eitzel, M.V., Cappadonna, J.L., Santos-Lang, C., Duerr, R.E., Virapongse, A., West, S.E., Kyba, C.C.M., Bowser, A., Cooper, C.B., Sforzi, A., Metcalfe, A.N., Harris, E.S., Thiel, M., Haklay, M., Ponciano, L., Roche, J., Ceccaroni, L., Shilling, F.M., Dörler, D., Heigl, F., Kiessling, T., Davis, B.Y., Jiang, Q., 2017. Citizen science terminology matters: exploring key terms. Citizen Sci. Theor. Pract. 2, 1.
Evans, K., Guariguata, M.R., 2008. Participatory Monitoring in Tropical Forest Management: A Review of Tools, Concepts and Lessons Learned. Center for International Forestry Research (CIFOR), Bogor, Indonesia.
Fernández-Llamazares, Á., Cabeza, M., 2017. Rediscovering the potential of indigenous storytelling for conservation practice. Conserv. Lett. 11, e12398.
Fragoso, J.M.V., Levi, T., Oliveira, L.F.B., Luzar, J.B., Overman, H., Read, J.M., Silvius, K.M., 2016. Line transect surveys underdetect terrestrial mammals: implications for the sustainability of subsistence hunting. PLoS One 11, e0152659.
Fundación Ciencia Ciudadana, 2017. Guía Para Conocer La Ciencia Ciudadana. Universidad Autónoma de Chile, Chile.
Funder, M., Ngaga, Y., Nielsen, M., Poulsen, M., Danielsen, F., 2013. Reshaping conservation: the social dynamics of participatory monitoring in Tanzania's community-managed forests. Conserv. Soc. 11, 218.
Gardiner, L.M., Bachman, S.P., 2016. The role of citizen science in a global assessment of extinction risk in palms (Arecaceae). Bot. J. Linn. Soc. 182, 543–550.

Geoghegan, H., Dyke, A., Pateman, R., West, S., Everett, G., 2016. Understanding Motivations for Citizen Science. Final Report on Behalf of UKEOF. UKEOF, Oxfordshire, UK.
Glasgow, H.B., Burkholder, J.M., Reed, R.E., Lewitus, A.J., Kleinman, J.E., 2004. Real-time remote monitoring of water quality: a review of current applications, and advancements in sensor, telemetry, and computing technologies. J. Exp. Mar. Biol. Ecol. 300, 409–448.
Gligo, N., 2016. Informe País: Estado Del Medio Ambiente En Chile. Universidad de Chile, Santiago, Chile.
Goldsmith, G.R., Morueta-Holme, N., Sandel, B., Fitz, E.D., Fitz, S.D., Boyle, B., Casler, N., Engemann, K., Jørgensen, P.M., Kraft, N.J.B., McGill, B., Peet, R.K., Piel, W.H., Spencer, N., Svenning, J.-C., Thiers, B.M., Violle, C., Wiser, S.K., Enquist, B.J., 2016. Plant-O-Matic: a dynamic and mobile guide to all plants of the Americas. Methods in Ecology and Evolution, 7, 960–965.
Goyder, D., Baker, W., Besnard, G., Dransfield, J., Gardiner, L., Moat, J., Rabehevitra, D., Rajaovelona, L., Rakotoarisoa, S., Rakotonasolo, F., Ralimanana, H., Randriamboavonjy, T., Razanatsoa, J., Sarasan, V., Vorontsova, M., Wilkin, P., Cable, S., 2017. Country focus—status of knowledge of Madagascan plants. In: Willis, K.J. (Ed.), State of the World'd Plants 2017. Royal Botanic Gardens, Kew, pp. 36–41. Report.
Graham, P.M., Dickens, C.W., Taylor, R.J., 2004. miniSASS—a novel technique for community participation in river health monitoring and management. Afr. J. Aquat. Sci. 29, 25–35.
Greenwood, J.J.D., 2007. Citizens, science and bird conservation. J. Ornithol. 148, 77–124.
Groom, Q., Weatherdon, L., Geijzendorffer, I.R., 2016. Is citizen science an open science in the case of biodiversity observations? J. Appl. Ecol. 54, 612–617.
Haklay, M., 2015. Citizen Science and Policy: A European Perspective. Woodrow Wilson International Center for Scholars, Washington, DC.
Handford, P., 2004. FORMAK Forest Monitoring Manual. FORMAK, Paekakariki, New Zealand.
Hannah, L., Dave, R., Lowry, P.P., Andelman, S., Andrianarisata, M., Andriamaro, L., Cameron, A., Hijmans, R., Kremen, C., MacKinnon, J., Randrianasolo, H.H., Andriambololonera, S., Razafimpahanana, A., Randriamahazo, H., Randrianarisoa, J., Razafinjatovo, P., Raxworthy, C., Schatz, G.E., Tadross, M., Wilme, L., 2008. Climate change adaptation for conservation in Madagascar. Biol. Lett. 4, 590–594.
Hardie-Boys, N., 2010. Valuing Community Group Contributions to Conservation. Science for Conservation. Report No. 299. Department of Conservation, Wellington, New Zealand.
Harper, G.J., Steininger, M.K., Tucker, C.J., Juhn, D., Hawkins, F., 2007. Fifty years of deforestation and forest fragmentation in Madagascar. Environ. Conserv. 34, 325–333.
He, Y., Tyson, E., 2017. Survey Results: Complexities & Overlaps in Existing Citizen Science Mosquito Projects. Woodrow Wilson International Center for Scholars Washington, D.C.
Hill, M.O., 2012. Local frequency as a key to interpreting species occurrence data when recording effort is not known. Methods Ecol. Evol. 3, 195–205.
Hochachka, W.M., Fink, D., Hutchinson, R.A., Sheldon, D., Wong, W.-K., Kelling, S., 2012. Data-intensive science applied to broad-scale citizen science. Trends Ecol. Evol. 27, 130–137.
Hortal, J., Jiménez-Valverde, A., Gómez, J.F., Lobo, J.M., Baselga, A., 2008. Historical bias in biodiversity inventories affects the observed environmental niche of the species. Oikos 117, 847–858.
Isaac, N.J.B., Pocock, M.J.O., 2015. Bias and information in biological records. Biol. J. Linn. Soc. 115, 522–531.

Isaac, N.J.B., van Strien, A.J., August, T.A., de Zeeuw, M.P., Roy, D.B., 2014. Statistics for citizen science: extracting signals of change from noisy ecological data. Methods Ecol. Evol. 5, 1052–1060.

IUCN, 2016. IUCN Red List of Threatened Species. Version 2016.3. IUCN, Gland, Switzerland.

Jenkins, C.N., Pimm, S.L., Joppa, L.N., 2013. Global patterns of terrestrial vertebrate diversity and conservation. Proc. Natl. Acad. Sci. U. S. A. 110, E2602–E2610.

Johnson, N., Alessa, L., Behe, C., Danielsen, F., Gearheard, S., Gofman-Wallingford, V., Kliskey, A., Krümmel, E.-M., Lynch, A., Mustonen, T., Pulsifer, P., Svoboda, M., 2015. The contributions of community-based monitoring and traditional knowledge to Arctic observing networks: reflections on the state of the field. Arctic 68, 28.

Johnson, C.N., Balmford, A., Brook, B.W., Buettel, J.C., Galetti, M., Guangchun, L., Wilmshurst, J.M., 2017. Biodiversity losses and conservation responses in the Anthropocene. Science 356, 270–275.

Jones, J.P.G., Andriamarovololona, M.M., Hockley, N., Gibbons, J.M., Milner-Gulland, E.J., 2008. Testing the use of interviews as a tool for monitoring trends in the harvesting of wild species. J. Appl. Ecol. 45, 1205–1212.

Joppa, L.N., O'Connor, B., Visconti, P., Smith, C., Geldmann, J., Hoffmann, M., Watson, J.E.M., Butchart, S.H.M., Virah-Sawmy, M., Halpern, B.S., Ahmed, S.E., Balmford, A., Sutherland, W.J., Harfoot, M., Hilton-Taylor, C., Foden, W., Minin, E.D., Pagad, S., Genovesi, P., Hutton, J., Burgess, N.D., 2016. Filling in biodiversity threat gaps. Science 352, 416–418.

Kennett, R., Danielsen, F., Silvius, K.M., 2015. Citizen science is not enough on its own. Nature 521, 161.

Keuskamp, J.A., Dingemans, B.J.J., Lehtinen, T., Sarneel, J.M., Hefting, M.M., 2013. Tea bag index: a novel approach to collect uniform decomposition data across ecosystems. Methods Ecol. Evol. 4, 1070–1075.

Kharas, H., 2017. In: The Unprecedented Expansion of the Global Middle Class: An Update. Global Economy & Development Working Paper 100. Brookings Institution, Washington, DC.

Kissling, W.D., Ahumada, J.A., Bowser, A., Fernandez, M., Fernández, N., García, E.A., Guralnick, R.P., Isaac, N.J.B., Kelling, S., Los, W., McRae, L., Mihoub, J.-B., Obst, M., Santamaria, M., Skidmore, A.K., Williams, K.J., Agosti, D., Amariles, D., Arvanitidis, C., Bastin, L., De Leo, F., Egloff, W., Elith, J., Hobern, D., Martin, D., Pereira, H.M., Pesole, G., Peterseil, J., Saarenmaa, H., Schigel, D., Schmeller, D.S., Segata, N., Turak, E., Uhlir, P.F., Wee, B., Hardisty, A.R., 2018. Building essential biodiversity variables (EBVs) of species distribution and abundance at a global scale. Biol. Rev. 93, 600–625.

Kosmala, M., Wiggins, A., Swanson, A., Simmons, B., 2016. Assessing data quality in citizen science. Front. Ecol. Environ. 14, 551–560.

Krasny, M.E., Russ, A., Tidball, K.G., Elmqvist, T., 2014. Civic ecology practices: participatory approaches to generating and measuring ecosystem services in cities. Ecosyst. Serv. 7, 177–186.

Krell, F.-T., 2004. Parataxonomy vs. taxonomy in biodiversity studies—pitfalls and applicability of 'morphospecies' sorting. Biodivers. Conserv. 13, 795–812.

Latombe, G., Pyšek, P., Jeschke, J.M., Blackburn, T.M., Bacher, S., Capinha, C., Costello, M.J., Fernández, M., Gregory, R.D., Hobern, D., Hui, C., Jetz, W., Kumschick, S., McGrannachan, C., Pergl, J., Roy, H.E., Scalera, R., Squires, Z.E., Wilson, J.R.U., Winter, M., Genovesi, P., McGeoch, M.A., 2017. A vision for global monitoring of biological invasions. Biol. Conserv. 213, 295–308.

Lawler, J.J., White, D., Sifneos, J.C., Master, L.L., 2003. Rare species and the use of indicator groups for conservation planning. Conserv. Biol. 17, 875–882.

Lawrence, A., 2006. 'No personal motive?' Volunteers, biodiversity, and the false dichotomies of participation. Ethics Place Environ. 9, 279–298.
LeBuhn, G., Droege, S., Connor, E.F., Gemmill-Herren, B., Potts, S.G., Minckley, R.L., Griswold, T., Jean, R., Kula, E., Roubik, D.W., Cane, J., Wright, K.W., Frankie, G., Parker, F., 2013. Detecting insect pollinator declines on regional and global scales. Conserv. Biol. 27, 113–120.
LeBuhn, G., Droege, S., Connor, E., Gemmill-Herren, B., Azzu, N., 2016. Protocol to Detect and Monitor Pollinator Communities: Guidance for Practitioners. FAO, Rome.
Lindenmayer, D.B., Likens, G.E., 2010. The science and application of ecological monitoring. Biol. Conserv. 143, 1317–1328.
Liu, H.-Y., Kobernus, M., Broday, D., Bartonova, A., 2014. A conceptual approach to a citizens' observatory—supporting community-based environmental governance. Environ. Health 13, 107.
Loftie-Eaton, M., 2015. Comparing reporting rates between the first and second southern African bird atlas projects. Ornithol. Observ. 6, 1–11.
Loiselle, S.A., Gasparini Fernandes Cunha, D., Shupe, S., Valiente, E., Rocha, L., Heasley, E., Belmont, P.P., Baruch, A., 2016. Micro and macroscale drivers of nutrient concentrations in urban streams in South, Central and North America. PLoS One 11, e0162684.
Loiselle, S.A., Frost, P.C., Turak, E., Thornhill, I., 2017. Citizen scientists supporting environmental research priorities. Sci. Total Environ. 598, 937.
Madagascar Catalogue, 2017. Catalogue of the Vascular Plants of Madagascar. Missouri Botanical Garden/Antananarivo, St. Louis, USA/Madagascar.
Maes, D., Isaac, N.J.B., Harrower, C.A., Collen, B., van Strien, A.J., Roy, D.B., 2015. The use of opportunistic data for IUCN Red List assessments. Biol. J. Linn. Soc. 115, 690–706.
Martinez-Harms, M.J., Bryan, B.A., Figueroa, E., Pliscoff, P., Runting, R.K., Wilson, K.A., 2017. Scenarios for land use and ecosystem services under global change. Ecosyst. Serv. 25, 56–68.
McGeoch, M.A., Squires, Z.E., 2015. An essential biodiversity variable approach to monitoring biological invasions: guide for countries. In: GEO BON Technical Series, vol. 2. GEO BON, Leipzig, Germany, 13 pp.
McGoff, E., Dunn, F., Cachazo, L.M., Williams, P., Biggs, J., Nicolet, P., Ewald, N.C., 2017. Finding clean water habitats in urban landscapes: professional researcher vs citizen science approaches. Sci. Total Environ. 581–582, 105–116.
McKinley, D.C., Miller-Rushing, A.J., Ballard, H.L., Bonney, R., Brown, H., Cook-Patton, S.C., Evans, D.M., French, R.A., Parrish, J.K., Phillips, T.B., Ryan, S.F., Shanley, L.A., Shirk, J.L., Stepenuck, K.F., Weltzin, J.F., Wiggins, A., Boyle, O.D., Briggs, R.D., Chapin, S.F., Hewitt, D.A., Preuss, P.W., Soukup, M.A., 2017. Citizen science can improve conservation science, natural resource management, and environmental protection. Biol. Conserv. 208, 15–28.
Meyer, C., Weigelt, P., Kreft, H., 2016. Multidimensional biases, gaps and uncertainties in global plant occurrence information. Ecol. Lett. 19, 992–1006.
Millenium Ecosystem Assessment, 2005. Ecosystems and Human Well-Being: A Framework for Assessment. Island Press.
Ministerio del Medio Ambiente, 2014. Quinto Informe de Biodiversidad de Chile. Ministerio del Medio Ambiente, Chile. http://portal.mma.gob.cl/wp-content/uploads/2017/08/Libro_Convenio_sobre_diversidad_Biologica.pdf.
Mittermeier, R.A., Turner, W.R., Larsen, F.W., Brooks, T.M., Gascon, C., 2011. Global biodiversity conservation: the critical role of hotspots. In: Biodiversity Hotspots. Springer Berlin Heidelberg, Berlin, Heidelberg, pp. 3–22.
Mustonen, T., 2015. Communal visual histories to detect environmental change in northern areas: examples of emerging north American and Eurasian practices. Ambio 44, 766–777.

Newson, S.E., Woodburn, R.J.W., Noble, D.G., Baillie, S.R., Gregory, R.D., 2005. Evaluating the Breeding Bird Survey for producing national population size and density estimates. Bird Study 52, 42–54.

Ortega-Álvarez, R., Sánchez-González, L.A., Rodríguez-Contreras, V., Vargas-Canales, V.M., Puebla-Olivares, F., Berlanga, H., 2012. Birding for and with people: integrating localparticipation in avian monitoring programs within high biodiversity areas in Southern Mexico. Sustainability 4, 1984–1998.

Overdevest, C., Stepenuck, K., 2004. Volunteer stream monitoring and local participation in natural resource issues. Hum. Ecol. Rev. 11, 177–185.

Parham, J., Crall, J., Stewart, C., Berger-Wolf, T., Rubenstein, D., 2017. In: Animal population censusing at scale with citizen science and photographic identification. AAAI 2017 Spring Symposium on AI for Social Good. Stanford, California.

Participatory Monitoring and Management Partnership (PMMP). (2015) Manaus Letter: recommendations for the participatory monitoring of biodiversity. In. International Seminar on Participatory Monitoring of Biodiversity for the Management of Natural Resources 2014 (eds P.A. Constantino, K.M. Silvius, J. Kleine Büning, P. Arroyo, F. Danielsen, C.C. Durigan, G. Estupinan, S. Hvalkof, M.K. Poulsen, & K.T. Ribeiro), PMMP, Manaus, Brazil.

Pearce-Higgins, J.W., Baillie, S.R., Boughey, K., Bourn, N.A.D., Foppen, R.P.B., Gillings, S., Gregory, R.D., Hunt, T., Jiguet, F., Lehikoinen, A., Musgrove, A.J., Robinson, R.A., Roy, D.B., Siriwardena, G.M., Walker, K.J., Wilson, J.D., 2018. Overcoming the challenges of public data archiving for citizen science biodiversity recording and monitoring schemes. J. Appl. Ecol. (in press).

Pereira, H.M., Ferrier, S., Walters, M., Geller, G.N., Jongman, R.H.G., Scholes, R.J., Bruford, M.W., Brummitt, N., Butchart, S.H.M., Cardoso, A.C., Coops, N.C., Dulloo, E., Faith, D.P., Freyhof, J., Gregory, R.D., Heip, C., Höft, R., Hurtt, G., Jetz, W., Karp, D.S., McGeoch, M.A., Obura, D., Onoda, Y., Pettorelli, N., Reyers, B., Sayre, R., Scharlemann, J.P.W., Stuart, S.N., Turak, E., Walpole, M., Wegmann, M., 2013. Essential biodiversity variables. Science 339, 277–278.

Peters, M.A., Hamilton, D., Eames, C., 2015. Action on the ground: a review of community environmental groups' restoration objectives, activities and partnerships in New Zealand. N. Z. J. Ecol. 39, 179–189.

Peters, M., Hamilton, D., Eames, C., Innes, J., Mason, N., 2016. The current state of community-based environmental monitoring in New Zealand. N. Z. J. Ecol. 40, 279–288.

Pettorelli, N., Laurance, W.F., O'Brien, T.G., Wegmann, M., Nagendra, H., Turner, W., 2014. Satellite remote sensing for applied ecologists: opportunities and challenges. J. Appl. Ecol. 51, 839–848.

Pew Research Center, 2016. Smartphone Ownership and Internet Usage Continues to Climb in Emerging Economies. Pew Research Center.

Pipek, P., Petrusková, T., Petrusek, A., Diblíková, L., Eaton, M.A., Pyšek, P., 2018. Dialects of an invasive songbird are preserved in its invaded but not native source range. Ecography 41, 245–254.

Pocock, M.J.O., Newson, S.E., Henderson, I.G., Peyton, J., Sutherland, W.J., Noble, D.G., Ball, S.G., Beckmann, B.C., Biggs, J., Brereton, T., Bullock, D.J., Buckland, S.T., Edwards, M., Eaton, M.A., Harvey, M.C., Hill, M.O., Horlock, M., Hubble, D.S., Julian, A.M., Mackey, E.C., Mann, D.J., Marshall, M.J., Medlock, J.M., O'Mahony, E.M., Pacheco, M., Porter, K., Prentice, S., Procter, D.A., Roy, H.E., Southway, S.E., Shortall, C.R., Stewart, A.J.A., Wembridge, D.E., Wright, M.A., Roy, D.B., 2015a. Developing and enhancing biodiversity monitoring programmes: a collaborative assessment of priorities. J. Appl. Ecol. 52, 686–695.

Pocock, M.J.O., Roy, H.E., Preston, C.D., Roy, D.B., 2015b. The Biological Records Centre: a pioneer of citizen science. Biol. J. Linn. Soc. 115, 475–493.

Pocock, M.J.O., Tweddle, J.C., Savage, J., Robinson, L.D., Roy, H.E., 2017. The diversity and evolution of ecological and environmental citizen science. PLoS One 12, e0172579.
Pocock, M.J.O., Roy, H.E., August, T., Kuria, A., Barasa, F., Bett, J., Githiru, M., Kairo, J., Kimani, J., Kinuthia, W., Kissui, B., Madindou, I., Mbogo, K., Mirembe, J., Mugo, P., Muniale, F.M., Njoroge, P., Njuguna, E.G., Olendo, M.I., Opige, M., Otieno, T.O., Ng'weno, C.C., Pallangyo, E., Thenya, T., Wanjiru, A., Trevelyan, R., 2018. Developing the global potential of citizen science: assessing opportunities that benefit people, society and the environment in East Africa. J. Appl. Ecol. https://doi.org/10.1111/1365-2664.13279.
Powney, G.D., Isaac, N.J.B., 2015. Beyond maps: a review of the applications of biological records. Biol. J. Linn. Soc. 115, 532–542.
Pretty, J., Smith, D., 2004. Social capital in biodiversity conservation and management. Conserv. Biol. 18, 631–638.
Proença, V., Martin, L.J., Pereira, H.M., Fernandez, M., McRae, L., Belnap, J., Böhm, M., Brummitt, N., García-Moreno, J., Gregory, R.D., Honrado, J.P., Jürgens, N., Opige, M., Schmeller, D.S., Tiago, P., van Swaay, C.A.M., 2017. Global biodiversity monitoring: from data sources to essential biodiversity variables. Biol. Conserv. 213, 256–263.
Pulsifer, P.L., Laidler, G.J., Taylor, D.R.F., Hayes, A., 2011. Towards an indigenist data management program: reflections on experiences developing an atlas of sea ice knowledge and use. Can. Geogr. 55, 108–124.
Raffaelli, D., 2007. Food webs, body size and the curse of the Latin binomial. In: Rooney, N., McCann, K.S., Noakes, D.L.G. (Eds.), From Energetics to Ecosystems: The Dynamics and Structure of Ecological Systems. Springer, The Netherlands, pp. 53–64.
Rakotoarinivo, M., Dransfield, J., 2012. Tahina spectabilis. The IUCN Red List of Threatened Species 2012: e.T195893A2430024. IUCN, Gland, Switzerland. https://doi.org/10.2305/IUCN.UK.2012.RLTS.T195893A2430024.en.
Rich, K.J., Ridealgh, M., West, S.E., Cinderby, S., Ashmore, M., 2015. Exploring the links between post-industrial landscape history and ecology through participatory methods. PLoS One 10, e0136522.
Robertson, G., Peters, M.A., 2006. Turning the Tide: An Estuaries Toolkit for New Zealand Communities. NZ Landcare Trust, Hamilton, New Zealand.
Robinson, O.J., Ruiz-Gutierrez, V., Fink, D., 2018. Correcting for bias in distribution modelling for rare species using citizen science data. Divers. Distrib. 24, 460–472.
Rotman, D., Preece, J., Hammock, J., Procita, K., Hansen, D., Parr, C., Lewis, D., Jacobs, D., 2012. In: Dynamic changes in motivation in collaborative citizen-science projects. Proceedings of the ACM 2012 Conference on Computer Supported Cooperative Work, p. 217.
Rotman, D., Hammock, J., Preece, J., Hansen, D., Boston, C., Bowser, A., He, Y., 2014. In: Motivations affecting initial and long-term participation in citizen science projects in three countries. iConference 2014 Proceedings. iSchools, pp. 110–124.
Roy, H.E., Pocock, M.J.O., Preston, C.D., Roy, D.B., Savage, J., Tweddle, J.C., Robinson, L.D., 2012. Understanding Citizen Science & Environmental Monitoring. Final Report on Behalf of UK-EOF. NERC Centre for Ecology & Hydrology, Wallingford, OxfordshireNERC,. Final Report on behalf of UK-EOF. & Museum., C. for E.& H. and N.H. (2012).
Roy, H.E., Baxter, E., Saunders, A., Pocock, M.J.O., 2016. Focal plant observations as a standardised method for pollinator monitoring: opportunities and limitations for mass participation citizen science. PLoS One 11, e0150794.
Sala, O.E., 2000. Global biodiversity scenarios for the year 2100. Science 287, 1770–1774.
SCBD, 2010. COP-10 Decision X/2. Secretariat of the Convention on Biological Diversity.

Schmeller, D.S., Henry, P.-Y., Julliard, R., Gruber, B., Clobert, J., Dziock, F., Lengyel, S., Nowicki, P., Déri, E., Budrys, E., Kull, T., Tali, K., Bauch, B., Settele, J., Van Swaay, C., Kobler, A., Babij, V., Papastergiadou, E., Henle, K., 2009. Advantages of volunteer-based biodiversity monitoring in Europe. Conserv. Biol. 23, 307–316.

Schmeller, D.S., Böhm, M., Arvanitidis, C., Barber-Meyer, S., Brummitt, N., Chandler, M., Chatzinikolaou, E., Costello, M.J., Ding, H., García-Moreno, J., Gill, M., Haase, P., Jones, M., Juillard, R., Magnusson, W.E., Martin, C.S., McGeoch, M., Mihoub, J.-B., Pettorelli, N., Proença, V., Peng, C., Regan, E., Schmiedel, U., Simaika, J.P., Weatherdon, L., Waterman, C., Xu, H., Belnap, J., 2017. Building capacity in biodiversity monitoring at the global scale. Biodivers. Conserv. 26, 2765–2790.

Schmiedel, U., Araya, Y., Bortolotto, M.I., Boeckenhoff, L., Hallwachs, W., Janzen, D., Kolipaka, S.S., Novotny, V., Palm, M., Parfondry, M., Smanis, A., Toko, P., 2016. Contributions of paraecologists and parataxonomists to research, conservation, and social development. Conserv. Biol. 30, 506–519.

Seed, L., Wolseley, P., Gosling, L., Davies, L., Power, S.A., 2013. Modelling relationships between lichen bioindicators, air quality and climate on a national scale: results from the UK OPAL air survey. Environ. Pollut. 182, 437–447.

Sharpe, A., Conrad, C., 2006. Community based ecological monitoring in Nova Scotia: challenges and opportunities. Environ. Monit. Assess. 113, 395–409.

Shirk, J.L., Ballard, H.L., Wilderman, C.C., Phillips, T., Wiggins, A., Jordan, R., McCallie, E., Minarchek, M., Lewenstein, B.V., Krasny, M.E., Bonney, R., 2012. Public participation in scientific research: a framework for deliberate design. Ecol. Soc. 17, 29–48.

Sinclair, A.J., Diduck, A.P., 2001. Public involvement in EA in Canada: a transformative learning perspective. Environ. Impact Assess. Rev. 21, 113–136.

Staddon, S.C., Nightingale, A., Shrestha, S.K., 2015. Exploring participation in ecological monitoring in Nepal's community forests. Environ. Conserv. 42, 268–277.

Stevens, M., Vitos, M., Lewis, J., Haklay, M., 2013. In: Participatory monitoring of poaching in the Congo basin. 21st GIS Research UK Conference, p. GISRUK 2013, Liverpool.

Sullivan, J.J., Molles, L.E., 2016. Biodiversity monitoring by community-based restoration groups in New Zealand. Ecol. Manage. Restor. 17, 210–217.

Sullivan, B.L., Wood, C.L., Iliff, M.J., Bonney, R.E., Fink, D., Kelling, S., 2009. eBird: a citizen-based bird observation network in the biological sciences. Biol. Conserv. 142, 2282–2292.

Sullivan, B.L., Aycrigg, J.L., Barry, J.H., Bonney, R.E., Bruns, N., Cooper, C.B., Damoulas, T., Dhondt, A.A., Dietterich, T., Farnsworth, A., Fink, D., Fitzpatrick, J.W., Fredericks, T., Gerbracht, J., Gomes, C., Hochachka, W.M., Iliff, M.J., Lagoze, C., La Sorte, F.A., Merrifield, M., Morris, W., Phillips, T.B., Reynolds, M., Rodewald, A.D., Rosenberg, K.V., Trautmann, N.M., Wiggins, A., Winkler, D.W., Wong, W.-K., Wood, C.L., Yu, J., Kelling, S., 2014. The eBird enterprise: an integrated approach to development and application of citizen science. Biol. Conserv. 169, 31–40.

Szabo, J.K., Vesk, P.A., Baxter, P.W.J., Possingham, H.P., 2010. Regional avian species declines estimated from volunteer-collected long-term data using list length analysis. Ecol. Appl. 20, 2157–2169.

Tengö, M., Hill, R., Malmer, P., Raymond, C.M., Spierenburg, M., Danielsen, F., Elmqvist, T., Folke, C., 2017. Weaving knowledge systems in IPBES, CBD and beyond—lessons learned for sustainability. Curr. Opin. Environ. Sustain. 26–27, 17–25.

Theobald, E.J., Ettinger, A.K., Burgess, H.K., DeBey, L.B., Schmidt, N.R., Froehlich, H.E., Wagner, C., HilleRisLambers, J., Tewksbury, J., Harsch, M.A., Parrish, J.K., 2015. Global change and local solutions: tapping the unrealized potential of citizen science for biodiversity research. Biol. Conserv. 181, 236–244.

Thessen, A., Patterson, D., 2011. Data issues in the life sciences. ZooKeys 150, 15–51.
Thornhill, I., Loiselle, S., Lind, K., Ophof, D., 2016. The citizen science opportunity for researchers and agencies. Bioscience 66, 720–721.
Thornhill, I., Ho, J.G., Zhang, Y., Li, H., Ho, K.C., Miguel-Chinchilla, L., Loiselle, S.A., 2017. Prioritising local action for water quality improvement using citizen science; a study across three major metropolitan areas of China. Sci. Total Environ. 584–585, 1268–1281.
Tipa, G., Teirney, L., 2003. A Cultural Health Index for Streams and Waterways: Indicators for Recognising and Expressing Maori Values. Report Prepared for the Ministry for the Environment ME475. Ministry for the Environment, Wellington, New Zealand.
Tittensor, D.P., Walpole, M., Hill, S.L.L., Boyce, D.G., Britten, G.L., Burgess, N.D., Butchart, S.H.M., Leadley, P.W., Regan, E.C., Alkemade, R., Baumung, R., Bellard, C., Bouwman, L., Bowles-Newark, N.J., Chenery, A.M., Cheung, W.W.L., Christensen, V., Cooper, H.D., Crowther, A.R., Dixon, M.J.R., Galli, A., Gaveau, V., Gregory, R.D., Gutierrez, N.L., Hirsch, T.L., Hoft, R., Januchowski-Hartley, S.R., Karmann, M., Krug, C.B., Leverington, F.J., Loh, J., Lojenga, R.K., Malsch, K., Marques, A., Morgan, D.H.W., Mumby, P.J., Newbold, T., Noonan-Mooney, K., Pagad, S.N., Parks, B.C., Pereira, H.M., Robertson, T., Rondinini, C., Santini, L., Scharlemann, J.P.W., Schindler, S., Sumaila, U.R., Teh, L.S.L., van Kolck, J., Visconti, P., Ye, Y., 2014. A mid-term analysis of progress toward international biodiversity targets. Science 346, 241–244.
Toivanen, T., Koponen, S., Kotovirta, V., Molinier, M., Chengyuan, P., 2013. Water quality analysis using an inexpensive device and a mobile phone. Environ. Syst. Res. 2, 9.
Toomey, A.H., Domroese, M.C., 2013. Can citizen science lead to positive conservation attitudes and behaviors? Hum. Ecol. Rev. 20, 50–62.
Topp-Jørgensen, E., Poulsen, M.K., Lund, J.F., Massao, J.F., 2005. Community-based monitoring of natural resource use and Forest quality in montane forests and Miombo Woodlands of Tanzania. Biodivers. Conserv. 14, 2653–2677.
Troudet, J., Grandcolas, P., Blin, A., Vignes-Lebbe, R., Legendre, F., 2017. Taxonomic bias in biodiversity data and societal preferences. Sci. Rep. 7, 9132.
Tulloch, A.I.T., Mustin, K., Possingham, H.P., Szabo, J.K., Wilson, K.A., 2013a. To boldly go where no volunteer has gone before: predicting volunteer activity to prioritize surveys at the landscape scale. Divers. Distrib. 19, 465–480.
Tulloch, A.I.T., Possingham, H.P., Joseph, L.N., Szabo, J., Martin, T.G., 2013b. Realising the full potential of citizen science monitoring programs. Biol. Conserv. 165, 128–138.
Ullmann, K., Vaughan, M., Kremen, C., Shih, T., Shepherd, M., 2010. California Pollinator Project: Citizen Scientist Pollinator Monitoring Guide. California Pollinator Project, Portland, Oregon.
Underhill, L.G., 2016. The fundamentals of the SABAP2 protocol. Biodivers. Observ. 7, 1–12.
UNGA, 2015. Transforming Our World: The 2030 Agenda for Sustainable Development (UNGA Resolution A/RES/70/1, 25 September 2015) ('2030 Agenda'). UNGA.
van der Wal, R., Sharma, N., Mellish, C., Robinson, A., Siddharthan, A., 2016. The role of automated feedback in training and retaining biological recorders for citizen science. Cons. Biol. 30, 550–561.
van Dyck, H., van Strien, A.J., Maes, D., van Swaay, C.A.M., 2009. Declines in common, widespread butterflies in a landscape under intense human use. Conserv. Biol. 23, 957–965.
van Strien, A.J., van Swaay, C.A.M., Termaat, T., 2013. Opportunistic citizen science data of animal species produce reliable estimates of distribution trends if analysed with occupancy models. J. Appl. Ecol. 50, 1450–1458.

van Swaay, C.A.M., Nowicki, P., Settele, J., van Strien, A.J., 2008. Butterfly monitoring in Europe: methods, applications and perspectives. Biodivers. Conserv. 17, 3455–3469.
Vianna, G.M.S., Meekan, M.G., Bornovski, T.H., Meeuwig, J.J., 2014. Acoustic telemetry validates a citizen science approach for monitoring sharks on coral reefs. PLoS One 9, e95565.
Vincent, A., Drag, N., Lyandres, O., Neville, S., Hoellein, T., 2017. Citizen science datasets reveal drivers of spatial and temporal variation for anthropogenic litter on Great Lakes beaches. Sci. Total Environ. 577, 105–112.
Vorontsova, M.S., Ratovonirina, G., Randriamboavonjy, T., 2013. Revision of Andropogon and Diectomis (Poaceae: Sacchareae) in Madagascar and the new Andropogon itremoensis from the Itremo Massif. Kew Bull. 68, 193–207.
Vörösmarty, C.J., McIntyre, P.B., Gessner, M.O., Dudgeon, D., Prusevich, A., Green, P., Glidden, S., Bunn, S.E., Sullivan, C.A., Liermann, C.R., Davies, P.M., 2010. Global threats to human water security and river biodiversity. Nature 468, 334.
Vörösmarty, C.J., Hoekstra, A.Y., Bunn, S.E., Conway, D., Gupta, J., 2015. Fresh water goes global. Science 349, 478–479.
Wachira, W., Jackson, C., Njoroge, P., 2015. Kenya Bird Map: an internet-based system for monitoring bird distribution and populations in Kenya. Scopus 34, 58–60.
West, S., Pateman, R., 2016. Recruiting and retaining participants in citizen science: what can be learned from the volunteering literature? Citizen Sci. Theor. Pract. 1, 15.
West, S., Pateman, R., 2017. How Could Citizen Science Support the Sustainable Development Goals? Stockholm Environment Institute, Sweden.
Wieczorek, J., Bloom, D., Guralnick, R., Blum, S., Döring, M., Giovanni, R., Robertson, T., Vieglais, D., 2012. Darwin core: an evolving community-developed biodiversity data standard. PLoS One 7, e29715.
Wilderman, C.C., Barron, A., Imgrund, L., 2004. In: Top down or bottom up? ALLARM's experience with two operational models for community science. Proceedings of the 4th National Water Quality Monitoring Council Conference, 17–20 May, Chattanooga, Tennessee.
Woodcock, B.A., Isaac, N.J.B., Bullock, J.M., Roy, D.B., Garthwaite, D.G., Crowe, A., Pywell, R.F., 2016. Impacts of neonicotinoid use on long-term population changes in wild bees in England. Nat. Commun. 7, 12459.
Wright, J.F., Furse, M.T., Moss, D., 1998. River classification using invertebrates: RIVPACS applications. Aquat. Conserv. 8, 617–631.
WWF, 2016. Living Planet Report. WWF, Gland, Switzerland.

CHAPTER SEVEN

A Replicated Network Approach to 'Big Data' in Ecology

Athen Ma*[,1], David A. Bohan[†], Elsa Canard[‡], Stéphane A.P. Derocles[†], Clare Gray[§,¶], Xueke Lu*[,||], Sarina Macfadyen[#], Gustavo Q. Romero, Pavel Kratina[§,1]**

*School of Electronic Engineering and Computer Science, Queen Mary University of London, London, United Kingdom
[†]UMR 1347 Agroécologie. AgroSup/UB/INRA, Pôle GESTion durable des ADventices, Dijon Cedex, France
[‡]INRA, Agrocampus-Ouest, Université de Rennes 1, UMR1349 IGEPP, Rennes Cedex, France
[§]School of Biological and Chemical Sciences, Queen Mary University of London, London, United Kingdom
[¶]Department of Life Sciences, Imperial College London, Ascot, Berkshire, United Kingdom
[||]School of Engineering, The University of Warwick, Coventry, United Kingdom
[#]CSIRO, Black Mountain, Acton, ACT, Australia
**Departamento de Biologia Animal, University of Campinas, Campinas, Brazil
[1]Corresponding authors: e-mail address: athen.ma@qmul.ac.uk; p.kratina@qmul.ac.uk

Contents

1. Introduction: A Need to Detect Ecosystem Change 226
   1.1 The Problem: Replication 228
   1.2 A Potential Solution: Replicated Networks Through Next-Generation Sequencing 231
2. Historical Perspective on Network Analysis 232
   2.1 Low-Level Network Properties: Common Metrics of Simple Networks 233
   2.2 Higher-Level Properties: Considerations for the Analysis of Complex Replicate Networks 238
   2.3 Dynamics: Assessing Disturbance in Complex Replicated Networks 240
3. Promising Future Avenues to 'Big Data', Network Analyses of Change 243
   3.1 Novel Food Web Profiling 244
   3.2 Null Models 246
   3.3 Weighted Networks 248
   3.4 Multilayer Networks 250
4. Conclusions 252
Acknowledgements 253
References 253
Further Reading 263

Advances in Ecological Research, Volume 59
ISSN 0065-2504
https://doi.org/10.1016/bs.aecr.2018.04.001

Abstract

Global environmental change is a pressing issue as evidenced by the rise of extreme weather conditions in many parts of the world, threatening the survival of vulnerable species and habitats. Effective monitoring of climatic and anthropogenic impacts is therefore critical to safeguarding ecosystems, and it would allow us to better understand their response to stressors and predict long-term impacts. Ecological networks provide a biomonitoring framework for examining the system-level response and functioning of an ecosystem, but have been, until recently, constrained by limited empirical data due to the laborious nature of their construction. Hence, most experimental designs have been confined to a single network or a small number of replicate networks, resulting in statistical uncertainty, low resolution, limited spatiotemporal scale and oversimplified assumptions.

Advances in data sampling and curation methodologies, such as next-generation sequencing (NGS) and the Internet 'Cloud', have facilitated the emergence of the 'Big Data' phenomenon in Ecology, enabling the construction of ecological networks to be carried out effectively and efficiently. This provides to ecologists an excellent opportunity to expand the way they study ecological networks. In particular, highly replicated networks are now within our grasp if new NGS technologies are combined with machine learning to develop network building methods. A replicated network approach will allow temporal and spatial variations embedded in the data to be taken into consideration, overcoming the limitations in the current 'single network' approach.

We are still at the embryonic stage in exploring replicated networks, and with these new opportunities we also face new challenges. In this chapter, we discuss some of these challenges and highlight potential approaches that will help us build and analyse replicated networks to better understand how complex ecosystems operate, and the services and functioning they provide, paving the way for deciphering ecological big data reliably in the future.

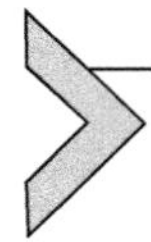

1. INTRODUCTION: A NEED TO DETECT ECOSYSTEM CHANGE

Large-scale monitoring of ecosystems has become ever more important in the face of global environmental change driven by the activity of man and changing climate. Current approaches to large-scale monitoring include the use of proxies or indicators, such as chemical indicators to evaluate the level of environmental pollution risks (e.g. Arshad and Martin, 2002; Schoenholtz et al., 2000) or biotic measures for assessing the condition of an ecosystem (e.g. Dale and Beyeler, 2001; Davies and Jackson, 2006). In the past 30 years, satellite- or aircraft-based remote sensing techniques have also led to radical changes to our understanding of the planet's environments, allowing large-scale and long-term processes, such as desertification, algal blooms and deforestation, to be followed (e.g. Lawton et al., 2001; Perennou et al., 2018; van der Werf et al., 2008). However, these two approaches cover the ecological extremes of what we would like to monitor. At one extreme, classical ecological biomonitoring is often limited to an individual species in a specific ecosystem; the data are therefore difficult to extrapolate to complex systems over large spatial scales. At the other extreme, remote sensing is explicitly large scale, covering many ecosystems and biomes simultaneously, but these data often lack ecological details and mechanisms that govern patterns of change.

Accelerating global and local changes prompt an urgent need to better understand responses of entire ecological networks and to predict the effects of future perturbations (Gray et al., 2014; Kratina et al., 2014). There are structural properties in ecological networks that reveal their underlying organisational rules and evolutionary and compensatory mechanisms in response to disturbance (Cohen, 1977; Lu et al., 2016; Woodward et al., 2010a,b). Some of these structures can be further linked to ecological functioning and services provided by an ecosystem (Thompson et al., 2012). Moreover, ecological network structures can be related to the long-term dynamics of an ecosystem, such as resilience and robustness (Dunne et al., 2002; Ledger et al., 2013; Memmott et al., 2004; Oliver et al., 2015). Network approaches to large-scale monitoring have the potential to provide forewarning of ecosystem degradation more informatively than existing biomonitoring and remote sensing (e.g. Dakos and Bascompte, 2014; Jiang et al., 2018).

In this middle ground between the ecological extremes of current biomonitoring and remote sensing approaches, a network-based approach to large scale, generic monitoring of ecosystems would be a valuable tool. We argue in support of networks as a generic analytical approach that can be applied in any domain of ecological research. Ecological networks can be examined for any ecosystem with little modification of either construction or analytical methodology. New molecular methods, for sampling species, and machine-learning approaches are starting to deliver the volumes of data necessary to monitor ecosystems. However, analytical methods that can be applied to such highly replicated network big data are little developed and their application lags behind the empirical advances. Here, we summarise the historic empirical research on ecological networks and the common analytical approaches used to analyse them. We highlight their assumptions and shortcomings in the face of emerging 'big' spatially and temporarily replicated network data. Finally, we review new promising avenues for future analyses, including the application of food web profiling, null models and weighted and multilayer networks. Overall we aim to summarise and highlight the new approaches to replicate big data in network ecology and their use in evaluating the changes and variation in network structure in order to detect and gauge environmental impact.

1.1 The Problem: Replication

The majority of past studies on ecological networks have been based on unreplicated sampling designs mainly because of the excessive research time and effort required to assemble complex, highly resolved networks. These 'single network' approaches have been used to study the structure and functioning of ecosystems, and much of our understanding of network topology has been characterised from such 'snapshot' data (Table 1). Networks were constructed from data aggregated across multiple time points and multiple sites. For example, Evans et al. (2013) quantified a network of multiple ecosystem services from a single organic farm in the United Kingdom. The farm network was comprised of 1502 unique interactions, and 560 taxa of plants and animals (Pocock et al., 2012). However, aggregation of data based on the presence or absence of species and expert knowledge of interactions can link taxa that do not occur at the same time and space and does not account for the fact that species pairs are located on a continuum of very weak to very strong interactions. Moreover, the assembly of a local community is often driven by local environmental conditions, species states, historical contingencies and stochastic assembly processes (Poisot et al., 2015). This means that an ecosystem can have multiple equilibrium states, with regional species pool (Ricklefs, 1987) and a diversity of interactions varying in time and space. Consequently, species cooccurrences in aggregated data do not always reveal an interaction between two species (Freilich et al., 2018; Olesen et al., 2011a,b) due to differences in phenology (Vázquez, 2005), ontogeny, consumer–resource body size ratios (Tsai et al., 2016) or low encounter rate (Canard et al., 2012). Some interactions may also be dependent on the presence of a third species or interaction between other species (Golubski and Abrams, 2011; Poisot et al., 2011a,b), or be limited to certain environmental conditions (Poisot et al., 2011a,b).

Replication of ecological networks is becoming more common, and it has the potential to improve the understanding of the magnitude of community changes following environmental perturbations (Tylianakis and Morris, 2017). There are recent examples of where replication has been used. For example, Tylianakis et al. (2007) pioneered the field by constructing

Table 1 Definitions of Common Terms Used in Network Analyses

Glossary	Definition
Background noise	Random natural variation in a network, which leads to the detection of false-positive connections. Replicated networks allow discriminating this background noise from the 'real' differences between networks caused by an ecosystem disturbance
Snapshot network	Single networks built on unreplicated sampling designs, often constructed from data aggregated across multiple time points and multiple sites
Network profiling	Techniques that search for patterns in networks. These methods characterise the frequency distribution of individual elements (e.g. network motifs) in order to identify their causes and consequences
Network backbone	Despite their typical link redundancy, networks seem to self-organise into robust, disassortative structures termed backbones. The backbone of a weighted network is composed from important individual nodes and their links and discriminates nodes and links (e.g. energy flux) that are insignificant in statistical sense
Core	A cohesive structure of closely interconnected nodes in a network. Large cores increase system redundancy by buffering external disturbance
Periphery	A structure of loosely connected nodes in a network
Higher-level system properties	General patterns, dynamics and properties of an ecological community that can be predicted without understanding all the underlying ecological details
Apparent competition	An increased abundance of one prey species negatively affects other prey species via increased abundance of shared predator

48 replicated networks along a gradient of habitat modification (from agricultural habitats to forests) and assessed how the disturbance and modification of habitats by humans affected the structure of species interactions as well as ecosystem function. They showed that species richness remained constant, but that the vulnerability of the networks decreased even while community evenness increased along the gradient of habitat modification. Morris et al. (2015) used a similar approach to construct 20 replicated networks of cavity-nesting Hymenoptera and their parasitoids and kleptoparasites along

an environmental gradient to demonstrate the effect of elevation on network structures. Dézerald et al. (2013) constructed 365 independent networks of metazoans inhabiting natural microecosystems (bromeliads) in tropical forests in French Guiana and showed that key metrics, such as linkage density and nestedness, varied across a gradient of canopy openness. Nielsen and Totland (2014) showed that the structural properties of mutualistic plant–pollinator networks in heavily managed boreal forests were conserved after habitat degradation, highlighting the resilience of species interactions to external disturbance.

Construction of replicated ecological networks is often constrained by the sampling effort required, as they are currently assembled in the same way as nonreplicated networks, but with replication across a range of environmental conditions (Morris et al., 2015; Nielsen and Totland, 2014; Tylianakis et al., 2007). This is highly resources costly and laborious. However, advances in computing, storage and processing that have occurred alongside the expansion of the Internet, and machine-learning approaches will increasingly result in more 'big data' becoming available (see Buttigieg et al., 2018). Big data is the 'catch-all' name given to very large databases that are accessed and analysed using bioinformatics approaches. In some cases, the analysis of biological big data has taken the form of simple data mining, looking for obvious correlations between biological variables. More recently, sophisticated approaches to hypothesising and testing relationships from big data have emerged. For example, statistical and logical machine-learning approaches have been used to build metagenomic and metabolomics interaction networks to identify genes or gene products associated with cancer or cell growth (e.g. Kourou et al., 2015). These techniques have also been applied to build ecological networks (e.g. Bohan et al., 2011a; Tamaddoni-Nezhad et al., 2013), both from classical ecological sample data and molecular ecological data, and will pave the way for generating replicated networks more quickly and economically.

Researchers have traditionally focussed on resolving networks that are relevant to the delivery of ecosystem services (e.g. pest control, pollination or nutrient cycling). In many communities, highly abundant species are key to the delivery of a service, and as a result, less attention has been paid to documenting rare species and their interactions. These rare species can fulfil essential ecosystem functions due to their rare trait combination (Mouillot et al., 2013). As we move towards more sophisticated approaches for gathering large amounts of empirical data (e.g. sequencing of environmental samples), it is envisioned that there will be higher levels of variation present in the data as more rare species and their interactions are recorded. With the emergence of big data in ecology, the research challenge will be to process the vast quantities of data that are being produced and to interpret their ecological significance (Woodward et al., 2014). This calls for more effective analytical tools for examining replicated ecological networks (e.g. Poisot et al., 2012) to complement the development and advances in ecological big data.

1.2 A Potential Solution: Replicated Networks Through Next-Generation Sequencing

A replicated network approach would allow a better understanding of how community structures, ecosystem functions and disturbance are interlinked, and we have already seen their advantages in a small number of studies. In 64 plant–pollinator networks, Kaiser-Bunbury et al. (2017) examined the reproductive performance of the 10 most abundant plant species to describe pollination across 4 restored and 4 unrestored mountaintop communities. The study showed that the estimates of interaction specialisation (Blüthgen et al., 2006) were lower in restored networks than in unrestored networks, suggesting that pollinators mediate less specialised interactions in the restored environments. More importantly, the differences in network structure reflect direct positive effects on pollination through fruit production of native plants. There are other ecological functions and associated ecosystem services that are more complex to examine than parasitism or pollination, such as the case of weed regulation by carabid beetles (Bohan et al., 2011b). This ecosystem service, as an example, relies on prey–predator interactions between weed seeds and carabid beetles that are very difficult to observe in nature and therefore would benefit from advanced molecular tools.

New molecular tools will provide an excellent opportunity to build ecological networks rapidly and efficiently, once methods are optimised and appropriate bioinformatics are developed and applied (Derocles et al., 2018; Evans et al., 2016). Most of the molecular methods available are based on high-throughput parallel sequencing, called 'next-generation sequencing' (NGS), where a whole community from various kinds of field samples is screened and analysed (e.g. environmental DNA obtained from samples of soil or water, mixture of cells or tissues from traps, such as arthropod pitfall traps). DNA-based approaches do not rely on taxonomic expertise, and some of these techniques can produce millions of DNA sequences within a relatively short time frame (Evans et al., 2016), and these approaches have advantages of being much faster than existing biomonitoring methods (Ji et al., 2013).

While researchers are starting to use NGS to construct replicated networks, current lack of replication means that it is not possible to identify or filter out the variation (noise) generated during the network assembly process, and this can potentially lead to an inaccurate interpretation of the results. In particular, NGS requires a bioinformatics pipeline for species identification (Kitson et al., 2016; Toju et al., 2013). Because accuracy of the identification strongly depends on the quality of DNA sequences and the completeness of the species database (Evans et al., 2016), any sample contamination can adversely affect the results (Piñol et al., 2014).

As we move towards gathering multiple datasets for a replicated network approach, utilising NGS will accelerate the generation of ecological big data and the construction of large numbers of networks, efficiently. A major ecological advantage of this is that it will be possible to discriminate the natural and structural variation encapsulated in replicates of a network from the variation due to disturbance, both allowing the 'real' differences among the ecological networks to be evaluated alongside ecosystem responses to perturbation. This is the first step to a next generation of biomonitoring that allows understanding and prediction of how environmental change affects ecological networks and ecosystems through time and space.

2. HISTORICAL PERSPECTIVE ON NETWORK ANALYSIS

A key objective in network ecology has been to relate network structure to the properties and dynamics of an ecological system so as to better understand the underlying ecological meanings and implications of environmental disturbances. For instance, changes in precipitation and temperature regimes alter the strength of biotic interactions and reorganise the structure of networks (Blois et al., 2013; Woodward et al., 2016). Flooding can also alter key network properties, reducing taxon richness, food-chain length, and increasing the proportion of species at basal trophic level (McLaughlin et al., 2013). Petchey et al. (2010) examined the effect of temperature on ecological networks by incorporating the relative activation energies of attack rates and handling times, and showed how diet breadth is linked to network connectance (i.e. the number of realised links as fraction of all possible). Networks can also make explicit indirect interactions that govern nonintuitive system behaviours (Rossberg, 2013) and alter the whole ecosystems (e.g. Montoya et al., 2009). A more mechanistic understanding of networks is then essential for successful biodiversity management and restoration.

Historically, mechanistic analyses have focused on simple networks, where all network parameters have been measured, analysed analytically and simulated. These simple networks have included isolated and remote ecosystems (e.g. Bear Island in the high Arctic), human-simplified networks with several dominant species (e.g. crop-pest) or networks that aggregated many species into several functional groups (Summerhayes and Elton, 1923). Big data has led to the recognition that ecological communities form complex networks with many sensitive parameters that can make deterministic modelling very difficult. Whereas for simple, unreplicated networks we can analytically track links among all species in a community (low-level properties), for complex and replicated networks require characterisation by what have been termed 'higher-level system properties'. This latter type of metric describes the general patterns, dynamics and properties of an ecological community that can be predicted without understanding all the underlying ecological details.

2.1 Low-Level Network Properties: Common Metrics of Simple Networks

There are numerous metrics for characterising networks (Costa et al., 2007) that have been applied to ecological networks (Bersier et al., 2002; Blüthgen et al., 2008; Ings et al., 2009; Thompson et al., 2012; Tylianakis et al., 2010; Vázquez et al., 2009). Linkage density describes the number of interaction links per node, and it indicates a diet specialisation averaged across the entire trophic network (Tylianakis et al., 2007). Connectance evaluates how many of the possible links in a network are present. Connectance has been found to increase with more generalised foraging in a community (Van Veen et al., 2008), and it has been linked to robustness of an ecosystem and the response of a network to perturbations (Blüthgen, 2010; Briand, 1983). The trophic similarity among nodes (e.g. species) and links (e.g. feeding pathways) has been used as a measure of redundancy (Cohen and Briand, 1984). The efficiency of a network characterises how closely are nodes connected to each other and the distribution of those connections in a network (Latora and Marchiori, 2001). The cumulative frequency (degree) distribution represents the probability of finding highly connected species (Ledger et al., 2012b, 2013; Montoya et al., 2009). The topological position of nodes, or groups of nodes sharing common biological traits, can be used to identify those that contribute most to network structure (e.g. Guimerà and Nunes Amaral, 2005; Olesen et al., 2007). These and other metrics have typically been used to characterise individual 'snapshot' networks. However, there are recent examples of the analysis of replicated networks using these metrics in order to identify the effect of disturbance (Box 1).

BOX 1 A Case Study Illustrating an Application of Common, Low-Level Food Web Metrics to Analyses of Spatially and Temporarily Replicated Network Data.

Current analysis of monitoring data obtained from replicated designs often relies on complex statistical methods to establish overall trends in structural responses (e.g. Gray et al., 2016; Lu et al., 2016; O'Gorman and Emmerson, 2010; O'Gorman et al., 2012). Gray et al. (2016) explored how the Upland Waters Monitoring Network (UWMN) of 23 lake and river sites distributed across the UK, recovered from acidification over 24 years. Considerable variation in the network response, among individual sites with differences in local environmental and climatic conditions and chemical and biological recovery rate, was found. Within individual sites there was noticeable variation in network structure from year to year, often obscuring overall trends. Large variation was found even among the sites that were not subject to changes in their acidity. Some sites that were not recovering from acidification showed significant monotonic changes in their network structure, while others experiencing dramatic changes in their acidity exhibited no observable monotonic change in their network structure.

Systematic changes in the network properties were only found when all the sites were analysed together, using appropriate analyses to account for pseudoreplication in the data. The full complexity of 13 acidity-related hydrochemical variables at each site in each year was reduced down to one acidity gradient through Principal Component Analysis, which was then used to investigate the presence of any trends in network structure. While considerable variation in network structure was present across all the food webs, significant trends did emerge (Fig. 1). There was an indication that biological recovery lagged behind chemical recovery, as most sites displayed either biological and chemical recovery or chemical but no biological recovery. Very few sites exhibited change in their food web structure in the absence of directional change in their acidity.

Although the large sample size of this analysis revealed trends across broad environmental gradients and across time, the food webs were limited by the manner in which they were constructed. They were built using 'inferred' feeding links harvested from the literature. The use of inferred feeding links in food web studies has been criticised on the basis that they might overestimate diet breadth and fail to detect behavioural differences between sites (Raffaelli, 2006), which might drive important changes to network structure. However, given the nature of

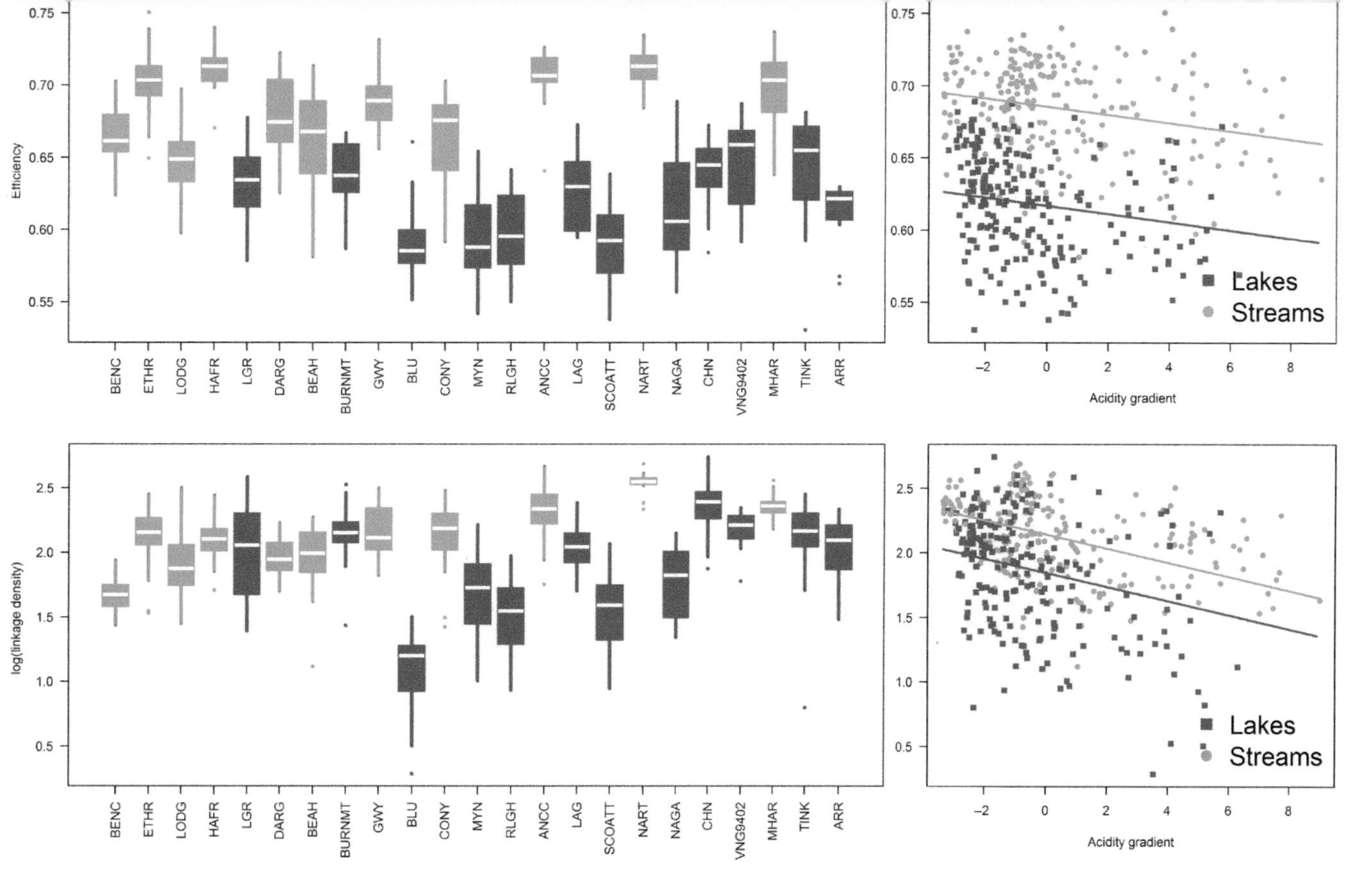

Fig. 1 The variation in network efficiency (*top panel*), linkage density (*middle panel*) and redundancy (*bottom panel*) across the sites in the UWMN. Sites are arranged by their initial acidified state, such that those sites with the lowest pH at the beginning of monitoring are to the *left*, while those with more circumneutral pH at the beginning of monitoring are to the *right*. Acidity among sites is represented by a general acidity gradient derived from the Principle Component Analysis (Gray et al., 2016).

(Continued)

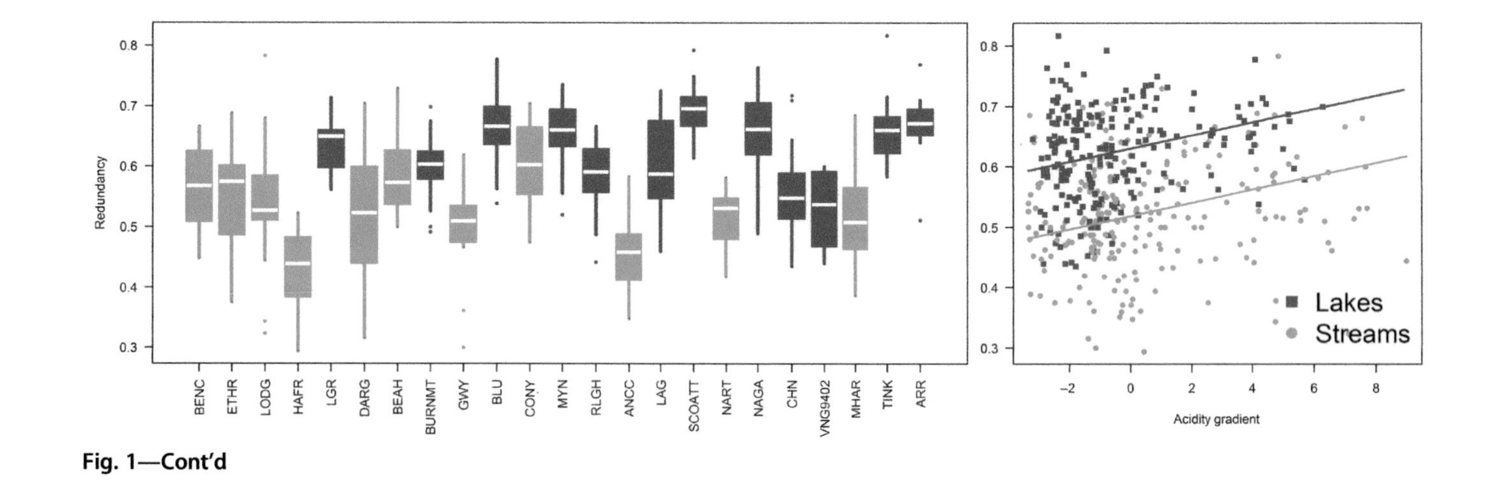
Redundancy
0.8
0.7
0.6
0.5
0.4
0.3
BENC
ETHR
LODG
HAFR
LGR
DARG
BEAH
BURNMT
GWY
BLU
CONY
MYN
RLGH
ANCC
LAG
SCOATT
NART
NAGA
CHN
VNG9402
MHAR
TINK
ARR
Lakes
Streams
−2
0
2
4
6
8
Acidity gradient

Fig. 1—Cont'd

building summary food webs in this manner, that they tend to overestimate interactions between species, they are more likely to be insensitive to environmental change rather than reveal erroneous trends. This is because it is more likely that the structure of summary food webs would be conserved, given that any changes in structure will be wholly or in part driven by changes in species composition rather than feeding behaviour. This means that the observed trends are likely to be broadly realistic. Such observational data are often well suited to the identification of trends in network or ecosystem change, in order to formulate formal hypotheses and design the accompanying experiments.

Given a known distribution of species and interactions within a network, the behaviour of the whole community can be described (Ings et al., 2009; Lewinsohn et al., 2006). 'Modular' structures, where groups of nodes interact among themselves more densely than with other nodes, have been reported for antagonistic networks. Modularity or compartmentalisation is common among abundant microorganisms that dominate energy transfer in soil ecosystems (Mulder et al., 2006). Composition of the modules and the distribution of 'redundant' links among modules would indicate the likelihood of cascading effects, whereby changes in one module or network propagate to the next in a 'domino-like' manner (Krause et al., 2003; Stouffer and Bascompte, 2011).

'Nested' structuring has often emerged from mutualistic networks, like plant–pollinator webs (Thebault and Fontaine, 2010). Nestedness describes networks where nodes with few connections tend to be linked to a subset of nodes interacting with more connected nodes (Bascompte et al., 2003). Consequently, most interactions appear asymmetric and organised around a group of highly interconnected nodes. However, both modular structure in mutualistic networks (Olesen et al., 2007) and nested structure in antagonistic networks (Cagnolo et al., 2011) have been reported. Profound nestedness has also been observed in randomly generated networks, and it has been argued that high nestedness therefore should not necessarily be taken as evidence for specialisation (Blüthgen, 2010; Blüthgen et al., 2007).

Common metrics and population dynamic models have been used in combination to examine the stability of simple networks (e.g. Layer et al., 2010, 2011). Nested patterns have been suggested to stabilise communities (Bastolla et al., 2009), though this approach has faced criticism of its underlying assumptions (Blüthgen, 2010), particularly those related to functional responses and interspecific competition. More recently, James et al. (2009) have shown that connectance, rather than nestedness, might better explain variation in stability. Despite this ongoing debate, the underlying logic is identical to that proposed in the 'insurance hypothesis' of positive biodiversity effects (Yachi and Loreau, 1999). Higher connectance is associated with higher functional redundancy, which may dampen the functional consequences of stochastic variations or loss of one or a few species (Reiss et al., 2009).

2.2 Higher-Level Properties: Considerations for the Analysis of Complex Replicate Networks

While analyses of simple 'snapshot' networks have provided a framework to assess the effects of environmental change or disturbance on ecosystem structure and function, such approaches rely on assumptions that are not always fully considered. First, the construction of ecological networks is largely carried out by pooling all interactions identified at several dates and in several locations into a single network, assuming that species interactions are invariant over time (at least for the duration of the study) and space. However, Derocles et al. (2014) showed that the structure of host–parasitoid food webs is much simpler at the local scale within a single field than at the global scale across the landscape, which included all crop fields of interest and uncultivated environments within the area. Despite this discrepancy, most analyses use aggregated data that overlooks the spatial variability in the landscape, which can significantly alter an ecological network structure (Tylianakis and Morris, 2017).

Second, pooling data across different habitats into a single network assumes the equal distribution of species and their interactions throughout the space. However, this may not always be true as there exist species that can have a disproportionally high number of interactions with other species only in some habitats (Evans et al., 2013). This assumption can therefore lead to inaccurate conclusions about ecological processes and the associated ecosystem services (e.g. natural pest control in Derocles et al., 2014). For instance, when two species share a common natural enemy, the presence of one species may negatively impact upon the dynamics of the other through an increase in predation risk or apparent competition (Holt, 1977; Morris et al., 2004; Muller and Godfray, 1997). In this context, an ecological network approach would allow the identification of potential cases of apparent competition through the detection of common natural enemies shared by prey species. This must be done cautiously as the focal food web might otherwise become an assemblage of multiple interactions (potentially disconnected either in time, space or both), rather than an accurate representation of the ecosystem. In fragmented ecosystems, in particular, at least some of the species studied may not be able to move freely between habitats. Barriers in the landscape, which impede free movement of individuals, would therefore be ignored when a single network is built using data from several different habitats.

Third, robustness analysis based on simulating species loss assumes that secondary extinctions are only driven by the direct negative effect of primary extinctions, which have bottom-up effects on a network (Barbosa et al., 2017; Sanders and van Veen, 2012). However, such cascading extinctions can also be the result of much more complex effects, sometimes driven by a top-down control (Sanders et al., 2015) or indirect effect such as apparent competition (Holt, 1977; Morris et al., 2004; Muller and Godfray, 1997). Such top-down control or indirect effects require information on species dynamics and so are less likely to be detected with 'snapshots' of the ecosystem structure that contain a single replicate network alone.

Empirical evidence for how ecological networks vary through time and space is still sparse. It is often unknown whether differences observed in network structure are due to an ecosystem disturbance or are a consequence of temporal and spatial dynamics. Evaluating such dynamics in network analyses would greatly enhance our understanding of ecosystem functioning (Heleno et al., 2014). While the robustness analysis performed on 'snapshot' networks relies on species extinction to identify cascading secondary extinction, it has been demonstrated that incorporating dynamics in ecological networks can have a significant impact on the robustness of the networks, as the ecological function of a species can disappear even before the species itself goes extinct (Säterberg et al., 2013). Therefore, cascading secondary extinctions may result from a simple decrease in the abundance of a given species because of the prey–predator dynamics constantly taking place in the ecosystem. Indeed, dynamical models tested both on theoretical and observed networks have shown that the species most likely to go extinct first are not those with increased mortality rates, i.e., those species that would be considered as 'primary extinction' in classical robustness analysis (Srinivasan et al., 2007). At the moment, the incorporation of spatial and temporal dynamics into the network ecology is still in its infancy (Tylianakis and Morris, 2017).

2.3 Dynamics: Assessing Disturbance in Complex Replicated Networks

Analyses of complex networks have been increasingly used to examine the community-wide impacts of external disturbance. Degree distribution, modularity and the inherently nested structure of mutualistic networks appear to confer resistance to perturbations (Fortuna and Bascompte, 2006). Analyses of pollination networks indicate a more rapid loss of interactions than of species, via extinction, following habitat destruction (Aizen et al., 2012). This would imply that rare plants can better resist habitat destruction than common plants that, in turn, support a large numbers of generalist pollinators (Aizen et al., 2012). However, despite there being this type of recurring response of networks to perturbation, the effects of global and regional changes on network structure are often variable and depend on local environmental conditions, and can include interactive effects of multiple stressors (Tylianakis et al., 2008). Moreover, the heterogeneous organisation of complex ecological networks is yet to be fully understood as novel network structures and their relationship with ecosystem stability and functioning are still emerging (Allesina and Pascual, 2008; Garcia-Domingo and Saldaña, 2008; Lu et al., 2016; Lurgi et al., 2012; Stouffer and Bascompte, 2011).

Certain network structures appear to be relatively little influenced by disturbance, due to nonrandom, scale-free distributions of links (Albert et al., 2000; Barabasi, 2009; Parrott, 2010). Such structured networks may include a few hubs, each having disproportionally high number of connections and potentially 'small-world' properties, where the mean path length between any two nodes is shorter than expected by chance (Watts, 1999; Watts and Strogatz, 1998). Scale-free and small-world mycorrhizal networks, as found in Douglas-fir forests (Beiler et al., 2010), are relatively resistant to random perturbations (removal or death of individual trees) because of the small probability of losing the whole hub. In contrast, the targeted removal of hub trees, via logging or pest infestation, would substantially fragment the mycorrhizal network and slowdown forest regeneration (Beiler et al., 2010; Bray, 2003).

Several studies have used complex networks, replicated in time and/or space, to determine the impacts of environmental disturbance (see Table 2 for a summary of approaches). The measurement of disturbance or management intervention is either explicitly incorporated into a survey design or is assessed across an implicit gradient. Commonly, network robustness analyses are used to determine how a given network attribute or property is

Table 2 Examples of the Different Ways That Disturbance Has Been Assessed in Networks, With Examples From Ecological Network Studies and Their Replication

Assessment	Examples	Type of Replication
Statistical testing of replicate networks collected in different states (sometimes due to an experimental treatment)	Pest-herbivore food webs on organic and conventional farms (Macfadyen et al., 2009) Networks from different habitats—plantation vs native forests in New Zealand (Peralta et al., 2017) Food webs in managed and plantation forests in New Zealand, before and after removal of a herbivore using a pesticide spray (Frost et al., 2016)	Statistical—multiple networks in the same state Usually temporally unreplicated
Statistical testing of networks collected across time (before and after a disturbance event)	Socioecological network of the West Cape ostrich industry before and after collapse due to avian influenza (Moore et al., 2016)	Usually spatially unreplicated
Introducing new species into ecosystems and assessing the network structural changes across time	Invasive species studies (Emer et al., 2016) Mutualistic interaction networks from paired temperate forests and paired oceanic islands which differ in the amount of alien species (Aizen et al., 2008)	Multiple sites with and without the introduced species (or paired sites)
Collecting networks across a gradient of habitats (e.g. land-use change) and inferring how these may have changed in the past in response to a disturbance	Food webs collected across a deforestation gradient or land-use intensity gradient, across five habitats (Laliberté and Tylianakis, 2010; Tylianakis and Morris, 2017; Tylianakis et al., 2007)	Multiple sites along a gradient sometimes replicate sites at each gradient level Usually temporally unreplicated, although there may be a temporal component to the gradient (e.g. succession)

Continued

Table 2 Examples of the Different Ways That Disturbance Has Been Assessed in Networks, With Examples From Ecological Network Studies and Their Replication—cont'd

Assessment	Examples	Type of Replication
Simulating disturbance on empirical networks by removing species or links	Functional extinctions in eight natural food webs by manipulating in silico species mortality rates (Säterberg et al., 2013) Simulated parasitoid species removal on farm food webs and examined herbivores released from control (Macfadyen et al., 2011) Four species extinction scenarios applied to detritus-based riverine food webs (Calizza et al., 2015) Extinction of links from alpine pollination networks (Santamaría et al., 2016) Species extinction and habitat loss scenarios applied to an agricultural network of ecological networks (Evans et al., 2013) Simulated species loss in freshwater food webs to examine the effect of drought (Lu et al., 2016)	Same as the underlying empirical dataset. Sometimes empirical data is related to a null model
Simulating disturbance on generated networks by removing species or links	Generated model food webs with 50 interacting species (Säterberg et al., 2013)	Replicated in a theoretical sense, usually by creating a null network through random restructuring of the rows and/or columns in a data matrix

altered as species or functional groups are removed from the network (e.g. Genini et al., 2010). Studies have examined the effects of primary species loss by measuring the rate of cascading secondary extinctions induced as consumers are left without resources and thus go extinct themselves, and so on (Memmott et al., 2004; Montoya et al., 2006; Pocock et al., 2012). Dunne et al. (2002) simulated cascading extinctions by removing species from 16 food webs and quantified the number of secondary extinctions that followed. The scenarios included random removal of species, removing species with the most links to other species or removing species with the least links to other species. When similar approaches were applied to food webs in detritus-based systems, it was perhaps unsurprising that the most connected species were the most important for food web resistance (Calizza et al., 2015).

The more recent development of meta-networks has started to incorporate the idea of spatial movement as a way of compensating for local species extinction within regional species pools that persist. For example, Mougi (2017) found that a moderate level of spatial coupling between habitats can attenuate some disturbance events due to species emigration. While the complexity and realism of network approaches have increased (e.g. Säterberg et al., 2013), some fundamental limitations still remain. In particular, much of the contribution of individual species dynamics to the network have been ignored, and we cannot yet predict either what roles species remaining in networks might assume or those functions might be maintained in the face of species extinctions (e.g. adaptive trophic behaviour, Valdovinos et al., 2010).

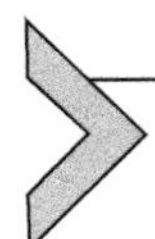

3. PROMISING FUTURE AVENUES TO 'BIG DATA', NETWORK ANALYSES OF CHANGE

The analysis of large ecological datasets inevitably comes with attendant challenges, such as sampling consistency, meta-data collection and curation (Raffaelli et al., 2014), and the choice of analytical method. However, much attention is required when dealing with large, replicated network datasets to avoid certain pitfalls. For example, given a sufficiently large sample, a statistical test will almost always exhibit a significant difference, unless the effect size is exactly zero. Very small differences between samples, however, even if significant, may not indicate meaningful change in ecosystem structure (Sullivan and Feinn, 2012). Trends found in large observational datasets may be slight, but may also be accompanied by large variation (e.g. Firbank et al., 2003). Unlike significance tests, effect size is independent of and therefore not confounded by sample size, all other things being equal (Sullivan and Feinn, 2012). Nevertheless, the appropriate use of statistical significance has allowed the establishment of trends in the structural responses of ecological networks (e.g. Gray et al., 2016; Lu et al., 2016; O'Gorman and Emmerson, 2010; O'Gorman et al., 2012).

Gauging structural variation across a collection of networks, in ecology and in other disciplines, is not always straightforward due to the complexity and scalability of the analysis. Methods do exist for ecologists to compare networks, with reference to both species composition and interactions (e.g. Poisot et al., 2012), and these approaches have notably been applied to plant–pollination networks (Carstensen et al., 2014; Trojelsgaard et al., 2015). However, these methods do not assess higher-level structural properties. In the following subsections, we highlight some advances in statistical and structural profiling techniques which have been used to examine variation in networks and that provide a potential roadmap to new tools for analysing ecological big data in the future.

3.1 Novel Food Web Profiling

The emergence of big data in ecology needs to be complemented by a comprehensive range of network characterisation metrics and profiling methods in order to gauge the nature and magnitude of change in structural properties in the face of external disturbance. Numerous studies have shown that simple food web measures, such as richness and connectance, are limited to capturing the overall global properties of ecological networks and are inadequate for reflecting the multiscale nature of their organisation (Ledger et al., 2012a; Melián and Bascompte, 2004; Srinivasan et al., 2007; Woodward et al., 2010a,b). The heterogeneous organisation of species interactions leads to the formation of structures within larger networks across multiple scales, which appear to be crucial to ecosystem response and functioning. Network motifs, which are small subnetworks of repeatable design, have been argued to be the fundamental building blocks of assembly of food webs (Bascompte and Stouffer, 2009; Milo, 2002), whereby their local stability can have a bottom-up effect to the overall stability of the networks (Allesina and Pascual, 2008; Garcia-Domingo and Saldaña, 2008). Similarly, the compartmentalisation of food webs has been found to be effective in isolating the spread of perturbations and hence help confer persistence and stability (e.g. Krause et al., 2003; Lu et al., 2016; Stouffer and Bascompte, 2011).

Complex networks are often highly variable in their structural organisation (Ahmed and Xing, 2009). There has been significant effort expended in trying to develop new techniques that extract from network variation meaningful properties that improve our understanding of network structure and dynamics. This effort has lead to the detection of characteristic motifs, clusters and backbones in interaction networks (Table 1). Profiling techniques characterise patterns in the frequency distribution of these elements in order to identify their causes and consequences. We have already seen success in the application of network metrics and profiling techniques from other scientific disciplines to ecological questions that then reveal novel patterns in species interactions across different scales (Bascompte and Stouffer, 2009; Leger et al., 2015), but the scope for cross-fertilisation is still enormous. Techniques based upon network efficiency (Latora and Marchiori, 2001), which can be defined as the reachability of nodes within a network computed from the shortest paths, might reflect the rate at which ecological information might diffuse across a whole network but have still to be applied widely in ecology.

Lu et al. (2016) examined the impact of drought on freshwater streams using the core profiling method (Ma and Mondragón, 2015). The analysis identified a substructure of highly connected species, a core profile, that potentially provides redundancy or buffering of the effects of perturbations to the networks. Less connected, peripheral species (periphery profile) were subject to higher levels of extinction in the event of drought. Importantly, link density within the core changed during this process, but the lower-level network property of overall connectance was unaffected. This result highlighted the different levels of response to perturbation at different structural levels within an ecosystem. Peripheral trophic specialists are more prone to extinction due to their inability to persist in harsher conditions. On the other hand, there is evidence that a more 'stable' part of the ecosystem, which consists of high-degree species (e.g. generalist consumers) with a wide niche selection, allows them to survive in less favourable environments (Chase, 2007; Chase and Leibold, 2003), undergoing a compensatory process and rebalancing the structural configuration of the ecosystem. The key question here is 'what would happen if the environmental conditions continue to degrade?' Would there be a critical point in which even the stable core of the ecosystem could no longer mitigate and withstand the effects of perturbations? These findings echo the view that there are more sophisticated topological features of ecological networks that have yet to be discovered, beyond low-level richness and connectance, which account for their responses to disturbance (Lurgi et al., 2012).

3.2 Null Models

Null models are increasingly being used to assess the significance of ecological patterns or to test specific ecological hypotheses (e.g. Toju et al., 2014; Ulrich and Gotelli, 2007, 2010). Broadly, there are two main approaches to a null model analysis that can be classed as either: (i) analyses of statistical significance when tested against a null hypothesis, defined by an appropriate statistical distribution or (ii) analyses that evaluate how much the pattern observed in the network data deviate from randomisations of the empirical data, with constraints imposed in order to preserve certain features of the empirical data. Significance can be measured either directly or indirectly. The direct method examines the proportion of the null distribution that is more extreme (on either tail) than the empirical value of the test statistic. Indirect significance testing compares (benchmarks) the empirical test statistic to the mean of the null distribution (Veech, 2012).

In network ecology, null models help to disentangle patterns of the network created by a mechanism of interest from patterns due to other mechanisms, often 'neutral' mechanisms—for instance relative abundance of species or sampling effect. Null model using randomisations of the empirical data are particularly appropriate to the study of ecological networks, as the impact of a unique mechanism on the complex network structure is often hard to anticipate a priori. The links of an empirical network are reshuffled to generate an ensemble of randomised networks (e.g. Solé et al., 2002; Vázquez et al., 2009), while conserving certain aspect of the empirical network. A null hypothesis is then used to examine how patterns of interaction exhibited by the empirical network differs from the random chance patterns, given the constraints determined by the researcher. The test of significance is performed by comparing test statistics from the empirical network to the mean value obtained from the ensemble of networks, usually given in the form of a z-score, which deviates incrementally from zero with statistical significance.

Numerous studies have used null models to benchmark the statistical significance of low-level structural properties, such as compartmentalisation, modularity and substructuring. For instance, Vázquez and Aizen (2003) examined the patterns of specialisation in pollination systems using a null model and randomisation. For all analyses, pollinator species were represented in rows and plant species in columns in a binary interaction matrix. During the randomisation of the empirical data, connectance was constrained, while the total number of interactions among the pollinator and plant species were randomly reshuffled. The results showed that the number of extreme specialists and extreme generalists in the data was significantly higher than expected under the null model. Stouffer and Bascompte (2011) applied a null model, in much the same way, to examine the effect of compartmentalisation on ecological networks by randomising the allocation of species to different compartments in the simulations, while conserving the number of compartments and their size in the network. Their analysis demonstrated that food web compartments reduce the propagation of extinctions throughout the community and that compartmentalisation enhances food web persistence (Stouffer and Bascompte, 2011). Also, some networks metrics themselves are very sensitive to the other properties, and null models are used to calculate a relative value of the metric. Nestedness, for example, is very sensitive to the size of the network, and the computation of a relative value is recommended for comparing networks (Bascompte et al., 2003).

Null models have been also used to assess the impact of drought on replicated freshwater food webs (Lu et al., 2016). Here, a null model was employed to identify functionally critical parts of the network (highly interconnected species that form the web core) that govern the ecosystem response. The link density in pairs of food webs, one undisturbed (control) and one drought-disturbed, was then compared to their null expectations (Fig. 2). The link density in the cores of both food webs differed significantly

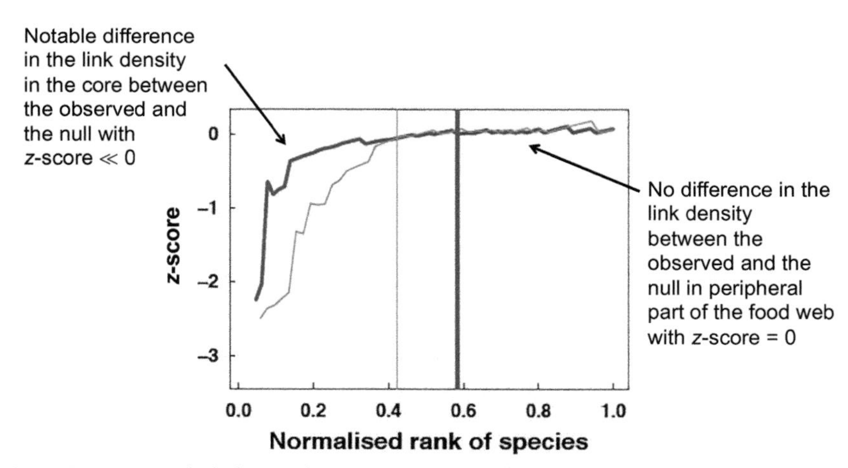

Fig. 2 Deviance in link density between a control (*dark thick line*) and drought (*light thin line*) web pair from their respective null models. Nodes were ordered by degree which were then *normalised* by the size of the network. Boundaries of the cores are marked by the respective *vertical lines*. The more negative *z*-scores in the cores reflect greater deviance from the null model.

from their respective null models (Fig. 2, left side), and the difference in link density between the web pair within the core was used to evaluate structural network change caused by drought. There were, however, no measureable changes in the link densities of the less connected species forming the periphery of the networks (Fig. 2, right side), which was not significantly different from a random process, suggesting both a high degree of variation and uncertainty in the network data.

3.3 Weighted Networks

Weighted networks (Barrat et al., 2004) have been used to characterise the 'strength' of individual interactions in ecological networks (Allesina et al., 2006; Bersier et al., 2002; Tylianakis et al., 2007; Ulanowicz et al., 2014). The magnitude of any link is specified by, but not limited to, biomass and energy fluxes (Thompson et al., 2012; Zhao et al., 2016). For example, one can represent nodes as species, edges as interactions and the weight of individual edges can be quantified by the frequency or likelihood of the given interaction across different replicates (Fig. 3A). Molecular data could then be used to provide information about the strength of individual interactions by pooling or aggregating individual samples into a weighted network. Building

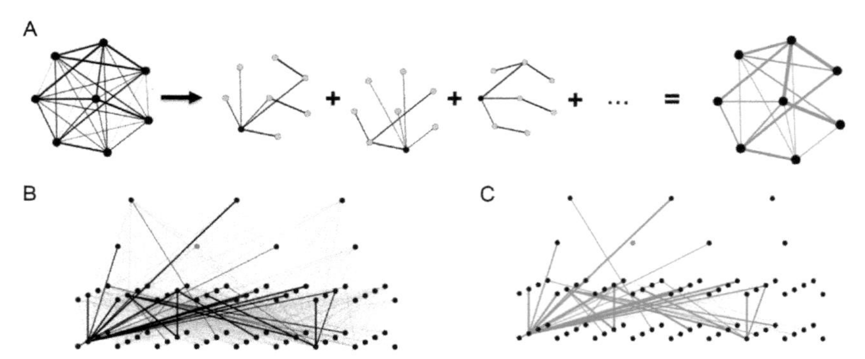

Fig. 3 (A) Devising the salient part of a weighted network whereby shortest path trees obtained from individual nodes are aggregated to define the salient links (in *red*). *Left*: A fully connected network with randomly assigned link weight. *Middle*: The shortest paths for each individual node to the rest of the network, where the thickness of each link represents the number of times it appears on each shortest path. *Right*: A consolidated graph by adding up each individual subplot. (B) A weighted network obtained by aggregated four (control) replicated webs from the Millstream dataset (Ledger et al., 2012a). The weight of an edge is proportional to the frequency that particular interaction has been observed across different replicates. (C) As in (B), but the weighted network now consists of the salient links (in *red*) only.

a network in this way means that the natural variation among the individual replicates is embedded in the resulting weighted network. Many simple but well-established metrics, such as centrality and clustering coefficients, which have already been applied to ecological networks (e.g. Dunne et al., 2002; Jordan, 2009; Stouffer et al., 2005), also have weighted counterparts (Barrat et al., 2004; Boccaletti et al., 2006), making their application to large scale, replicated networks relatively straightforward. In addition, many profiling techniques have been developed to depict high-level structural properties in weighted networks, including substructures (Csermely et al., 2013; Ma and Mondragón, 2015; Rombach et al., 2014; Rossa et al., 2013) and network backbones (Grady et al., 2012; Serrano et al., 2009).

The backbone of a weighted network can be used to discriminate nodes and links that are insignificant, in a statistical sense (Grady et al., 2012; Serrano et al., 2009). For the analysis of replicated ecological networks, this could help identify those critical nodes and links that are key for the transfer of fluxes from other parts of the network from those that are indistinguishable from noise. These methods would also lead to a fuller understanding of an ecosystem response by differentiating ecologically important changes from temporary fluctuations. For example, local environmental conditions in different sites can vary naturally and phenotypically plastic species might colonise certain sites where these conditions are in their favour. This can result in multiple stable states in an ecosystem due to a combination of environmental variation, the stochastic nature of community assembly and the invasion history in the locality (Chase, 2003, 2007). When the environment becomes harsher, these flexible species may then go extinct, but given that their occurrence and interactions within the rest of the ecosystem are sporadic, the expectation is that opportunistic species should fall outside the backbone of a network.

One way to define the backbone of a weighted network is to consider both the global and local importance of individual nodes and their links (Serrano et al., 2009). Here, the global property of a node is determined by its degree, and the local property is based on a node's connection strength to its immediate neighbours. For a given node with a degree k, the method tests the distribution of its local connection strengths against a null hypothesis for randomly selected nodes of the same degree and therefore discriminates those links that are statically insignificant. Grady et al. (2012) proposed a different backbone method that identifies the salient or important links in a weighted network (Fig. 3B). For a given node r, the shortest paths from node r to the other nodes in the network are calculated, and the method constructs a shortest path tree for node *r* by summarising the most effective routes from node *r* to the rest of the network. This is repeated for all nodes, and the salient links are those that are most commonly found across all shortest path trees (see Fig. 3C for an example). This method benefits from requiring no parameterisation or reference to a null model, as the distribution of link salience was found to be bimodal. These explicitly network-based, backbone methods depict the efficiency of information flows, which in ecology may therefore be interpreted as the transfer of energy fluxes in ecological networks, which in turn is related to trophic length (Levine, 1980).

3.4 Multilayer Networks

Multilayer networks are one of the more recent and promising approaches that can be applied to replicated complex 'big data'. Until recently, studies on ecological networks primarily focused on a single type of interaction, such as predation in food webs (but see Pocock et al., 2012 for an integrated network analysis), operating over a fixed 'snapshot' of time or space. Multilayer networks were developed as a mathematical framework for representing complex systems (Kivela et al., 2014), where different interaction types among nodes or networks operating at different times or spatial locations can be encoded into individual layers within the framework. If we consider a transportation system with many stations that are linked by more than one type of transportation mode, then the way in which stations are connected by a given transportation mode can be encoded as a single layer. Individual layers can then be linked together via common interchanges, forming a multilayer transportation network. The temporal and spatial structure of ecological networks can similarly be represented using multilayer networks. Individual snapshot networks, which vary in time or space, can form a single layer (Pilosof et al., 2017), with different snapshot networks forming further layers. This allows the analysis of complex datasets with explicit connections within and among multiple layers (Boccaletti et al., 2014; Kivela et al., 2014; Pilosof et al., 2017). Ecological layers can include multiple, interconnected processes such as dispersal, colonisation or extinction, opening a new opportunity to advance the analysis and explanation of complex replicated network data, particularly when subject to change.

In a set of replicated food webs, each individual food web could be represented as a network of species and their trophic relationships using a single layer within a multilayer network, with the links within this layer being referred to as intralayer edges. Individual layers can then be bound together

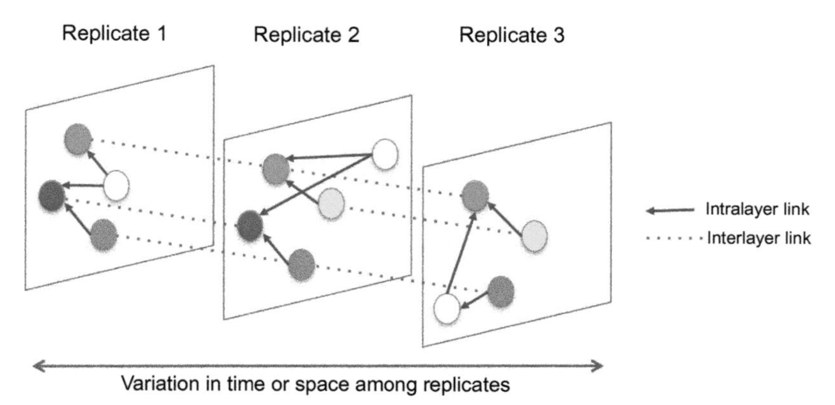

Fig. 4 An example of a multilayer network with three replicates of food webs. Nodes within an individual replicate are connected in the usual way based on the predation relationships between predators and prey, forming the intralayer links. Interlayer connections are inserted between common species among layers which are denoted by the coloured nodes.

to form a multilayer network by introducing interlayer edges between species that are common to more than one layer (Fig. 4). Collating replicated networks into a multilayer network provides an opportunity to uncover characteristics among the species interactions and how they are altered by perturbations. More importantly, this approach allows us to track how responses persist across the different replicates and allows us to gain an insight into their relative significance.

There exists a range of metrics and analytical techniques to profile multilayer networks (see Boccaletti et al., 2014 for a review). Pilosof et al. (2017) highlighted module profiling in multilayer networks in an ecological context using an example which links a plant–herbivore network with a plant-root-parasite network. This multilayer network illustrated how parasitism impacts on herbivory. In the case of replicated networks, detecting modular structures within a multilayer framework might enable us to establish rules of module size and composition and their persistence across the different replicates. This would potentially shed light onto the coevolutionary processes that underlie species and community stability. One example, particularly pertinent to biomonitoring, is that species which belong to more than one module could be the key hubs or keystones within the ecosystem. The evaluation of robustness to the effects of perturbation might further help identify those species that are central to the ecosystem stability. Within a multilayer network architecture, the impact of species removal within a single, local layer can be quantified globally by looking at the cross-layer effects. Multilayer analyses may therefore become an invaluable tool for determining whether a network response to disturbance is a localised phenomenon or an effect that cascades across layers.

Multilayer network approaches have the potential to provide a methodology for addressing issues of natural variation among replicated networks, through a process of complexity reduction that aggregates layers that are similar or nondistinguishable from one other (De Domenico et al., 2015). 'Distinguishability' of layers is defined by a distance function, based on the proportion of overlapping elements between two network layers. This approach allows us to gauge the degree of variation across different replicated networks as layers. High distinguishability between two layers would indicate a high level of variation between either the species composition or interactions of the two network layers, or both, and suggest at the presence of multiple equilibriums within the ecosystem. If two layers are nondistinguishable, it should then be possible to identify the key constituent parts that are common to the two layers, such as the core (Lu et al., 2016; Ma and Mondragón, 2015; Rombach et al., 2014) or backbone (Grady et al., 2012; Serrano et al., 2009), which describe critical changes or responses to perturbation.

4. CONCLUSIONS

We are entering the 'Big Data' era in ecology. Recent advances in diet tracing techniques and NGS will potentially generate large amounts of good quality and highly resolved ecological network data. This opens up an enormous opportunity for sophisticated, replicated experimental and biomonitoring designs that can overcome the shortcomings of current 'snapshot' approach, including incomplete sampling, limited spatiotemporal scale and oversimplistic assembly assumptions. If we are all to capitalise on this big data phenomenon in ecology, there is a real need for more open data and tools, shared among ecologists, rather than duplicating effort in curating data and implementing code (Hampton et al., 2013). Whereas DNA sequences are being deposited in an appropriate online repositories, such as GenBank (Benson et al., 2013), there is now an urgent priority to catalogue and deposit ecological data of species compositions, abundances, cooccurrences or dietary biomarker in similar open access repositories (Buttigieg et al., 2018; Nielsen et al., 2018; Pauli et al., 2017).

New opportunities associated with the emergence of large-scale datasets also present new challenges. In this chapter, we consider the question 'How do we study ecological big data reliably in the future to detect change?'. Whereas the historical approaches are well suited to characterise simple, nonreplicated networks, these low-level methods do not incorporate natural variation, change and spatiotemporal dynamics of highly replicated complex networks. New multiscale approaches to network ecology based on food web profiling, null models, weighted networks and multilayer networks allow us to do this by taking into account the high replication brought by big ecological data. Analytical frameworks already exist in network science for examining temporal and spatial variation among networks which can now be applied in ecology (e.g. Barrat et al., 2004; Blonder et al., 2012; Gray et al., 2016; Kivela et al., 2014; Lu et al., 2016; Pilosof et al., 2017). We envision that this cross-fertilisation between the different disciplines will continue and help identify novel structural properties that would enable us to better predict the impact of environmental drivers and threats. These approaches build on the current standard biomonitoring methods and promise to provide a new framework to understanding how biodiversity, species interactions and ecosystem functions respond to a range of natural and anthropogenic disturbance events (Thompson et al., 2012).

ACKNOWLEDGEMENTS

We would like to thank Mark Ledger for sharing the Millstream dataset, Axel G. Rossberg and Alfred Burian for constructive discussion and comments on the manuscript. X.L. and C.G. were supported by Queen Mary University of London. X.L. was additionally supported by the Chinese Scholarship Council, and C.G. was additionally supported by the Freshwater Biological Association. D.A.B. and E.C. were supported by the project ANR-17-CE32-0011, and D.A.B. and S.D. also received support from the projects FACCE SURPLUS PREAR and FACCE C-IPM BioAWARE.

REFERENCES

Ahmed, A., Xing, E.P., 2009. Recovering time-varying networks of dependencies in social and biological studies. Proc. Natl. Acad. Sci. U.S.A. 106, 11878–11883. https://doi.org/10.1073/pnas.0901910106.

Aizen, M.A., Morales, C.L., Morales, J.M., 2008. Invasive mutualists erode native pollination webs. PLoS Biol. 6, e31. San Franc. https://doi.org/10.1371/journal.pbio.0060031.

Aizen, M.A., Sabatino, M., Tylianakis, J.M., 2012. Specialization and rarity predict nonrandom loss of interactions from mutualist networks. Science 335, 1486–1489. https://doi.org/10.1126/science.1215320.

Albert, R., Jeong, H., Barabasi, A.-L., 2000. Error and attack tolerance of complex networks. Nature 406, 378–382. https://doi.org/10.1038/35019019.

Allesina, S., Pascual, M., 2008. Network structure, predator–prey modules, and stability in large food webs. Theor. Ecol. 1, 55–64. https://doi.org/10.1007/s12080-007-0007-8.

Allesina, S., Bodini, A., Bondavalli, C., 2006. Secondary extinctions in ecological networks: bottlenecks unveiled. Ecol. Model. 194, 150–161. https://doi.org/10.1016/j.ecolmodel.2005.10.016.
Arshad, M.A., Martin, S., 2002. Identifying critical limits for soil quality indicators in agro-ecosystems. Agric. Ecosyst. Environ. 88, 153–160. https://doi.org/10.1016/S0167-8809(01)00252-3.
Barabasi, A.-L., 2009. Scale-free networks: a decade and beyond. Science 325, 412–413. https://doi.org/10.1126/science.1173299.
Barbosa, M., Fernandes, G.W., Lewis, O.T., Morris, R.J., 2017. Experimentally reducing species abundance indirectly affects food web structure and robustness. J. Anim. Ecol. 86, 327–336. https://doi.org/10.1111/1365-2656.12626.
Barrat, A., Barthelemy, M., Pastor-Satorras, R., Vespignani, A., 2004. The architecture of complex weighted networks. Proc. Natl. Acad. Sci. U.S.A. 101, 3747–3752. https://doi.org/10.1073/pnas.0400087101.
Bascompte, J., Stouffer, D.B., 2009. The assembly and disassembly of ecological networks. Philos. Trans. R. Soc. Lond. B Biol. Sci. 364, 1781–1787. https://doi.org/10.1098/rstb.2008.0226.
Bascompte, J., Jordano, P., Melián, C.J., Olesen, J.M., 2003. The nested assembly of plant-animal mutualistic networks. Proc. Natl. Acad. Sci. U.S.A. 100, 9383–9387. https://doi.org/10.1073/pnas.1633576100.
Bastolla, U., Fortuna, M.A., Pascual-Garcia, A., Ferrera, A., Luque, B., Bascompte, J., 2009. The architecture of mutualistic networks minimizes competition and increases biodiversity. Nature 458, 1018–1020.
Beiler, K.J., Durall, D.M., Simard, S.W., Maxwell, S.A., Kretzer, A.M., 2010. Architecture of the wood-wide web: Rhizopogon spp. genets link multiple Douglas-fir cohorts. New Phytol. 185, 543–553. https://doi.org/10.1111/j.1469-8137.2009.03069.x.
Benson, D.A., Cavanaugh, M., Clark, K., Karsch-Mizrachi, I., Lipman, D.J., Ostell, J., Sayers, E.W., 2013. GenBank. Nucleic Acids Res. 41, D36–D42.
Bersier, L.-F., Banasek-Richter, C., Cattin, M.-F., 2002. Quantitative descriptors of food-web matrices. Ecology 83, 2394–2407. https://doi.org/10.1890/0012-9658(2002)083[2394:QDOFWM]2.0.CO;2.
Blois, J.L., Zarnetske, P.L., Fitzpatrick, M.C., Finnegan, S., 2013. Climate change and the past, present, and future of biotic interactions. Science 341, 499–504. https://doi.org/10.1126/science.1237184.
Blonder, B., Wey, T.W., Dornhaus, A., James, R., Sih, A., 2012. Temporal dynamics and network analysis. Methods Ecol. Evol. 3, 958–972. https://doi.org/10.1111/j.2041-210X.2012.00236.x.
Blüthgen, N., 2010. Why network analysis is often disconnected from community ecology: a critique and an ecologist's guide. Basic Appl. Ecol. 11, 185–195. https://doi.org/10.1016/j.baae.2010.01.001.
Blüthgen, N., Menzel, F., Blüthgen, N., 2006. Measuring specialization in species interaction networks. BMC Ecol. 6, 9.
Blüthgen, N., Menzel, F., Hovestadt, T., Fiala, B., Blüthgen, N., 2007. Specialization, constraints, and conflicting interests in mutualistic networks. Curr. Biol. 17, 341–346. https://doi.org/10.1016/j.cub.2006.12.039.
Blüthgen, N., Fründ, J., Vázquez, D.P., Menzel, F., 2008. What do interaction network metrics tell us about specialization and biological traits. Ecology 89, 3387–3399. https://doi.org/10.1890/07-2121.1.
Boccaletti, S., Latora, V., Moreno, Y., Chavez, M., Hwang, D., 2006. Complex networks: structure and dynamics. Phys. Rep. 424, 175–308. https://doi.org/10.1016/j.physrep.2005.10.009.

Boccaletti, S., Bianconi, G., Criado, R., del Genio, C.I., Gómez-Gardeñes, J., Romance, M., Sendiña-Nadal, I., Wang, Z., Zanin, M., 2014. The structure and dynamics of multilayer networks. Phys. Rep. 544, 1–122. https://doi.org/10.1016/j.physrep.2014.07.001.

Bohan, D.A., Caron-Lormier, G., Muggleton, S., Raybould, A., Tamaddoni-Nezhad, A., 2011a. Automated discovery of food webs from ecological data using logic-based machine learning. PLoS One 6, e29028. https://doi.org/10.1371/journal.pone.0029028.

Bohan, D.A., Powers, S.J., Champion, G., Haughton, A.J., Hawes, C., Squire, G., Cussans, J., Mertens, S.K., 2011b. Modelling rotations: can crop sequences explain arable weed seedbank abundance?: crop sequence effects on seedbanks. Weed Res. 51, 422–432. https://doi.org/10.1111/j.1365-3180.2011.00860.x.

Bray, D., 2003. Molecular networks: the top-down view. Science 301, 1864–1865. https://doi.org/10.1126/science.1089118.

Briand, F., 1983. Environmental control of food web structure. Ecology 64, 253–263. https://doi.org/10.2307/1937073.

Buttigieg, P.L., Fadeev, E., Bienhold, C., Hehemann, L., Offre, P., Boetius, A., 2018. Marine microbes in 4D—using time series observation to assess the dynamics of the ocean microbiome and its links to ocean health. Curr. Opin. Microbiol. 43, 169–185.

Cagnolo, L., Salvo, A., Valladares, G., 2011. Network topology: patterns and mechanisms in plant-herbivore and host-parasitoid food webs: patterns and determinants of food web topology. J. Anim. Ecol. 80, 342–351. https://doi.org/10.1111/j.1365-2656.2010.01778.x.

Calizza, E., Costantini, M.L., Rossi, L., 2015. Effect of multiple disturbances on food web vulnerability to biodiversity loss in detritus-based systems. Ecosphere 6, 1–20. https://doi.org/10.1890/ES14-00489.1.

Canard, E., Mouquet, N., Marescot, L., Gaston, K.J., Gravel, D., Mouillot, D., 2012. Emergence of structural patterns in neutral trophic networks. PLoS One 7, e38295. https://doi.org/10.1371/journal.pone.0038295.

Carstensen, D.W., Sabatino, M., Trøjelsgaard, K., Morellato, L.P.C., 2014. Beta diversity of plant-pollinator networks and the spatial turnover of pairwise interactions. PLoS One 9, e112903. https://doi.org/10.1371/journal.pone.0112903.

Chase, J.M., 2003. Community assembly: when should history matter? Oecologia 136, 489–498. https://doi.org/10.1007/s00442-003-1311-7.

Chase, J.M., 2007. Drought mediates the importance of stochastic community assembly. Proc. Natl. Acad. Sci. U.S.A. 104, 17430–17434. https://doi.org/10.1073/pnas.0704350104.

Chase, J.M., Leibold, M.A., 2003. Ecological Niches: Linking Classical and Contemporary Approaches. University of Chicago Press, Chicago.

Cohen, J.E., 1977. Food webs and the dimensionality of trophic niche space. Proc. Natl. Acad. Sci. U.S.A. 74, 4533–4536.

Cohen, J.E., Briand, F., 1984. Trophic links of community food webs. Proc. Natl. Acad. Sci. U.S.A. 81, 4105–4109.

Costa, L.d.F., Rodrigues, F.A., Travieso, G., Villas Boas, P.R., 2007. Characterization of complex networks: a survey of measurements. Adv. Phys. 56, 167–242. https://doi.org/10.1080/00018730601170527.

Csermely, P., London, A., Wu, L.-Y., Uzzi, B., 2013. Structure and dynamics of core/periphery networks. J. Complex Networks 1, 93–123. https://doi.org/10.1093/comnet/cnt016.

Dakos, V., Bascompte, J., 2014. Critical slowing down as early warning for the onset of collapse in mutualistic communities. Proc. Natl. Acad. Sci. U.S.A. 111, 17546–17551.

Dale, V.H., Beyeler, S.C., 2001. Challenges in the development and use of ecological indicators. Ecol. Indic. 1, 3–10. https://doi.org/10.1016/S1470-160X(01)00003-6.

Davies, S.P., Jackson, S.K., 2006. The biological condition gradient: a descriptive model for interpreting change in aquatic ecosystems. Ecol. Appl. 16, 1251–1266. https://doi.org/10.1890/1051-0761.

De Domenico, M., Nicosia, V., Arenas, A., Latora, V., 2015. Structural reducibility of multilayer networks. Nat. Commun. 6, 6864. https://doi.org/10.1038/ncomms7864.

Derocles, S.A.P., Le Ralec, A., Besson, M.M., Maret, M., Walton, A., Evans, D.M., Plantegenest, M., 2014. Molecular analysis reveals high compartmentalization in aphid-primary parasitoid networks and low parasitoid sharing between crop and noncrop habitats. Mol. Ecol. 23, 3900–3911. https://doi.org/10.1111/mec.12701.

Derocles, S.A.P., Bohan, D.A., Dumbrell, A.J., Kitson, J.J.N., Massol, F., Pauvert, C., Plantegenest, M., Vacher, C., Evans, D.M., 2018. Biomonitoring for the 21st century: integrating next-generation sequencing into ecological network analysis. Adv. Ecol. Res. 58, 1–62.

Dézerald, O., Leroy, C., Corbara, B., Carrias, J.-F., Pélozuelo, L., Dejean, A., Céréghino, R., 2013. Food-web structure in relation to environmental gradients and predator-prey ratios in tank-bromeliad ecosystems. PLoS One 8, e71735. https://doi.org/10.1371/journal.pone.0071735.

Dunne, J.A., Williams, R.J., Martinez, N.D., 2002. Network structure and biodiversity loss in food webs: robustness increases with connectance. Ecol. Lett. 5, 558–567. https://doi.org/10.1046/j.1461-0248.2002.00354.x.

Emer, C., Memmott, J., Vaughan, I.P., Montoya, D., Tylianakis, J.M., 2016. Species roles in plant–pollinator communities are conserved across native and alien ranges. Divers. Distrib. 22, 841–852. https://doi.org/10.1111/ddi.12458.

Evans, D.M., Pocock, M.J.O., Memmott, J., 2013. The robustness of a network of ecological networks to habitat loss. Ecol. Lett. 16, 844–852. https://doi.org/10.1111/ele.12117.

Evans, D.M., Kitson, J.J.N., Lunt, D.H., Straw, N.A., Pocock, M.J.O., 2016. Merging DNA metabarcoding and ecological network analysis to understand and build resilient terrestrial ecosystems. Funct. Ecol. 30, 1904–1916. https://doi.org/10.1111/1365-2435.12659.

Firbank, L.G., Heard, M.S., Woiwod, I.P., Hawes, C., Haughton, A.J., Champion, G.T., Scott, R.J., Hill, M.O., Dewar, A.M., Squire, G.R., May, M.J., Brooks, D.R., Bohan, D.A., Daniels, R.E., Osborne, J.L., Roy, D.B., Black, H.I.J., Rothery, P., Perry, J.N., 2003. An introduction to the farm-scale evaluations of genetically modified herbicide-tolerant crops. J. Appl. Ecol. 40, 2–16. https://doi.org/10.1046/j.1365-2664.2003.00787.x.

Fortuna, M.A., Bascompte, J., 2006. Habitat loss and the structure of plant-animal mutualistic networks: mutualistic networks and habitat loss. Ecol. Lett. 9, 281–286. https://doi.org/10.1111/j.1461-0248.2005.00868.x.

Freilich, M.A., Wieters, E., Broitman, B.R., Marquet, P.A., Navarrete, S.A., 2018. Species co-occurrence networks: can they reveal trophic and non-trophic interactions in ecological communities? Ecology 99, 690–699. https://doi.org/10.1002/ecy.2142.

Frost, C.M., Peralta, G., Rand, T.A., Didham, R.K., Varsani, A., Tylianakis, J.M., 2016. Apparent competition drives community-wide parasitism rates and changes in host abundance across ecosystem boundaries. Nat. Commun. 7, 12644. https://doi.org/10.1038/ncomms12644.

Garcia-Domingo, J.L., Saldaña, J., 2008. Effects of heterogeneous interaction strengths on food web complexity. Oikos 117, 336–343. https://doi.org/10.1111/j.2007.0030-1299.16261.x.

Genini, J., Morellato, L.P.C., Guimaraes, P.R., Olesen, J.M., 2010. Cheaters in mutualism networks. Biol. Lett. 6, 494–497. https://doi.org/10.1098/rsbl.2009.1021.

Golubski, A.J., Abrams, P.A., 2011. Modifying modifiers: what happens when interspecific interactions interact? J. Anim. Ecol. 80, 1097–1108. https://doi.org/10.1111/j.1365-2656.2011.01852.x.

Grady, D., Thiemann, C., Brockmann, D., 2012. Robust classification of salient links in complex networks. Nat. Commun. 3, 864. https://doi.org/10.1038/ncomms1847.

Gray, C., Baird, D.J., Baumgartner, S., Jacob, U., Jenkins, G.B., O'Gorman, E.J., Lu, X., Ma, A., Pocock, M.J.O., Schuwirth, N., Thompson, M., Woodward, G., 2014. FORUM: ecological networks: the missing links in biomonitoring science. J. Appl. Ecol. 51, 1444–1449. https://doi.org/10.1111/1365-2664.12300.

Gray, C., Hildrew, A.G., Lu, X., Ma, A., McElroy, D., Monteith, D., O'Gorman, E., Shilland, E., Woodward, G., 2016. Recovery and nonrecovery of freshwater food webs from the effects of acidification. Adv. Ecol. Res. 55, 475–534. https://doi.org/10.1016/bs.aecr.2016.08.009.

Guimerà, R., Nunes Amaral, L.A., 2005. Functional cartography of complex metabolic networks. Nature 433, 895–900. https://doi.org/10.1038/nature03288.

Hampton, S.E., Strasser, C.A., Tewksbury, J.J., Gram, W.K., Budden, A.E., Batcheller, A.L., Duke, C.S., Porter, J.H., 2013. Big data and the future of ecology. Front. Ecol. Environ. 11, 156–162. https://doi.org/10.1890/120103.

Heleno, R., Garcia, C., Jordano, P., Traveset, A., Gomez, J.M., Bluthgen, N., Memmott, J., Moora, M., Cerdeira, J., Rodriguez-Echeverria, S., Freitas, H., Olesen, J.M., 2014. Ecological networks: delving into the architecture of biodiversity. Biol. Lett. 10, 20131000. –20131000. https://doi.org/10.1098/rsbl.2013.1000.

Holt, R.D., 1977. Predation, apparent competition, and the structure of prey communities. Theor. Popul. Biol. 12, 197–229. https://doi.org/10.1016/0040-5809(77)90042-9.

Ings, T.C., Montoya, J.M., Bascompte, J., Blüthgen, N., Brown, L., Dormann, C.F., Edwards, F., Figueroa, D., Jacob, U., Jones, J.I., Lauridsen, R.B., Ledger, M.E., Lewis, H.M., Olesen, J.M., van Veen, F.J.F., Warren, P.H., Woodward, G., 2009. Review: ecological networks—beyond food webs. J. Anim. Ecol. 78, 253–269. https://doi.org/10.1111/j.1365-2656.2008.01460.x.

James, R., Croft, D.P., Krause, J., 2009. Potential banana skins in animal social network analysis. Behav. Ecol. Sociobiol. 63, 989–997. https://doi.org/10.1007/s00265-009-0742-5.

Ji, Y., Ashton, L., Pedley, S.M., Edwards, D.P., Tang, Y., Nakamura, A., Kitching, R., Dolman, P.M., Woodcock, P., Edwards, F.A., Larsen, T.H., Hsu, W.W., Benedick, S., Hamer, K.C., Wilcove, D.S., Bruce, C., Wang, X., Levi, T., Lott, M., Emerson, B.C., Yu, D.W., 2013. Reliable, verifiable and efficient monitoring of biodiversity via metabarcoding. Ecol. Lett. 16, 1245–1257. https://doi.org/10.1111/ele.12162.

Jiang, J., Huang, Z., Seager, T.P., Lin, W., Grebogi, C., Hastings, A., Lai, Y., 2018. Predicting tipping points in mutualistic networks through dimension reduction. PNAS 115 (4), E639–E647.

Jordan, F., 2009. Keystone species and food webs. Philos. Trans. R. Soc. Lond. B Biol. Sci. 364, 1733–1741. https://doi.org/10.1098/rstb.2008.0335.

Kaiser-Bunbury, C.N., Mougal, J., Whittington, A.E., Valentin, T., Gabriel, R., Olesen, J.M., Blüthgen, N., 2017. Ecosystem restoration strengthens pollination network resilience and function. Nature 542, 223–227. https://doi.org/10.1038/nature21071.

Kitson, J.J.N., Hahn, C., Sands, R.J., Straw, N.A., Evans, D.M., Lunt, D.H., 2016. Nested metabarcode tagging: a robust tool for studying species interactions in ecology and evolution. bioRxiv 1–24. https://doi.org/10.1101/035071.

Kivela, M., Arenas, A., Barthelemy, M., Gleeson, J.P., Moreno, Y., Porter, M.A., 2014. Multilayer networks. J. Complex Networks 2, 203–271. https://doi.org/10.1093/comnet/cnu016.

Kourou, K., Exarchos, T.P., Exarchos, K.P., Karamouzis, M.V., Fotiadis, D.I., 2015. Machine learning applications in cancer prognosis and prediction. Comput. Struct. Biotechnol. J. 13, 8–17. https://doi.org/10.1016/j.csbj.2014.11.005.

Kratina, P., Mac Nally, R., Kimmerer, W.J., Thomson, J.R., Winder, M., 2014. Human-induced biotic invasions and changes in plankton interaction networks. J. Appl. Ecol. 51, 1066–1074. https://doi.org/10.1111/1365-2664.12266.

Krause, A.E., Frank, K.A., Mason, D.M., Ulanowicz, R.E., Taylor, W.W., 2003. Compartments revealed in food-web structure. Nature 426, 282–285. https://doi.org/10.1038/nature02115.

Laliberté, E., Tylianakis, J.M., 2010. Deforestation homogenizes tropical parasitoid—host networks. Ecology 91, 1740–1747.

Latora, V., Marchiori, M., 2001. Efficient behavior of small-world networks. Phys. Rev. Lett. 87, 198701. https://doi.org/10.1103/PhysRevLett.87.198701.

Lawton, R.O., Nair, U.S., Pielke Sr., R.A., Welch, R.M., 2001. Climatic impact of tropical lowland deforestation on nearby montane cloud forests. Science 294, 584–587.

Layer, K., Riede, J.O., Hildrew, A.G., Woodward, G., 2010. Food web structure and stability in 20 streams across a wide pH gradient. Adv. Ecol. Res. 42, 265–299.

Layer, K., Hildrew, A.G., Jenkins, G.B., Riede, J.O., Rossiter, S.J., Townsend, C.R., Woodward, G., 2011. Long-term dynamics of a well-characterised food web: four decades of acidification and recovery in the Broadstone Stream model system. Adv. Ecol. Res. 44, 69–117.

Ledger, M.E., Brown, L.E., Edwards, F.K., Milner, A.M., Woodward, G., 2012a. Drought alters the structure and functioning of complex food webs. Nat. Clim. Chang. 3, 223–227. https://doi.org/10.1038/nclimate1684.

Ledger, M.E., Harris, R.M.L., Armitage, P.D., Milner, A.M., 2012b. Climate change impacts on community resilience. Adv. Ecol. Res. 46, 211–258. Elsevier. https://doi.org/10.1016/B978-0-12-396992-7.00003-4.

Ledger, M.E., Brown, L.E., Edwards, F.K., Hudson, L.N., Milner, A.M., Woodward, G., 2013. Extreme climatic events alter aquatic food webs. Adv. Ecol. Res. 48, 343–395. Elsevier. https://doi.org/10.1016/B978-0-12-417199-2.00006-9.

Leger, J.-B., Daudin, J.-J., Vacher, C., 2015. Clustering methods differ in their ability to detect patterns in ecological networks. Methods Ecol. Evol. 6, 474–481. https://doi.org/10.1111/2041-210X.12334.

Levine, S., 1980. Several measures of trophic structure applicable to complex food webs. J. Theor. Biol. 83, 195–207. https://doi.org/10.1016/0022-5193(80)90288-X.

Lewinsohn, T.M., Inácio Prado, P., Jordano, P., Bascompte, J., Olesen, J.M., 2006. Structure in plant-animal interaction assemblages. Oikos 113, 174–184. https://doi.org/10.1111/j.0030-1299.2006.14583.x.

Lu, X., Gray, C., Brown, L.E., Ledger, M.E., Milner, A.M., Mondragón, R.J., Woodward, G., Ma, A., 2016. Drought rewires the cores of food webs. Nat. Clim. Chang. 6, 875–878. https://doi.org/10.1038/nclimate3002.

Lurgi, M., Lopez, B.C., Montoya, J.M., 2012. Novel communities from climate change. Philos. Trans. R. Soc. B Biol. Sci. 367, 2913–2922. https://doi.org/10.1098/rstb.2012.0238.

Ma, A., Mondragón, R.J., 2015. Rich-cores in networks. PLoS One 10, e0119678. https://doi.org/10.1371/journal.pone.0119678.

Macfadyen, S., Gibson, R., Polaszek, A., Morris, R.J., Craze, P.G., Planqué, R., Symondson, W.O.C., Memmott, J., 2009. Do differences in food web structure between organic and conventional farms affect the ecosystem service of pest control? Ecol. Lett. 12, 229–238. https://doi.org/10.1111/j.1461-0248.2008.01279.x.

Macfadyen, S., Craze, P.G., Polaszek, A., van Achterberg, K., Memmott, J., 2011. Parasitoid diversity reduces the variability in pest control services across time on farms. Proc. Biol. Sci. 278, 3387–3394. https://doi.org/10.1098/rspb.2010.2673.

McLaughlin, Ó.B., Emmerson, M.C., O'Gorman, E.J., 2013. Habitat isolation reduces the temporal stability of island ecosystems in the face of flood disturbance. Adv. Ecol. Res. 48, 225–284. Elsevier. https://doi.org/10.1016/B978-0-12-417199-2.00004-5.
Melián, C.J., Bascompte, J., 2004. Food web cohesion. Ecology 85, 352–358. https://doi.org/10.1890/02-0638.
Memmott, J., Waser, N.M., Price, M.V., 2004. Tolerance of pollination networks to species extinctions. Proc. Biol. Sci. 271, 2605–2611. https://doi.org/10.1098/rspb.2004.2909.
Milo, R., 2002. Network motifs: simple building blocks of complex networks. Science 298, 824–827. https://doi.org/10.1126/science.298.5594.824.
Montoya, J.M., Pimm, S.L., Solé, R.V., 2006. Ecological networks and their fragility. Nature 442, 259–264. https://doi.org/10.1038/nature04927.
Montoya, J., Woodward, G., Emmerson, M.C., Solé, R.V., 2009. Press perturbations and indirect effects in real food webs. Ecology 90, 2426–2433. https://doi.org/10.1890/08-0657.1.
Moore, C., Grewar, J., Cumming, G.S., 2016. Quantifying network resilience: comparison before and after a major perturbation shows strengths and limitations of network metrics. J. Appl. Ecol. 53, 636–645. https://doi.org/10.1111/1365-2664.12486.
Morris, R.J., Lewis, O.T., Godfray, H.C.J., 2004. Experimental evidence for apparent competition in a tropical forest food web. Nature 428, 310–313. https://doi.org/10.1038/nature02394.
Morris, R.J., Sinclair, F.H., Burwell, C.J., 2015. Food web structure changes with elevation but not rainforest stratum. Ecography 38, 792–802. https://doi.org/10.1111/ecog.01078.
Mougi, A., 2017. Spatial complexity enhances predictability in food webs. Sci. Rep. 7, 43440. https://doi.org/10.1038/srep43440.
Mouillot, D., Bellwood, D.R., Baraloto, C., Chave, J., Galzin, R., Harmelin-Vivien, M., Kulbicki, M., Lavergne, S., Lavorel, S., Mouquet, N., Paine, T., Renaud, J., Thuiller, W., 2013. Rare species support vulnerable functions in high-diversity ecosystems. PLoS Biol. San Franc. 11, e1001569. https://doi.org/10.1371/journal.pbio.1001569.
Mulder, C., den Hollander, H., Schouten, T., Rutgers, M., 2006. Allometry, biocomplexity, and web topology of hundred agro-environments in the Netherlands. Ecol. Complex. 3, 219–230. https://doi.org/10.1016/j.ecocom.2006.05.004.
Muller, C.B., Godfray, H.C.J., 1997. Apparent competition between two aphid species. J. Anim. Ecol. 66 (1), 57–64. https://doi.org/10.2307/5965.
Nielsen, A., Totland, Ø., 2014. Structural properties of mutualistic networks withstand habitat degradation while species functional roles might change. Oikos 123, 323–333. https://doi.org/10.1111/j.1600-0706.2013.00644.x.
Nielsen, M.J., Clare, L.E., Hayden, B., Brett, T.M., Kratina, P., 2018. Diet tracing in ecology: method comparison and selection. Methods Ecol. Evol. 9, 278–291. https://doi.org/10.1111/2041-210X.12869.
O'Gorman, E.J., Emmerson, M.C., 2010. Manipulating interaction strengths and the consequences for trivariate patterns in a marine food web. Adv. Ecol. Res. 42, 301–419. Elsevier. https://doi.org/10.1016/B978-0-12-381363-3.00006-X.
O'Gorman, E.J., Fitch, J.E., Crowe, T.P., 2012. Multiple anthropogenic stressors and the structural properties of food webs. Ecology 93, 441–448. https://doi.org/10.1890/11-0982.1.
Olesen, J.M., Bascompte, J., Dupont, Y.L., Jordano, P., 2007. The modularity of pollination networks. Proc. Natl. Acad. Sci. U.S.A. 104, 19891–19896. https://doi.org/10.1073/pnas.0706375104.
Olesen, J.M., Bascompte, J., Dupont, Y.L., Elberling, H., Rasmussen, C., Jordano, P., 2011a. Missing and forbidden links in mutualistic networks. Proc. R. Soc. B Biol. Sci. 278, 725–732. https://doi.org/10.1098/rspb.2010.1371.

Olesen, J.M., Stefanescu, C., Traveset, A., 2011b. Strong, long-term temporal dynamics of an ecological network. PLoS One 6, e26455. https://doi.org/10.1371/journal.pone.0026455.

Oliver, T.H., et al., 2015. Biodiversity and resilience of ecosystem functions. Trends Ecol. Evol. 30, 673–684.

Parrott, L., 2010. Measuring ecological complexity. Ecol. Indic. 10, 1069–1076. https://doi.org/10.1016/j.ecolind.2010.03.014.

Pauli, J.N., Newsome, S.D., Cook, J.A., Harrod, C., Steffan, S.A., Baker, C.J., … Cerling, T.E., 2017. Opinion: why we need a centralized repository for isotopic data. Proc. Natl. Acad. Sci. U.S.A. 114, 2997–3001.

Peralta, G., Frost, C.M., Didham, R.K., Rand, T.A., Tylianakis, J.M., 2017. Non-random food-web assembly at habitat edges increases connectivity and functional redundancy. Ecology 98, 995–1005. https://doi.org/10.1002/ecy.1656.

Perennou, C., Guelmami, A., Paganini, M., Philipson, P., Poulin, B., Strauch, A., Tottrup, C., Truckenbrodt, J., Geijzendorffer, I.R., 2018. Mapping Mediterranean wetlands with remote sensing: a good-looking map is not always a good map. Adv. Ecol. Res. 58, 243–277.

Petchey, O.L., Brose, U., Rall, B.C., 2010. Predicting the effects of temperature on food web connectance. Philos. Trans. R. Soc. Lond. B Biol. Sci. 365, 2081–2091. https://doi.org/10.1098/rstb.2010.0011.

Pilosof, S., Porter, M.A., Pascual, M., Kéfi, S., 2017. The multilayer nature of ecological networks. Nat. Ecol. Evol. 1, 101. https://doi.org/10.1038/s41559-017-0101.

Piñol, J., San Andrés, V., Clare, E.L., Mir, G., Symondson, W.O.C., 2014. A pragmatic approach to the analysis of diets of generalist predators: the use of next-generation sequencing with no blocking probes. Mol. Ecol. Resour. 14, 18–26. https://doi.org/10.1111/1755-0998.12156.

Pocock, M.J.O., Evans, D.M., Memmott, J., 2012. The robustness and restoration of a network of ecological networks. Science 335, 973–977. https://doi.org/10.1126/science.1214915.

Poisot, T., Bever, J.D., Nemri, A., Thrall, P.H., Hochberg, M.E., 2011a. A conceptual framework for the evolution of ecological specialisation: evolution of ecological specialisation. Ecol. Lett. 14, 841–851. https://doi.org/10.1111/j.1461-0248.2011.01645.x.

Poisot, T., Lepennetier, G., Martinez, E., Ramsayer, J., Hochberg, M.E., 2011b. Resource availability affects the structure of a natural bacteria-bacteriophage community. Biol. Lett. 7, 201–204. https://doi.org/10.1098/rsbl.2010.0774.

Poisot, T., Canard, E., Mouillot, D., Mouquet, N., Gravel, D., 2012. The dissimilarity of species interaction networks. Ecol. Lett. 15, 1353–1361. https://doi.org/10.1111/ele.12002.

Poisot, T., Stouffer, D.B., Gravel, D., 2015. Beyond species: why ecological interaction networks vary through space and time. Oikos 124, 243–251. https://doi.org/10.1111/oik.01719.

Raffaelli, D., 2006. Food webs, body size and the curse of the Latin binomial. In: Rooney, N., McCann, K.S., Noakes, D.L.G. (Eds.), From Energetics to Ecosystems: The Dynamics and Structure of Ecological Systems. Springer, Netherlands, pp. 53–64. https://doi.org/10.1007/978-1-4020-5337-5_3.

Raffaelli, D., Bullock, J.M., Cinderby, S., Durance, I., Emmett, B., Harris, J., Hicks, K., Oliver, T.H., Paterson, D., White, P.C.L., 2014. Big data and ecosystem research programmes. Adv. Ecol. Res. 51, 41–77. Elsevier. https://doi.org/10.1016/B978-0-08-099970-8.00004-X.

Reiss, J., Bridle, J.R., Montoya, J.M., Woodward, G., 2009. Emerging horizons in biodiversity and ecosystem functioning research. Trends Ecol. Evol. 24, 505–514.

Ricklefs, R.E., 1987. Community diversity: relative roles of local and regional processes. Science 235, 167–171. https://doi.org/10.1126/science.235.4785.167.
Rombach, M.P., Porter, M.A., Fowler, J.H., Mucha, P.J., 2014. Core-periphery structure in networks. SIAM J. Appl. Math. 74, 167–190. https://doi.org/10.1137/120881683.
Rossa, F.D., Dercole, F., Piccardi, C., 2013. Profiling core-periphery network structure by random walkers. Sci. Rep. 3. https://doi.org/10.1038/srep01467.
Rossberg, A.G., 2013. Food Webs and Biodiversity: Foundations, Models, Data. Wiley.
Sanders, D., van Veen, F.J.F., 2012. Indirect commensalism promotes persistence of secondary consumer species. Biol. Lett. 8, 960–963. https://doi.org/10.1098/rsbl.2012.0572.
Sanders, D., Kehoe, R., van Veen, F.J.F., 2015. Experimental evidence for the population-dynamic mechanisms underlying extinction cascades of carnivores. Curr. Biol. 25, 3106–3109. https://doi.org/10.1016/j.cub.2015.10.017.
Santamaría, S., Galeano, J., Pastor, J.M., Méndez, M., 2016. Removing interactions, rather than species, casts doubt on the high robustness of pollination networks. Oikos 125, 526–534. https://doi.org/10.1111/oik.02921.
Säterberg, T., Sellman, S., Ebenman, B., 2013. High frequency of functional extinctions in ecological networks. Nature 499, 468–470. https://doi.org/10.1038/nature12277.
Schoenholtz, S.H., Miegroet, H.V., Burger, J.A., 2000. A review of chemical and physical properties as indicators of forest soil quality: challenges and opportunities. For. Ecol. Manage. 138, 335–356. https://doi.org/10.1016/S0378-1127(00)00423-0.
Serrano, M.A., Boguna, M., Vespignani, A., 2009. Extracting the multiscale backbone of complex weighted networks. Proc. Natl. Acad. Sci. U.S.A. 106, 6483–6488. https://doi.org/10.1073/pnas.0808904106.
Solé, R.V., Alonso, D., McKane, A., 2002. Self–organized instability in complex ecosystems. Philos. Trans. R. Soc. Lond. B. Biol. Sci. 357, 667–681. https://doi.org/10.1098/rstb.2001.0992.
Srinivasan, U.T., Dunne, J.A., Harte, J., Martinez, N.D., 2007. Response of complex food webs to realistic extinction sequences. Ecology 88, 671–682. https://doi.org/10.1890/06-0971.
Stouffer, D.B., Bascompte, J., 2011. Compartmentalization increases food-web persistence. Proc. Natl. Acad. Sci. U.S.A. 108, 3648–3652. https://doi.org/10.1073/pnas.1014353108.
Stouffer, D.B., Camacho, J., Guimerà, R., Ng, C.A., Nunes Amaral, L.A., 2005. Quantitative patterns in the structure of model and empirical food webs. Ecology 86, 1301–1311. https://doi.org/10.1890/04-0957.
Sullivan, G.M., Feinn, R., 2012. Using effect size—or why the *P* value is not enough. J. Grad. Med. Educ. 4, 279–282. https://doi.org/10.4300/JGME-D-12-00156.1.
Summerhayes, V.S., Elton, C.S., 1923. Contributions to the ecology of Spitsbergen and Bear Island. J. Ecol. 11, 214–286.
Tamaddoni-Nezhad, A., Milani, G.A., Raybould, A., Muggleton, S., Bohan, D.A., 2013. Construction and validation of food webs using logic-based machine learning and text mining. Adv. Ecol. Res. 49, 225–289. Elsevier. https://doi.org/10.1016/B978-0-12-420002-9.00004-4.
Thebault, E., Fontaine, C., 2010. Stability of ecological communities and the architecture of mutualistic and trophic networks. Science 329, 853–856. https://doi.org/10.1126/science.1188321.
Thompson, R.M., Brose, U., Dunne, J.A., Hall, R.O., Hladyz, S., Kitching, R.L., Martinez, N.D., Rantala, H., Romanuk, T.N., Stouffer, D.B., Tylianakis, J.M., 2012. Food webs: reconciling the structure and function of biodiversity. Trends Ecol. Evol. 27, 689–697. https://doi.org/10.1016/j.tree.2012.08.005.

Toju, H., Sato, H., Yamamoto, S., Kadowaki, K., Tanabe, A.S., Yazawa, S., Nishimura, O., Agata, K., 2013. How are plant and fungal communities linked to each other in below-ground ecosystems? A massively parallel pyrosequencing analysis of the association specificity of root-associated fungi and their host plants. Ecol. Evol. 3, 3112–3124. https://doi.org/10.1002/ece3.706.

Toju, H., Guimarães, P.R., Olesen, J.M., Thompson, J.N., 2014. Assembly of complex plant–fungus networks. Nat. Commun. 5, 5273. https://doi.org/10.1038/ncomms6273.

Trojelsgaard, K., Jordano, P., Carstensen, D.W., Olesen, J.M., 2015. Geographical variation in mutualistic networks: similarity, turnover and partner fidelity. Proc. Biol. Sci. 282, 20142925–20142925. https://doi.org/10.1098/rspb.2014.2925.

Tsai, C.-H., Hsieh, C., Nakazawa, T., 2016. Predator-prey mass ratio revisited: does preference of relative prey body size depend on individual predator size? Funct. Ecol. 30, 1979–1987. https://doi.org/10.1111/1365-2435.12680.

Tylianakis, J.M., Morris, R.J., 2017. Ecological networks across environmental gradients. Annu. Rev. Ecol. Evol. Syst. 48, 25–48. https://doi.org/10.1146/annurev-ecolsys-110316-022821.

Tylianakis, J.M., Tscharntke, T., Lewis, O.T., 2007. Habitat modification alters the structure of tropical host-parasitoid food webs. Nature 445, 202–205. https://doi.org/10.1038/nature05429.

Tylianakis, J.M., Didham, R.K., Bascompte, J., Wardle, D.A., 2008. Global change and species interactions in terrestrial ecosystems. Ecol. Lett. 11, 1351–1363. https://doi.org/10.1111/j.1461-0248.2008.01250.x.

Tylianakis, J.M., Laliberté, E., Nielsen, A., Bascompte, J., 2010. Conservation of species interaction networks. Biol. Conserv. 143, 2270–2279. https://doi.org/10.1016/j.biocon.2009.12.004.

Ulanowicz, R.E., Holt, R.D., Barfield, M., 2014. Limits on ecosystem trophic complexity: insights from ecological network analysis. Ecol. Lett. 17, 127–136. https://doi.org/10.1111/ele.12216.

Ulrich, W., Gotelli, N.J., 2007. Null model analysis of species nestedness patterns. Ecology 88, 1824–1831. https://doi.org/10.1890/06-1208.1.

Ulrich, W., Gotelli, N.J., 2010. Null model analysis of species associations using abundance data. Ecology 91, 3384–3397. https://doi.org/10.1890/09-2157.1.

Valdovinos, F.S., Ramos-Jiliberto, R., Garay-Narváez, L., Urbani, P., Dunne, J.A., 2010. Consequences of adaptive behaviour for the structure and dynamics of food webs. Ecol. Lett. 13, 1546–1559. https://doi.org/10.1111/j.1461-0248.2010.01535.x.

van der Werf, G.R., Dempewolf, J., Trigg, S.N., Randerson, J.T., Kasibhatla, P.S., Giglio, L., Murdiyarso, D., Peters, W., Morton, D.C., Collatz, G.J., Dolman, A.J., DeFries, R.S., 2008. Climate regulation of fire emissions and deforestation in equatorial Asia. Proc. Natl. Acad. Sci. U.S.A. 105, 20350–20355.

Van Veen, F.J.F., Muller, C.B., Pell, J.K., Godfray, H.C.J., 2008. Food web structure of three guilds of natural enemies: predators, parasitoids and pathogens of aphids. J. Anim. Ecol. 77, 191–200.

Vázquez, D.P., 2005. Degree distribution in plant-animal mutualistic networks: forbidden links or random interactions? Oikos 108, 421–426. https://doi.org/10.1111/j.0030-1299.2005.13619.x.

Vázquez, D.P., Aizen, M.A., 2003. Null model analyses of specialization in plant-pollinator interactions. Ecology 84, 2493–2501. https://doi.org/10.1890/02-0587.

Vázquez, D.P., Chacoff, N.P., Cagnolo, L., 2009. Evaluating multiple determinants of the structure of plant-animal mutualistic networks. Ecology 90, 2039–2046. https://doi.org/10.1890/08-1837.1.

Veech, J.A., 2012. Significance testing in ecological null models. Theor. Ecol. 5, 611–616. https://doi.org/10.1007/s12080-012-0159-z.
Watts, D.J., 1999. Networks, dynamics, and the small-world phenomenon. Am. J. Sociol. 105, 493–527. https://doi.org/10.1086/210318.
Watts, D.J., Strogatz, S.H., 1998. Collective dynamics of "small-world" networks. Nature 393, 440–442.
Woodward, G., Perkins, D.M., Brown, L.E., 2010a. Climate change and freshwater ecosystems: impacts across multiple levels of organization. Philos. Trans. R. Soc. Lond. B Biol. Sci. 365, 2093–2106. https://doi.org/10.1098/rstb.2010.0055.
Woodward, G., et al., 2010b. Ecological networks in a changing climate. Adv. Ecol. Res. 42, 72–138.
Woodward, G., Dumbrell, A.J., Baird, D.J., Hajibabaei, M., 2014. Preface. Adv. Ecol. Res. 51, ix–xiii. Elsevier. https://doi.org/10.1016/B978-0-08-099970-8.09985-1.
Woodward, G., Bonada, N., Brown, L.E., Death, R.G., Durance, I., Gray, C., Hladyz, S., Ledger, M.E., Milner, A.M., Ormerod, S.J., Thompson, R.M., Pawar, S., 2016. The effects of climatic fluctuations and extreme events on running water ecosystems. Philos. Trans. R. Soc. Lond. B Biol. Sci. 371, 20150274. https://doi.org/10.1098/rstb.2015.0274.
Yachi, S., Loreau, M., 1999. Biodiversity and ecosystem productivity in a fluctuating environment: the insurance hypothesis. Proc. Natl Acad. Sci. U.S.A. 96, 1463–1468.
Zhao, L., Zhang, H., O'Gorman, E.J., Tian, W., Ma, A., Moore, J.C., Borrett, S.R., Woodward, G., 2016. Weighting and indirect effects identify keystone species in food webs. Ecol. Lett. 19, 1032–1040. https://doi.org/10.1111/ele.12638.

FURTHER READING

Allesina, S., Alonso, D., Pascual, M., 2008. A general model for food web structure. Science 320, 658–661. https://doi.org/10.1126/science.1156269.
Baiser, B., Gotelli, N.J., Buckley, H.L., Miller, T.E., Ellison, A.M., 2012. Geographic variation in network structure of a nearctic aquatic food web: network structure in an aquatic food web. Glob. Ecol. Biogeogr. 21, 579–591. https://doi.org/10.1111/j.1466-8238.2011.00705.x.
Bascompte, J., 2010. Structure and dynamics of ecological networks. Science 329, 765–766. https://doi.org/10.1126/science.1194255.
Bascompte, J., Jordano, P., Olesen, J.M., 2006. Asymmetric coevolutionary networks facilitate biodiversity maintenance. Science 312, 431–433. https://doi.org/10.1126/science.1123412.
Bohan, D.A., Boursault, A., Brooks, D.R., Petit, S., 2011c. National-scale regulation of the weed seedbank by carabid predators: carabid seed predation. J. Appl. Ecol. 48, 888–898. https://doi.org/10.1111/j.1365-2664.2011.02008.x.
Carnicer, J., Jordano, P., Melián, C.J., 2009. The temporal dynamics of resource use by frugivorous birds: a network approach. Ecology 90, 1958–1970. https://doi.org/10.1890/07-1939.1.
Dalsgaard, B., Magård, E., Fjeldså, J., Martín González, A.M., Rahbek, C., Olesen, J.M., Ollerton, J., Alarcón, R., Cardoso Araujo, A., Cotton, P.A., Lara, C., Machado, C.G., Sazima, I., Sazima, M., Timmermann, A., Watts, S., Sandel, B., Sutherland, W.J., Svenning, J.-C., 2011. Specialization in plant-hummingbird networks is associated with species richness, contemporary precipitation and quaternary climate-change velocity. PLoS One 6, e25891. https://doi.org/10.1371/journal.pone.0025891.
Gabor, C., Nepusz, T., 2006. The igraph software package for complex network research. InterJournal Complex Syst. 1695, 1–9.

Gray, C., 2016. Towards a Networks Based Approach to Biomonitoring. Queen Mary University of London, London, UK.
Hudson, L.N., Emerson, R., Jenkins, G.B., Layer, K., Ledger, M.E., Pichler, D.E., Thompson, M.S.A., O'Gorman, E.J., Woodward, G., Reuman, D.C., 2013. Cheddar: analysis and visualisation of ecological communities in R. Methods Ecol. Evol. 4, 99–104. https://doi.org/10.1111/2041-210X.12005.
Jordano, P., 1987. Patterns of mutualistic interactions in pollination and seed dispersal: connectance, dependence asymmetries, and coevolution. Am. Nat. 129, 657–677. https://doi.org/10.1086/284665.
Lewinsohn, T.M., Roslin, T., 2008. Four ways towards tropical herbivore megadiversity. Ecol. Lett. 11, 398–416. https://doi.org/10.1111/j.1461-0248.2008.01155.x.
Morris, R.J., Gripenberg, S., Lewis, O.T., Roslin, T., 2014. Antagonistic interaction networks are structured independently of latitude and host guild. Ecol. Lett. 17, 340–349. https://doi.org/10.1111/ele.12235.
Olesen, J.M., Bascompte, J., Elberling, H., Jordano, P., 2008. Temporal dynamics in a pollination network. Ecology 89, 1573–1582. https://doi.org/10.1890/07-0451.1.
Petanidou, T., Potts, S.G., 2006. Mutual use of resources in Mediterranean plant–pollinator communities: how specialized are pollination webs. In: Plant–Pollinator Interactions: From Specialization to Generalization. University of Chicago Press, Chicago, IL, pp. 220–244.
Sala, O.E., Chapin, F.S., Armesto, J.J., Berlow, E., 2000. Global biodiversity scenarios for the year 2100. Science 287, 1770–1774. https://doi.org/10.1126/science.287.5459.1770.
Schleuning, M., Blüthgen, N., Flörchinger, M., Braun, J., Schaefer, H.M., Böhning-Gaese, K., 2011. Specialization and interaction strength in a tropical plant-frugivore network differ among forest strata. Ecology 92, 26–36. https://doi.org/10.1890/09-1842.1.
Schmid-Araya, J.M., Hildrew, A.G., Robertson, A., Schmid, P.E., Winterbottom, J., 2002. The importance of meiofauna in food webs: evidence from an acid stream. Ecology 83, 1271–1285. https://doi.org/10.1890/0012-9658(2002)083[1271:TIOMIF]2.0.CO;2.
Sih, A., Hanser, S.F., McHugh, K.A., 2009. Social network theory: new insights and issues for behavioral ecologists. Behav. Ecol. Sociobiol. 63, 975–988. https://doi.org/10.1007/s00265-009-0725-6.
Tavares-Cromar, A.F., Williams, D.D., 1996. The importance of temporal resolution in food web analysis: evidence from a detritus-based stream. Ecol. Monogr. 66, 91–113. https://doi.org/10.2307/2963482.
Thompson, R.M., Townsend, C.R., 1999. The effect of seasonal variation on the community structure and food-web attributes of two streams: implications for food-web science. Oikos 87, 75. https://doi.org/10.2307/3546998.
Wood, S.A., Russell, R., Hanson, D., Williams, R.J., Dunne, J.A., 2015. Effects of spatial scale of sampling on food web structure. Ecol. Evol. 5, 3769–3782. https://doi.org/10.1002/ece3.1640.

ADVANCES IN ECOLOGICAL RESEARCH VOLUME 1–59

CUMULATIVE LIST OF TITLES

Aerial heavy metal pollution and terrestrial ecosystems, **11**, 218
Advances in monitoring and modelling climate at ecologically relevant scales, **58**, 101
Age determination and growth of Baikal seals (*Phoca sibirica*), **31**, 449
Age-related decline in forest productivity: pattern and process, **27**, 213
Allometry of body size and abundance in 166 food webs, **41**, 1
Analysis and interpretation of long-term studies investigating responses to climate change, **35**, 111
Analysis of processes involved in the natural control of insects, **2**, 1
Ancient Lake Pennon and its endemic molluscan faun (Central Europe; Mio-Pliocene), **31**, 463
Ant-plant-homopteran interactions, **16**, 53
Anthropogenic impacts on litter decomposition and soil organic matter, **38**, 263
Arctic climate and climate change with a focus on Greenland, **40**, 13
Arrival and departure dates, **35**, 1
Assessing the contribution of micro-organisms and macrofauna to biodiversity-ecosystem functioning relationships in freshwater microcosms, **43**, 151
A belowground perspective on Dutch agroecosystems: how soil organisms interact to support ecosystem services, **44**, 277
The benthic invertebrates of Lake Khubsugul, Mongolia, **31**, 97
Big data and ecosystem research programmes, **51**, 41
Biodiversity, species interactions and ecological networks in a fragmented world **46**, 89
Biogeography and species diversity of diatoms in the northern basin of Lake Tanganyika, **31**, 115
Bioinformatics for biomonitoring: species detection and diversity estimates across next-generation sequencing platforms, **59**, 1
Biological strategies of nutrient cycling in soil systems, **13**, 1
Biomanipulation as a restoration tool to combat eutrophication: recent advances and future challenges, **47**, 411

Biomonitoring of human impacts in freshwater ecosystems: the good, the bad and the ugly, **44**, 1
Biomonitoring for the 21st century: integrating next-generation sequencing into ecological network analysis, **58**, 1
Bray-Curtis ordination: an effective strategy for analysis of multivariate ecological data, **14**, 1
Body size, life history and the structure of host-parasitoid networks, **45**, 135
Breeding dates and reproductive performance, **35**, 69
Can a general hypothesis explain population cycles of forest Lepidoptera? **18**, 179
Carbon allocation in trees; a review of concepts for modeling, **25**, 60
Catchment properties and the transport of major elements to estuaries, **29**, 1
A century of evolution in *Spartina anglica*, **21**, 1
Changes in substrate composition and rate-regulating factors during decomposition, **38**, 101
The challenge of future research on climate change and avian biology, **35**, 237
The challenges of linking ecosystem services to biodiversity: lessons from a large-scale freshwater study, **54**, 87
Challenges with inferring how land-use affects terrestrial biodiversity: study design, time, space and synthesis, **58**, 163
The colne estuary: a long-term microbial ecology observatory, **55**, 227
Climate change and eco-evolutionary dynamics in food webs, **47**, 1
Climate change impacts on community resilience: evidence from a drought disturbance experiment **46**, 211
Climate change influences on species interrelationships and distributions in high-Arctic Greenland, **40**, 81
Climate-driven range shifts within benthic habitats across a marine biogeographic transition zone, **55**, 325
Climate influences on avian population dynamics, **35**, 185
Climatic and geographic patterns in decomposition, **38**, 227
Climatic background to past and future floods in Australia, **39**, 13
The climatic response to greenhouse gases, **22**, 1
Coevolution of mycorrhizal symbionts and their hosts to metal-contaminated environment, **30**, 69
Community genetic and competition effects in a model pea aphid system, **50**, 239
Communities of parasitoids associated with leafhoppers and planthoppers in Europe, **17**, 282

Community structure and interaction webs in shallow marine hardbottom communities: tests of an environmental stress model, **19**, 189
A complete analytic theory for structure and dynamics of populations and communities spanning wide ranges in body size, **46**, 427
Complexity, evolution, and persistence in host-parasitoid experimental systems with *Callosobruchus* beetles as the host, **37**, 37
Connecting the green and brown worlds: Allometric and stoichiometric predictability of above- and below-ground networks, **49**, 69
Conservation of the endemic cichlid fishes of Lake Tanganyika; implications from population-level studies based on mitochondrial DNA, **31**, 539
Constructing nature: laboratory models as necessary tools for investigating complex ecological communities, **37**, 333
Construction and validation of food webs using logic-based machine learning and text mining, **49**, 225
The contribution of laboratory experiments on protists to understanding population and metapopulation dynamics, **37**, 245
The cost of living: field metabolic rates of small mammals, **30**, 177
Cross-scale approaches to forecasting biogeographic responses to climate change, **55**, 371
Decomposers: soil microorganisms and animals, **38**, 73
The decomposition of emergent macrophytes in fresh water, **14**, 115
Delays, demography and cycles; a forensic study, **28**, 127
Dendroecology; a tool for evaluating variations in past and present forest environments, **19**, 111
Determinants of density-body size scaling within food webs and tools for their detection, **45**, 1
Detrital dynamics and cascading effects on supporting ecosystem services, **53**, 97
The development of regional climate scenarios and the ecological impact of green-house gas warming, **22**, 33
Developments in ecophysiological research on soil invertebrates, **16**, 175
The direct effects of increase in the global atmospheric CO_2 concentration on natural and commercial temperate trees and forests, **19**, 2; **34**, 1
Disentangling the pathways and effects of ecosystem service co-production, **54**, 245
Distributional (In)congruence of biodiversity—ecosystem functioning, **46**, 1
The distribution and abundance of lake dwelling Triclads-towards a hypothesis, **3**, 1

DNA metabarcoding meets experimental ecotoxicology: Advancing knowledge on the ecological effects of Copper in freshwater ecosystems, **51**, 79
Do eco-evo feedbacks help us understand nature? Answers from studies of the trinidadian guppy, **50**, 1
The dynamics of aquatic ecosystems, **6**, 1
The dynamics of endemic diversification: molecular phylogeny suggests an explosive origin of the Thiarid Gastropods of Lake Tanganyika, **31**, 331
The dynamics of field population of the pine looper, *Bupalis piniarius* L. (Lep, Geom.), **3**, 207
Earthworm biotechnology and global biogeochemistry, **15**, 369
Ecological aspects of fishery research, **7**, 114
Eco-evolutionary dynamics of agricultural networks: implications for sustainable management, **49**, 339
Eco-evolutionary dynamics: experiments in a model system, **50**, 167
Eco-evolutionary dynamics of individual-based food webs, **45**, 225
Eco-evolutionary dynamics in a three-species food web with intraguild predation: intriguingly complex, **50**, 41
Eco-evolutionary dynamics of plant–insect communities facing disturbances: implications for community maintenance and agricultural management, **52**, 91
Eco-evolutionary interactions as a consequence of selection on a secondary sexual trait, **50**, 143
Eco-evolutionary spatial dynamics: rapid evolution and isolation explain food web persistence, **50**, 75
Ecological conditions affecting the production of wild herbivorous mammals on grasslands, **6**, 137
Ecological networks in a changing climate, **42**, 71
Ecological and evolutionary dynamics of experimental plankton communities, **37**, 221
Ecological implications of dividing plants into groups with distinct photosynthetic production capabilities, **7**, 87
Ecological implications of specificity between plants and rhizosphere microorganisms, **31**, 122
Ecological interactions among an Orestiid (Pisces: Cyprinodontidae) species flock in the littoral zone of Lake Titicaca, **31**, 399
Ecological studies at Lough Ine, **4**, 198
Ecological studies at Lough Hyne, **17**, 115
Ecology of mushroom-feeding Drosophilidae, **20**, 225

The ecology of the Cinnabar moth, **12**, 1
Ecology of coarse woody debris in temperate ecosystems, **15**, 133; **34**, 59
Ecology of estuarine macrobenthos, **29**, 195
Ecology, evolution and energetics: a study in metabolic adaptation, **10**, 1
Ecology of fire in grasslands, **5**, 209
The ecology of pierid butterflies: dynamics and interactions, **15**, 51
The ecology of root lifespan, **27**, 1
The ecology of serpentine soils, **9**, 225
Ecology, systematics and evolution of Australian frogs, **5**, 37
Ecophysiology of trees of seasonally dry Tropics: comparison among phonologies, **32**, 113
Ecosystems and their services in a changing world: an ecological perspective, **48**, 1
Effect of flooding on the occurrence of infectious disease, **39**, 107
Effects of food availability, snow, and predation on breeding performance of waders at Zackenberg, **40**, 325
Effect of hydrological cycles on planktonic primary production in Lake Malawi Niassa, **31**, 421
Effects of climatic change on the population dynamics of crop pests, **22**, 117
Effects of floods on distribution and reproduction of aquatic birds, **39**, 63
The effects of invasive species on the decline in species richness: a global meta-analysis, **56**, 61
The effects of modern agriculture nest predation and game management on the population ecology of partridges (*Perdix perdix* and *Alectoris rufa*), **11**, 2
Effective river restoration in the 21st century: From trial and error to novel evidence-based approaches, **55**, 529
El Niño effects on Southern California kelp forest communities, **17**, 243
Empirically characterising trophic networks: What emerging DNA-based methods, stable isotope and fatty acid analyses can offer, **49**, 177
Empirical evidences of density-dependence in populations of large herbivores, **41**, 313
Endemism in the Ponto-Caspian fauna, with special emphasis on the Oncychopoda (Crustacea), **31**, 179
Energetics, terrestrial field studies and animal productivity, **3**, 73
Energy in animal ecology, **1**, 69
Environmental warming in shallow lakes: a review of potential changes in community structure as evidenced from space-for-time substitution approaches, **46**, 259

Environmental warming and biodiversity-ecosystem functioning in freshwater microcosms: partitioning the effects of species identity, richness and metabolism, **43**, 177
Estimates of the annual net carbon and water exchange of forests: the EUROFLUX methodology, **30**, 113
Estimating forest growth and efficiency in relation to canopy leaf area, **13**, 327
Estimating relative energy fluxes using the food web, species abundance, and body size, **36**, 137
Evolution and endemism in Lake Biwa, with special reference to its gastropod mollusc fauna, **31**, 149
Evolutionary and ecophysiological responses of mountain plants to the growing season environment, **20**, 60
The evolutionary ecology of carnivorous plants, **33**, 1
Evolutionary inferences from the scale morphology of Malawian Cichlid fishes, **31**, 377
Explosive speciation rates and unusual species richness in haplochromine cichlid fishes: effects of sexual selection, **31**, 235
Extreme climatic events alter aquatic food webs: a synthesis of evidence from a mesocosm drought experiment, **48**, 343
The evolutionary consequences of interspecific competition, **12**, 127
The exchange of ammonia between the atmosphere and plant communities, **26**, 302
Faster, higher and stronger? The Pros and Cons of molecular faunal data for assessing ecosystem condition, **51**, 1
Faunal activities and processes: adaptive strategies that determine ecosystem function, **27**, 92
Fire frequency models, methods and interpretations, **25**, 239
Floods down rivers: from damaging to replenishing forces, **39**, 41
Food webs, body size, and species abundance in ecological community description, **36**, 1
Food webs: theory and reality, **26**, 187
Food web structure and stability in 20 streams across a wide pH gradient, **42**, 267
Forty years of genecology, **2**, 159
Foraging in plants: the role of morphological plasticity in resource acquisitions, **25**, 160
Fossil pollen analysis and the reconstruction of plant invasions, **26**, 67
Fractal properties of habitat and patch structure in benthic ecosystems, **30**, 339

Free air carbon dioxide enrichment (FACE) in global change research: a review, **28**, 1
From Broadstone to Zackenberg: space, time and hierarchies in ecological networks, **42**, 1
From natural to degraded rivers and back again: a test of restoration ecology theory and practice, **44**, 119
Functional traits and trait-mediated interactions: connecting community-level interactions with ecosystem functioning, **52**, 319
The general biology and thermal balance of penguins, **4**, 131
General ecological principles which are illustrated by population studies of Uropodid mites, **19**, 304
Generalist predators, interactions strength and food web stability, **28**, 93
Genetic correlations in multi-species plant/herbivore interactions at multiple genetic scales: implications for eco-evolutionary dynamics, **50**, 263
Genetic and phenotypic aspects of life-history evolution in animals, **21**, 63
Geochemical monitoring of atmospheric heavy metal pollution: theory and applications, **18**, 65
Global climate change leads to mistimed avian reproduction, **35**, 89
Global persistence despite local extinction in acarine predator-prey systems: lessons from experimental and mathematical exercises, **37**, 183
Habitat isolation reduces the temporal stability of island ecosystems in the face of flood disturbance, **48**, 225
Heavy metal tolerance in plants, **7**, 2
Herbivores and plant tannins, **19**, 263
High-Arctic plant–herbivore interactions under climate influence, **40**, 275
High-Arctic soil CO_2 and CH_4 production controlled by temperature, water, freezing, and snow, **40**, 441
Historical changes in environment of Lake Titicaca: evidence from Ostracod ecology and evolution, **31**, 497
How agricultural intensification affects biodiversity and ecosystem services **55**, 43
How well known is the ichthyodiversity of the large East African lakes? **31**, 17
Human and environmental factors influence soil faunal abundance-mass allometry and structure, **41**, 45
Human ecology is an interdisciplinary concept: a critical inquiry, **8**, 2
Hutchinson reversed, or why there need to be so many species, **43**, 1
Hydrology and transport of sediment and solutes at Zackenberg, **40**, 197

The Ichthyofauna of Lake Baikal, with special reference to its zoogeographical relations, **31**, 81
Impact of climate change on fishes in complex Antarctic ecosystems, **46**, 351
Impacts of invasive species on food webs: a review of empirical data, **56**, 1
Impacts of warming on the structure and functioning of aquatic communities: individual- to ecosystem-level responses, **47**, 81
Implications of phylogeny reconstruction for Ostracod speciation modes in Lake Tanganyika, **31**, 301
Importance of climate change for the ranges, communities and conservation of birds, **35**, 211
Importance of microorganisms to macroorganisms invasions: is the essential invisible to the eye? (The Little Prince, A. de Saint-Exupéry, 1943), **57**, 99
Increased stream productivity with warming supports higher trophic levels, **48**, 285
Individual-based food webs: species identity, body size and sampling effects, **43**, 211
Individual variability: the missing component to our understanding of predator–prey interactions, **52**, 19
Individual variation decreases interference competition but increases species persistence, **52**, 45
Industrial melanism and the urban environment, **11**, 373
Individual trait variation and diversity in food webs, **50**, 203
Inherent variation in growth rate between higher plants: a search for physiological causes and ecological consequences, **23**, 188; **34**, 283
Insect herbivory below ground, **20**, 1
Insights into the mechanism of speciation in Gammarid crustaceans of Lake Baikal using a population-genetic approach, **31**, 219
Interaction networks in agricultural landscape mosaics, **49**, 291
Integrated coastal management: sustaining estuarine natural resources, **29**, 241
Integration, identity and stability in the plant association, **6**, 84
Intrinsic and extrinsic factors driving match–mismatch dynamics during the early life history of marine fishes, **47**, 177
Inter-annual variability and controls of plant phenology and productivity at Zackenberg, **40**, 249
Introduction, **38**, 1
Introduction, **39**, 1
Introduction, **40**, 1
Invasions toolkit: Current methods for tracking the spread and impact of invasive species, **56**, 85

Invasions of host-associated microbiome networks, **57**, 201
Island Biogeography of Food Webs, **56**, 183
Isopods and their terrestrial environment, **17**, 188
Lake Biwa as a topical ancient lake, **31**, 571
Lake flora and fauna in relation to ice-melt, water temperature, and chemistry at Zackenberg, **40**, 371
The landscape context of flooding in the Murray–Darling basin, **39**, 85
Landscape ecology as an emerging branch of human ecosystem science, **12**, 189
Late quaternary environmental and cultural changes in the Wollaston Forland region, Northeast Greenland, **40**, 45
Learning ecological networks from next-generation sequencing data, **54**, 1
Linking biodiversity, ecosystem functioning and services, and ecological resilience: Towards an integrative framework for improved management, **53**, 55
Linking DNA metabarcoding and text mining to create network-based biomonitoring tools: a case study on boreal wetland macroinvertebrate communities, **59**, 33
Linking spatial and temporal change in the diversity structure of ancient lakes: examples from the ecology and palaeoecology of the Tanganyikan Ostracods, **31**, 521
Litter decomposition as an indicator of stream ecosystem functioning at local-to-continental scales: insights from the European *RivFunction* project, **55**, 99
Litter fall, **38**, 19
Litter production in forests of the world, **2**, 101
Locally extreme environments as natural long-term experiments in ecology, **55**, 283
Long-term changes in Lake Balaton and its fish populations, **31**, 601
Long-term dynamics of a well-characterised food web: four decades of acidification and recovery in the broadstone stream model system, **44**, 69
Macrodistribution, swarming behaviour and production estimates of the lakefly *Chaoborus edulis* (Diptera: Chaoboridae) in Lake Malawi, **31**, 431
Making waves: the repeated colonization of fresh water by Copepod crustaceans, **31**, 61
Manipulating interaction strengths and the consequences for trivariate patterns in a marine food web, **42**, 303
Manipulative field experiments in animal ecology: do they promise more than they can deliver? **30**, 299

Mapping mediterranean wetlands with remote sensing: a good-looking map is not always a good map, **58**, 243
Marine ecosystem regime shifts induced by climate and overfishing: a review for the Northern Hemisphere, **47**, 303
Massively introduced managed species and their consequences for plant–pollinator interactions, **57**, 147
Mathematical model building with an application to determine the distribution of Durshan® insecticide added to a simulated ecosystem, **9**, 133
Mechanisms of microthropod-microbial interactions in soil, **23**, 1
Mechanisms of primary succession: insights resulting from the eruption of Mount St Helens, **26**, 1
Mesocosm experiments as a tool for ecological climate-change research, **48**, 71
Methods in studies of organic matter decay, **38**, 291
The method of successive approximation in descriptive ecology, **1**, 35
Meta-analysis in ecology, **32**, 199
Microbial experimental systems in ecology, **37**, 273
Microevolutionary response to climatic change, **35**, 151
Migratory fuelling and global climate change, **35**, 33
The mineral nutrition of wild plants revisited: a re-evaluation of processes and patterns, **30**, 1
Modelling and projecting the response of local terrestrial biodiversity worldwide to land use and related pressures: the PREDICTS project, **58**, 201
Modelling interaction networks for enhanced ecosystem services in agroecosystems, **49**, 437
Modelling terrestrial carbon exchange and storage: evidence and implications of functional convergence in light-use efficiency, **28**, 57
Modelling the potential response of vegetation to global climate change, **22**, 93
Module and metamer dynamics and virtual plants, **25**, 105
Modeling individual animal histories with multistate capture–recapture models, **41**, 87
Mutualistic interactions in freshwater modular systems with molluscan components, **20**, 126
Mycorrhizal links between plants: their functioning and ecological significances, **18**, 243
Mycorrhizas in natural ecosystems, **21**, 171
The nature of species in ancient lakes: perspectives from the fishes of Lake Malawi, **31**, 39

Networking agroecology: Integrating the diversity of agroecosystem interactions, **49**, 1
A network-based method to detect patterns of local crop biodiversity: Validation at the species and infra-species levels, **53**, 259
Nitrogen dynamics in decomposing litter, **38**, 157
Nocturnal insect migration: effects of local winds, **27**, 61
Noninvasive analysis of the soil microbiome: biomonitoring strategies using the volatilome, community analysis, and environmental data, **59**, 93
Nonlinear stochastic population dynamics: the flour beetle *Tribolium* as an effective tool of discovery, **37**, 101
Novel and disrupted trophic links following invasion in freshwater ecosystems, **57**, 55
Nutrient cycles and H^+ budgets of forest ecosystems, **16**, 1
Nutrients in estuaries, **29**, 43
On the evolutionary pathways resulting in C_4 photosynthesis and crassulacean acid metabolism (CAM), **19**, 58
Origin and structure of secondary organic matter and sequestration of C and N, **38**, 185
Oxygen availability as an ecological limit to plant distribution, **23**, 93
Parasitism between co-infecting bacteriophages, **37**, 309
Persistence of plants and pollinators in the face of habitat loss: Insights from trait-based metacommunity models, **53**, 201
Robustness trade-offs in model food webs: invasion probability decreases while invasion consequences increase with connectance, **56**, 263
Scaling-up trait variation from individuals to ecosystems, **52**, 1
Temporal variability in predator–prey relationships of a forest floor food web, **42**, 173
The past as a key to the future: the use of palaeoenvironmental understanding to predict the effects of man on the biosphere, **22**, 257
Towards an integration of biodiversity–ecosystem functioning and food web theory to evaluate relationships between multiple ecosystem services, **53**, 161
Pattern and process of competition, **4**, 11
Parasites and biological invasions: predicting ecological alterations at levels from individual hosts to whole networks, **57**, 1
Permafrost and periglacial geomorphology at Zackenberg, **40**, 151
Perturbing a marine food web: consequences for food web structure and trivariate patterns, **47**, 349
Phenetic analysis, tropic specialization and habitat partitioning in the Baikal Amphipod genus *Eulimnogammarus* (Crustacea), **31**, 355

Photoperiodic response and the adaptability of avian life cycles to environmental change, **35**, 131
Phylogeny of a gastropod species flock: exploring speciation in Lake Tanganyika in a molecular framework, **31**, 273
Phenology of high-Arctic arthropods: effects of climate on spatial, seasonal, and inter-annual variation, **40**, 299
Phytophages of xylem and phloem: a comparison of animal and plant sapfeeders, **13**, 135
Population and community body size structure across a complex environmental gradient, **52**, 115
The population biology and Turbellaria with special reference to the freshwater triclads of the British Isles, **13**, 235
Population cycles in birds of the Grouse family (Tetraonidae), **32**, 53
Population cycles in small mammals, **8**, 268
Population dynamical responses to climate change, **40**, 391
Population dynamics, life history, and demography: lessons from *Drosophila*, **37**, 77
Population dynamics in a noisy world: lessons from a mite experimental system, **37**, 143
Population regulation in animals with complex life-histories: formulation and analysis of damselfly model, **17**, 1
Positive-feedback switches in plant communities, **23**, 264
The potential effect of climatic changes on agriculture and land use, **22**, 63
Predation and population stability, **9**, 1
Predicted effects of behavioural movement and passive transport on individual growth and community size structure in marine ecosystems, **45**, 41
Predicting the responses of the coastal zone to global change, **22**, 212
Predictors of individual variation in movement in a natural population of threespine stickleback (*Gasterosteus aculeatus*), **52**, 65
Present-day climate at Zackenberg, **40**, 111
The pressure chamber as an instrument for ecological research, **9**, 165
Primary production by phytoplankton and microphytobenthos in estuaries, **29**, 93
Principles of predator-prey interaction in theoretical experimental and natural population systems, **16**, 249
The production of marine plankton, **3**, 117
Production, turnover, and nutrient dynamics of above and below ground detritus of world forests, **15**, 303

Protecting an ecosystem service: approaches to understanding and mitigating threats to wild insect pollinators, **54**, 135
Quantification and resolution of a complex, size-structured food web, **36**, 85
Quantifying the biodiversity value of repeatedly logged rainforests: gradient and comparative approaches from borneo, **48**, 183
Quantitative ecology and the woodland ecosystem concept, **1**, 103
14 Questions for invasion in ecological networks, **56**, 293
Realistic models in population ecology, **8**, 200
Recovery and nonrecovery of freshwater food webs from the effects of acidification, **55**, 469
Recommendations for the next generation of global freshwater biological monitoring tools, **55**, 609
References, **38**, 377
The relationship between animal abundance and body size: a review of the mechanisms, **28**, 181
Relative risks of microbial rot for fleshy fruits: significance with respect to dispersal and selection for secondary defence, **23**, 35
Renewable energy from plants: bypassing fossilization, **14**, 57
A replicated network approach to 'big data' in ecology, **59**, 225
Responses of soils to climate change, **22**, 163
Rodent long distance orientation ("homing"), **10**, 63
The role of body size in complex food webs: a cold case, **45**, 181
The role of body size variation in community assembly, **52**, 201
Scale effects and extrapolation in ecological experiments, **33**, 161
Scale dependence of predator-prey mass ratio: determinants and applications, **45**, 269
Scaling of food-web properties with diversity and complexity across ecosystems, **42**, 141
Scaling from traits to ecosystems: developing a general trait driver theory via integrating trait-based and metabolic scaling theories, **52**, 249
Secondary production in inland waters, **10**, 91
Seeing double: size-based and taxonomic views of food web structure, **45**, 67
The self-thinning rule, **14**, 167
Shifting impacts of climate change: Long-term patterns of plant response to elevated CO_2, drought, and warming across ecosystems, **55**, 437
Shifts in the Diversity and Composition of Consumer Traits Constrain the Effects of Land Use on Stream Ecosystem Functioning, **52**, 169
A simulation model of animal movement patterns, **6**, 185
Snow and snow-cover in central Northeast Greenland, **40**, 175

Soil and plant community characteristics and dynamics at Zackenberg, **40**, 223
Soil arthropod sampling, **1**, 1
Soil diversity in the Tropics, **21**, 316
Soil fertility and nature conservation in Europe: theoretical considerations and practical management solutions, **26**, 242
Solar ultraviolet-b radiation at Zackenberg: the impact on higher plants and soil microbial communities, **40**, 421
Some economics of floods, **39**, 125
Spatial and inter-annual variability of trace gas fluxes in a heterogeneous high-Arctic landscape, **40**, 473
Spatial root segregation: are plants territorials? **28**, 145
Species abundance patterns and community structure, **26**, 112
Stochastic demography and conservation of an endangered perennial plant (*Lomatium bradshawii*) in a dynamic fire regime, **32**, 1
Stomatal control of transpiration: scaling up from leaf to regions, **15**, 1
Stream ecosystem functioning in an agricultural landscape: the importance of terrestrial–aquatic linkages, **44**, 211
Structure and function of microphytic soil crusts in wildland ecosystems of arid to semiarid regions, **20**, 180
Studies on the cereal ecosystems, **8**, 108
Studies on grassland leafhoppers (Auchenorrhbyncha, Homoptera) and their natural enemies, **11**, 82
Studies on the insect fauna on Scotch Broom *Sarothamnus scoparius* (L.) Wimmer, **5**, 88
Sustained research on stream communities: a model system and the comparative approach, **41**, 175
Systems biology for ecology: from molecules to ecosystems, **43**, 87
The study area at Zackenberg, **40**, 101
Sunflecks and their importance to forest understorey plants, **18**, 1
A synopsis of the pesticide problem, **4**, 75
The temperature dependence of the carbon cycle in aquatic ecosystems, **43**, 267
Temperature and organism size – a biological law for ecotherms? **25**, 1
Terrestrial plant ecology and ^{15}N natural abundance: the present limits to interpretation for uncultivated systems with original data from a Scottish old field, **27**, 133
Theories dealing with the ecology of landbirds on islands, **11**, 329
A theory of gradient analysis, **18**, 271; **34**, 235
Throughfall and stemflow in the forest nutrient cycle, **13**, 57

Tiddalik's travels: the making and remaking of an aboriginal flood myth, **39**, 139
Towards understanding ecosystems, **5**, 1
Tradeoffs and compatibilities among ecosystem services: biological, physical and economic drivers of multifunctionality, **54**, 207
Trends in the evolution of Baikal amphipods and evolutionary parallels with some marine Malacostracan faunas, **31**, 195
Trophic interactions in population cycles of voles and lemmings: a model-based synthesis **33**, 75
The use of perturbation as a natural experiment: effects of predator introduction on the community structure of zooplanktivorous fish in Lake Victoria, **31**, 553
The unique contribution of rothamsted to ecological research at large temporal scales, **55**, 3
The use of statistics in phytosociology, **2**, 59
Unanticipated diversity: the discovery and biological exploration of Africa's ancient lakes, **31**, 1
Understanding ecological concepts: the role of laboratory systems, **37**, 1
Understanding the social impacts of floods in Southeastern Australia, **39**, 159
Unravelling the impacts of micropollutants in aquatic ecosystems: interdisciplinary studies at the interface of large-scale ecology, **55**, 183
Using fish taphonomy to reconstruct the environment of ancient Lake Shanwang, **31**, 483
Using large-scale data from ringed birds for the investigation of effects of climate change on migrating birds: pitfalls and prospects, **35**, 49
Using social media for biomonitoring: how Facebook, Twitter, Flickr and other social networking platforms can provide large-scale biodiversity data, **59**, 133
Vegetation, fire and herbivore interactions in heathland, **16**, 87
Vegetational distribution, tree growth and crop success in relation to recent climate change, **7**, 177
Vertebrate predator–prey interactions in a seasonal environment, **40**, 345
A vision for global biodiversity monitoring with citizen science, **59**, 169
The visualisation of ecological networks, and their use as a tool for engagement, advocacy and management, **54**, 41
Volatile biomarkers for aquatic ecological research, **59**, 75
Water flow, sediment dynamics and benthic biology, **29**, 155
When ranges collide: evolutionary history, phylogenetic community interactions, global change factors, and range size differentially affect plant productivity, **50**, 293

When microscopic organisms inform general ecological theory, **43**, 45
Why we need sustainable networks bridging countries, disciplines, cultures and generations for aquatic biomonitoring 2.0: a perspective derived from the *DNAqua-Net* COST action, **58**, 63
10 years later: Revisiting priorities for science and society a decade after the millennium ecosystem assessment, **53**, 1
Zackenberg in a circumpolar context, **40**, 499
The zonation of plants in freshwater lakes, **12**, 37.

CPI Antony Rowe
Chippenham, UK
2018-10-31 11:24